Principles and Applications in Speed Sensing and Energy Harvesting for Smart Roads

Luay Taha
The Pennsylvania State University, Altoona, USA

Sohail Anwar
The Pennsylvania State University, Altoona, USA

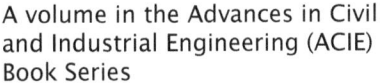

A volume in the Advances in Civil and Industrial Engineering (ACIE) Book Series

Published in the United States of America by
IGI Global
Engineering Science Reference (an imprint of IGI Global)
701 E. Chocolate Avenue
Hershey PA, USA 17033
Tel: 717-533-8845
Fax: 717-533-8661
E-mail: cust@igi-global.com
Web site: http://www.igi-global.com

Copyright © 2024 by IGI Global. All rights reserved. No part of this publication may be reproduced, stored or distributed in any form or by any means, electronic or mechanical, including photocopying, without written permission from the publisher.
Product or company names used in this set are for identification purposes only. Inclusion of the names of the products or companies does not indicate a claim of ownership by IGI Global of the trademark or registered trademark.

Library of Congress Cataloging-in-Publication Data

CIP DATA IN PROCESS

Principles and Applications in Speed Sensing and Energy Harvesting for Smart Roads
Luay Taha, Sohail Anwar
2024 Engineering Science Reference
ISBN: 9781668492147(hc) | ISBN: 9781668492154(sc) | eISBN: 9781668492161

This book is published in the IGI Global book series Advances in Civil and Industrial Engineering (ACIE) (ISSN: 2326-6139; eISSN: 2326-6155)

British Cataloguing in Publication Data
A Cataloguing in Publication record for this book is available from the British Library.

All work contributed to this book is new, previously-unpublished material.
The views expressed in this book are those of the authors, but not necessarily of the publisher.

For electronic access to this publication, please contact: eresources@igi-global.com.

Advances in Civil and Industrial Engineering (ACIE) Book Series

Ioan Constantin Dima
University Valahia of Târgovişte, Romania

ISSN:2326-6139
EISSN:2326-6155

MISSION

Private and public sector infrastructures begin to age, or require change in the face of developing technologies, the fields of civil and industrial engineering have become increasingly important as a method to mitigate and manage these changes. As governments and the public at large begin to grapple with climate change and growing populations, civil engineering has become more interdisciplinary and the need for publications that discuss the rapid changes and advancements in the field have become more in-demand. Additionally, private corporations and companies are facing similar changes and challenges, with the pressure for new and innovative methods being placed on those involved in industrial engineering.

The **Advances in Civil and Industrial Engineering (ACIE) Book Series** aims to present research and methodology that will provide solutions and discussions to meet such needs. The latest methodologies, applications, tools, and analysis will be published through the books included in **ACIE** in order to keep the available research in civil and industrial engineering as current and timely as possible.

COVERAGE

- Materials Management
- Transportation Engineering
- Optimization Techniques
- Construction Engineering
- Operations Research
- Quality Engineering
- Structural Engineering
- Urban Engineering
- Ergonomics
- Hydraulic Engineering

IGI Global is currently accepting manuscripts for publication within this series. To submit a proposal for a volume in this series, please contact our Acquisition Editors at Acquisitions@igi-global.com or visit: http://www.igi-global.com/publish/.

The Advances in Civil and Industrial Engineering (ACIE) Book Series (ISSN 2326-6139) is published by IGI Global, 701 E. Chocolate Avenue, Hershey, PA 17033-1240, USA, www.igi-global.com. This series is composed of titles available for purchase individually; each title is edited to be contextually exclusive from any other title within the series. For pricing and ordering information please visit http://www.igi-global.com/book-series/advances-civil-industrial-engineering/73673. Postmaster: Send all address changes to above address. Copyright © 2024 IGI Global. All rights, including translation in other languages reserved by the publisher. No part of this series may be reproduced or used in any form or by any means – graphics, electronic, or mechanical, including photocopying, recording, taping, or information and retrieval systems – without written permission from the publisher, except for non commercial, educational use, including classroom teaching purposes. The views expressed in this series are those of the authors, but not necessarily of IGI Global.

Titles in this Series

For a list of additional titles in this series, please visit:
http://www.igi-global.com/book-series/advances-civil-industrial-engineering/73673

Emerging Engineering Technologies and Industrial Applications
Younes El Kacimi (Ibn Tofail University, Morocco) and Khaoula Alaoui (Ibn Tofail University, Morocco)
Engineering Science Reference • © 2024 • 426pp • H/C (ISBN: 9798369313350) • US $300.00

Emerging Materials, Technologies, and Solutions for Energy Harvesting
Shilpa Mehta (Auckland University of Technology, New Zealand) Arij Naser Abougreen (University of Tripoli, Libya) and Shashi Kant Gupta (Eudoxia Research University, USA)
Engineering Science Reference • © 2024 • 368pp • H/C (ISBN: 9798369320037) • US $315.00

Innovations in Machine Learning and IoT for Water Management
Abhishek Kumar (Chandigarh University, India) Arun Lal Srivastav (Chitkara University, India) Ashutosh Kumar Dubey (University of Castilla-La Mancha, Spain) Vishal Dutt (AVN Innovations Pvt. Ltd., India) and Narayan Vyas (AVN Innovations Pvt. Ltd., India)
Engineering Science Reference • © 2024 • 312pp • H/C (ISBN: 9798369311943) • US $300.00

Intelligent Engineering Applications and Applied Sciences for Sustainability
Brojo Kishore Mishra (NIST Institute of Science and Technology (Autonomous), India & Saveetha College of Liberal Arts and Sciences, SIMATS, Chennai, India)
Engineering Science Reference • © 2023 • 542pp • H/C (ISBN: 9798369300442) • US $270.00

Global Science's Cooperation Opportunities, Challenges, and Good Practices
Mohamed Moussaoui (National School of Applied Sciences of Tangier (ENSAT), Abdelmalek Essaadi University, Morocco)
Engineering Science Reference • © 2023 • 357pp • H/C (ISBN: 9781668478745) • US $245.00

701 East Chocolate Avenue, Hershey, PA 17033, USA
Tel: 717-533-8845 x100 • Fax: 717-533-8661
E-Mail: cust@igi-global.com • www.igi-global.com

Table of Contents

Preface xiii

Chapter 1
The Current Status and Scope of Smart Highways in India 1
    Aditya Singh, Amrita Vishwa Vidyapeetham, India

Chapter 2
Renewable Energy-Integrated Electric Vehicle Charging Infrastructure
Across Cold Region Roads 42
    Dharmbir Prasad, Asansol Engineering College, India
    Rudra Pratap Singh, Asansol Engineering College, India
    Tanuja Tiwary, Asansol Engineering College, India
    Ranadip Roy, Sanaka Educational Trust's Group of Institutions, India
    Md. Irfan Khan, IAC Electricals Pvt. Ltd., India

Chapter 3
Intelligent Gas Detection Systems: Leveraging IoT and Machine Learning for
Early Warning of Hazardous Gas and LPG Leakage 76
    Kondireddy Muni Sankar, Department of Information Technology, Vel's
        University (VISTAS), Chennai, India
    B. Booba, Department of Information Technology, Vel's University
        (VISTAS), Chennai, India

Chapter 4
The SWIPT-Enabled Cooperative Full-Duplex Relaying Communication for
6G Radio Networks 108
    Rajeev Kumar, Indian Institute of Information Technology, Sri City, India

Chapter 5
Energy Harvesting and Smart Highways for Sustainable Transportation
Infrastructure: Revolutionizing Roads Using Nanotechnology 136
> *Mohanraj Gopal, School of Computer Science Engineering and
>   Information Systems, Vellore Institute of Technology, Vellore, India*
> *J. Lurdhumary, Department of Electronics and Communication
>   Engineering, DMI College of Engineering, Chennai, India*
> *S. Bathrinath, Department of Mechanical Engineering, Kalasalingam
>   Academy of Research and Education, Krishnankoil, India*
> *A. Parvathi Priya, Department of Chemistry, R.M.K. Engineering
>   College, Chennai, India*
> *Atul Sarojwal, Department of Electrical Engineering, MJP Rohilkhand
>   University, Bareilly, India*
> *S. Boopathi, Department of Mechanical Engineering, Muthayammal
>   Engineering College, Namakkal, India*

Chapter 6
AI and ML Adaptive Smart-Grid Energy Management Systems: Exploring
Advanced Innovations 166
> *S. Saravanan, Department of Electrical and Electronics Engineering,
>   B.V. Raju Institute of Technology, Narsapur, India*
> *Richa Khare, Amity School of Applied Sciences, Amity University,
>   Lucknow, India*
> *K. Umamaheswari, Department of Electrical and Electronics
>   Engineering, Dr. Mahalingam College of Engineering and
>   Technology, Pollachi, India*
> *Smriti Khare, Amity School of Applied Sciences, Amity University,
>   Lucknow, India*
> *B. S. Krishne Gowda, Department of Commerce, Government College
>   for Women, Chintamani, India*
> *Sampath Boopathi, Department of Mechanical Engineering,
>   Muthayammal Engineering College, Namakkal, India*

Chapter 7
Smart Highways-Based Piezoelectric Vehicle Speed Sensor 197
   Luay Y. Taha, Pennsylvania State University, Altoona, USA

Chapter 8
The Principles and Applications of Electrostatic Transducers 241
   Rita Tareq Aljadiri, Higher Colleges of Technology, UAE

Compilation of References 275

About the Contributors 300

Index 303

Detailed Table of Contents

Preface xiii

Chapter 1
The Current Status and Scope of Smart Highways in India 1
  Aditya Singh, Amrita Vishwa Vidyapeetham, India

Due to the advancement in technology, electronic technologies are being implemented on roads as well as on highways. This helps in the improvement of traffic engineering and transportation facility. In this chapter, smart highway will be explained and how they are important in the present scenario. A number of research and scientific publications will be covered to find out the gaps in the current research in regard to the topic. Some smart highway projects in the country India will be covered along with smart highway's present status in the country. Some advantages and challenges will be talked about in this chapter. Then, data from different sources will be collected to perform graphical analyses in order to understand the future of smart highways in the country.

Chapter 2
Renewable Energy-Integrated Electric Vehicle Charging Infrastructure
Across Cold Region Roads 42
  Dharmbir Prasad, Asansol Engineering College, India
  Rudra Pratap Singh, Asansol Engineering College, India
  Tanuja Tiwary, Asansol Engineering College, India
  Ranadip Roy, Sanaka Educational Trust's Group of Institutions, India
  Md. Irfan Khan, IAC Electricals Pvt. Ltd., India

The growing popularity of electric vehicles and the need for environmentally friendly transportation have made it necessary to build smart roadways with charging station equipped by energy harvesting and management systems. In order to collect and transform ambient energy into electrical power, the suggested system makes use of a variety of energy harvesting technologies embedded in the road surface, including solar panels, wind energy, etc. A complex energy management system integrated into the smart road infrastructure effectively stores, distributes, and transfers the captured energy in real time to electric vehicle charging stations or to the vehicles themselves. The suggested system comprises of essential elements such as intelligent inverters, decentralized energy storage units, and a central management system that manages the entire process. In the proposed renewable energy integrated charging system having energy production of 3,00,233 kWh/year with energy storage of 161 kWh.

Chapter 3
Intelligent Gas Detection Systems: Leveraging IoT and Machine Learning for Early Warning of Hazardous Gas and LPG Leakage..76
    Kondireddy Muni Sankar, Department of Information Technology, Vel's University (VISTAS), Chennai, India
    B. Booba, Department of Information Technology, Vel's University (VISTAS), Chennai, India

The chapter explores the integration of internet of things (IoT) and machine learning (ML) algorithms in hazardous gas detection systems, specifically LPG leakage. The gas detection systems industry is undergoing a significant transformation due to technological advancements, innovative sensor technologies, and intelligent solutions. This shift is crucial for safety in residential and industrial environments. The integration of gas sensing technologies with IoT devices has led to real-time monitoring, data analytics, and smart automation. Smart homes and industrial facilities benefit from interconnected gas detection systems that detect anomalies and trigger automated responses. Regulatory bodies are implementing standards to integrate advanced gas detection technologies into the automobile sector, focusing on sensor capabilities, data analytics, and automation integration. This integration will create a safer environment, prioritizing safety in connected living and working spaces. The chapter also discusses real-world case studies using automobiles.

Chapter 4
The SWIPT-Enabled Cooperative Full-Duplex Relaying Communication for 6G Radio Networks...108
    Rajeev Kumar, Indian Institute of Information Technology, Sri City, India

This chapter presents the simultaneously wireless information and power transfer (SWIPT)-enabled cooperative communication for 6th generation (6G) radio networks. These networks integrate many applications and use tiny internet-of-thing (IoT) devices that need large amounts of power to maintain energy threshold level of the networks for a long time. Therefore, the energy harvesting (EH) is a promising technique to provide sufficient power to small IoT devices and energy constrained user terminals in the 6G radio networks. For this purpose, the authors consider two-way full-duplex (FD) relaying network to derive closed-form expressions of end-to-end ergodic capacity by exploiting power splitting (PS) and time switching (TS) protocols for Rayleigh fading channels. From the numerical results, they observe that the proposed FD-based EH policies perform better as compared to half-duplex EH policies. Finally, this chapter provides future research directions.

Chapter 5
Energy Harvesting and Smart Highways for Sustainable Transportation Infrastructure: Revolutionizing Roads Using Nanotechnology 136
    Mohanraj Gopal, School of Computer Science Engineering and
        Information Systems, Vellore Institute of Technology, Vellore, India
    J. Lurdhumary, Department of Electronics and Communication
        Engineering, DMI College of Engineering, Chennai, India
    S. Bathrinath, Department of Mechanical Engineering, Kalasalingam
        Academy of Research and Education, Krishnankoil, India
    A. Parvathi Priya, Department of Chemistry, R.M.K. Engineering
        College, Chennai, India
    Atul Sarojwal, Department of Electrical Engineering, MJP Rohilkhand
        University, Bareilly, India
    S. Boopathi, Department of Mechanical Engineering, Muthayammal
        Engineering College, Namakkal, India

The chapter explores the integration of nanotechnology, energy harvesting, and smart highways into global transportation infrastructure, aiming to create sustainable and efficient systems. Nanotechnology enhances road surface durability and functionality, offering increased strength, resilience, and self-healing properties. Energy harvesting techniques, such as piezoelectric and solar technologies, harness kinetic and solar energy from vehicular motion and sunlight, powering infrastructure, streetlights, and even the grid. Smart highways, enabled by interconnected sensors and communication systems, monitor traffic flow, adjust speed limits, provide real-time updates, and autonomously manage transportation systems. These innovations not only promise a sustainable transportation ecosystem but also catalyze economic growth, environmental preservation, and enhanced quality of life for communities worldwide.

Chapter 6
AI and ML Adaptive Smart-Grid Energy Management Systems: Exploring
Advanced Innovations 166
    S. Saravanan, Department of Electrical and Electronics Engineering,
        B.V. Raju Institute of Technology, Narsapur, India
    Richa Khare, Amity School of Applied Sciences, Amity University,
        Lucknow, India
    K. Umamaheswari, Department of Electrical and Electronics
        Engineering, Dr. Mahalingam College of Engineering and
        Technology, Pollachi, India
    Smriti Khare, Amity School of Applied Sciences, Amity University,
        Lucknow, India
    B. S. Krishne Gowda, Department of Commerce, Government College
        for Women, Chintamani, India
    Sampath Boopathi, Department of Mechanical Engineering,
        Muthayammal Engineering College, Namakkal, India

The chapter explores the transformative role of artificial intelligence (AI) and machine learning (ML) in shaping smart energy management systems (SEMS) and predicts innovations by 2030. It discusses AI principles in energy optimization, predictive analytics in smart grids, and renewable energy integration through AI-driven strategies. The chapter also addresses critical aspects like predictive maintenance, consumer-centric solutions, cybersecurity challenges, ethical considerations, and regulatory frameworks for responsible AI implementation. By examining challenges and prospects, it provides insights into the dynamic future of energy management driven by AI and ML advancements.

Chapter 7
Smart Highways-Based Piezoelectric Vehicle Speed Sensor 197
    Luay Y. Taha, Pennsylvania State University, Altoona, USA

In this chapter, the authors proposed a piezoelectric based vehicle speed sensor. The sensors produce pulses when stressed by a vehicle's front and back wheels, passed over them. Due to the vehicle physical dimension, the sensor pulse, due to the back wheel stressing, has some delay time as compared with the front wheel sensing pulse, due to the front wheel stressing. The time delay between these pulses is estimated using microcontroller program. Then, the program computes the vehicle speed from the distance to time delay ratio.

Chapter 8
The Principles and Applications of Electrostatic Transducers 241
Rita Tareq Aljadiri, Higher Colleges of Technology, UAE

This chapter provides an overview of electrostatic transducers, describing the fundamental principles of converting mechanical energy into electrical energy using variable capacitors. It explains the operation principle of electrostatic transducers, emphasizing the structure types and conversion mechanisms. The chapter outlines the variable capacitor factors, design considerations, and implementation requirements, followed by an analysis of electrostatic conversion mechanisms. A comparative analysis of capacitor structures and power processing circuits highlights the optimal design choices, considering efficiency, power output, and scalability. Furthermore, the chapter explores the applications of electrostatic harvesters, focusing on integrating them into smart road infrastructure. A case study on smart road development in the UAE showcases the prospects of using electrostatic transducers to enhance road connectivity, efficiency, and sustainability.

Compilation of References 275

About the Contributors 300

Index 303

Preface

In recent years, smart roads and road energy harvesting have attracted great attention from industry and academia, as it can provide traffic information and harvest energy to power traffic ancillary facilities and low-power wireless sensor devices to support car networking and intelligent transportation. The system is a kind of transport infrastructure which utilizes data analytics in addition to sensors, to enhance safety as well as optimize traffic. The system was first executed in 1997 in the country Netherlands, which further laid the foundation for more than one thousand systems which are presently active globally. Smart Highways or Smart Roads have used information as well as communication technologies to build a present-day Internet of Things (IOT) infrastructure, which not only gathers real time traffic but also analyses it. So, utilizing this data, analysts design systems in addition to algorithms to improve daily traffic conditions. This helps in understanding the weather conditions, condition of the road surface, and traffic patterns in real time, and accordingly appropriate adjustments are made as needed.

Smart Highway technologies (SRT) involves traffic monitoring, real time monitoring, active transportation, remote monitoring, autonomously driving cars, 3D Printing, traffic signal management, road sign recognition, vehicle tracking, pedestrian safety, Laser scanner, high-tech cameras, traffic congestion sensors, IoT connectivity, traffic management networks, traffic lights optimization, solar powered roadways, glow in the dark roads, interactive lights, electric priority lanes for charging Electric Vehicles (EV's), weather detection, and the roads that honk system.

Smart highways technologies have been implemented all over the world. In the USA, four million miles of roadways spanning the continent. To prevent any further deterioration, the region is investing heavily in adopting SRT. Hundred billion dollars were invested to allocate over to repair dilapidated roads and highways to incorporate intelligent road systems, EV charging stations, and to provide broadband connectivity to improve internet access and to build a framework for intelligent transportation systems (ITSs) (Analytics Vidhya, 2023). The US Department of Transportation's ITS Joint Program Office (JPO) has spurred the development and use of ITS to move people and goods more safely and efficiently. (Smith, E. (2022).

An example of a smart road in the USA is the Virginia Smart Road. Different types of sensors were installed during the construction of the road, such as strain gauges, pressure cells, thermocouples, time domain reflectometry (TDR) probes, and resistivity probes (Al-Qadi et al., 2004). In Europe, the continent has been prompt in adopting SRT, given it houses a vast network of roadways and several globally renowned highways. Several projects have been undertaken to make the network more advanced and maintained. For example, the Ursa Major neo project aims to achieve high-grade traffic management and recuperate traveler experiences. URSA MAJOR addresses first and foremost freight transport on the TEN-T road network in a corridor linking the most important economic regions between Rotterdam and Sicily. To accomplish this, 35 Road, Inland Terminal, and Port Operators from 3 EU Member States (Italy, Germany, The Netherlands) have formed a consortium and incorporated Switzerland and Austria as transit countries (Nocera et al., 2018). In Asia, China already has numerous intelligent road systems in place and is extending the network to more provinces. For example, the highway from Chengdu to Yibin (seetao, 2021) is a 157km long smart highway equipped with sensors, radars, and IoT devices to collect data. Japan has formed unified standards throughout the country and employed electronic toll collection system (ETC) and vehicle information and communication system (VICS) (Liu et al., 2021). In India, it implemented more than 1500 kilometers of ITS throughout the country. For instance, EFKON India Pvt. Ltd is a major company that provides facilities for Smart Highways, like ITS, Toll Management System (TMS), Highway Traffic Management System, Electronic Enforcement System, Smart City solutions, and Tunnel automation works. Another example is the FASTag in India, which is an electronic toll collection system, operated by the National Highways Authority of India (NHAI). It employs Radio Frequency Identification (RFID) technology for making toll payments directly from the prepaid or savings account linked to it or directly toll owner (Sontakke et al., 2019).

The major elements in smart highway infrastructure are sensors, ITS, energy harvesting system, and data analytics as well as management system.

The sensing technologies include the use of piezoelectric sensors, image sensors, three point-laser sensors, distributed fiber-optic acoustic sensors, three-axis digital magnetic sensors, wireless magnetic sensor network, micro-electromechanical system (MEMS) SmatRock sensors, camera, and GPS and MEMS gyro. The sensing signals are used for different purposes such as monitoring, speed estimation, and traffic control. The vehicle speed is estimated using different estimation algorithms such as adaptive signature cropping, pulse extraction, feature extraction, frame difference method, Image motion detection, laser change of Hight triggering, visual servoing controller, wavelet-denoising algorithm, dual-threshold algorithm, cross correlation, sum of absolute differences, circular convolution, signals centers of mass, dynamic time warping, normalized cross-correlation, smoothed coherence transform, phase

Preface

transform, playback speed, and Kalman filtering. These algorithms can be classified as a time domain approach, frequency domain approach, or a combination between the two. Most of these algorithms detect the vehicle, estimate the speed, and classify the vehicle type.

The ITS is a system which takes the assistance of advanced technologies to utilize real time data to help in the guidance of automobiles through a given traffic on the highway, by enhancing safety as well as decreasing traffic congestion. Some of these technologies include calling for emergency services when an accident occurs, using cameras to enforce traffic laws or signs that mark speed limit changes depending on conditions.

The paramount quality of smart highways is the ability to implement energy harvesting technologies, where mainly sensors as well as cameras decrease the use of external surveillance of the highways. Different energy harvesting techniques have been implemented in smart highways, such as piezoelectric generator, asphalt solar collector (ASC), asphalt pavements with solar cells, thermoelectric generator (TEG), and Induction Heating. These technologies address the growing demand for sustainable energy sources in transportation infrastructure. Piezoelectric materials embedded in roadways can convert the kinetic energy generated by vehicular motion into electrical power. This harvested energy can then be utilized to power streetlights, traffic management systems, and other infrastructure components (Kokkinopoulos et al., 2014; Hill et al., 2014). Similarly, solar technologies integrated into smart highways harness sunlight to generate renewable energy. This dual approach of harvesting energy from both vehicular motion and sunlight not only reduces the dependence on traditional power sources but also contributes to the overall sustainability of transportation infrastructure, paving the way for cleaner and more eco-friendly systems (Andriopoulou, 2012). The thermoelectric generator (TEGs) is another way to harvest energy from smart roads. The TEG is based on the Seebeck effect and works by converting geothermal energy into electricity. Geothermal energy refers to the energy generated by the difference in heat between the surface of the road and the layers underneath. The higher the temperature differential, the larger the amount of electrical energy generated. This makes this kind of solution especially suitable for regions that have extremely hot weather year-round. Energy harvesting Asphalt solar collector combined with piping system is also another way of energy source. The radiation from the sun and atmosphere is absorbed in the pavement through an increase in warmth which is captured by water piping system and stored in the ground or other storage reservoirs over summertime. The stored energy could then be used for supporting nearby facilities for district heating and cooling, electricity, recharging or de-icing the roads (e.g. via a hydronic system. Induction Heating is a method of introducing conductive particles in the asphalt mixture, with the objective of heating it via induction. This offers an alternative way of energy harvesting through pavements (Wu et al., 2011).

Data analytics as well as management system is required to facilitate historical traffic data needed in constructing smart highways. In ITS, data can be obtained from diverse sources, such as GPS, smart cards, sensors, video detectors, and social media. Using accurate and effective data analytics of disorganized data can provide better service for ITS. As the required volume of data is extremely huge, from Trillion-byte level to Petabyte, traditional data processing systems are inefficient, and cannot meet the data analytics requirement (Zhu et al., 2018). Big Data analytics, on the other hand, provides ITS with a new technical method using data mining, artificial intelligence, machine learning, data fusion, social networks and so on (Bello-Orgaz, et al., 2016). The merits of implementing Big Data analytics, such as Apache Hadoop and Spark, in ITS can resolve three problems: data storage, data analysis and data management. It can also improve the ITS operation efficiency and give information or provide decisions to manage traffic. Improving the ITS safety level is another important feature of Big Data analytics. The real time transportation information collected by advanced sensor and detection techniques, can be utilized, Through Big Data analytics, to predict the occurrence of traffic accident quickly; thus, improving the emergency rescue ability.

THE CHALLENGES AND SOLUTIONS

As explained, Smart Roads can be used for wide range of applications:

- Innovation (facilitation of autonomous vehicles operations),
- User service quality (traffic monitoring, communication, safety),
- Infrastructure monitoring (damages detection),
- Energy source (solar panels, piezoelectric materials, …etc).

These applications require implementation. Several challenges are encountered to build smart highway infrastructure. These challenges can be classified into three categories:

- The challenge of the cost of materials, sensor technology, communication technology, and construction (Trubia et al., 2020).
- The challenge of safe autonomous and semi-autonomous driving vehicle operation such as the accuracy and decision of car sensing algorithm to make accurate steering decisions (Wiegand, 2019).

Preface

- The challenges of transportation sector such as high fuel prices, high levels of CO_2 emissions, increasing traffic congestion, and improved road safety (Guerrero-Ibanez et al., 2015).

Numerous studies (for a summary and review see Aditya's chapter in this book, besides Taha and Anwar 'chapter) have indicated that there are different factors that must be considered in smart highways infrastructure.

The cost of Road pavements is high due to the use of heavy materials whose construction requires high costs of heavy vehicles, skilled workers, required storage, and material movement. These aspects increase pollution caused by emissions during the transportation and production phase. Benevolo et al., 2016, proposed utilizing new low-cost materials that brings advantages from an energetic, operational, and qualitative point of view.

The cost of sensor technology is another factor that must be considered in smart highways construction. Smart materials, such as piezoelectric materials, are good candidates for low-cost materials used in highway sensor technology. These materials have the features of real time monitoring, and the ability to identify current condition of materials and structures (Goulias, 2000).

Modern low power Communication technologies are used nowadays to minimize the cost of communication in smart highways. These key technologies include WLAN (WiFi), WiMAX, LTE, LTE-A, Bluetooth, Zigbee, Z-Wave, and LoRaWAN. They can be used in most smart city applications, such as smart grids and metering, smart street lighting, smart health monitoring, and smart transportation (Yaqoob et al., 2017).

Challenges related to constructing smart highways are the pavement construction. Different types of sensors and energy harvesters require paving underneath the asphalt layer. This adds complexity and cost to the smart highway's construction. Shi, 2020, explained different approaches used for constructing the smart pavement. This field is open for further research and developments.

Different studies were reported to address the safety challenges encountered in autonomous driving vehicle operation. If the algorithm receives wrongful information, the car might make a decision that might result in an accident or probably undesired behavior. Thus, high sensing accuracy is required. Also, the classified and identified vehicles on the road must utilize the captured information precisely so that to make accurate steering decisions. Also, the delay problem between the steering decision of the teleoperator and the movement of the car must be minimized (Georg et al., 2018; Wiegand, 2019).

In recent years, modern society has been facing challenges related to traffic congestion, fuel high prices, and increased pollution due to CO_2 emissions. Thus, developing a sustainable intelligent transportation system, using emerging technologies such as connected vehicles, cloud computing, and the IOT, is vital. Several projects were implemented all over the world to address these challenges, such as USDOT Connected Vehicle Research Program (USA) (Birdsall, 2013), COMSafety European Union-USA (COMSafety, 2014), COMPANYON Project (Sweden) (European Commission, 2014), Instant Mobility Project (2011), IoT6 Project (Ziegler et al., 2013), BMW and Siemens trial system (Narla, 2013), and Smart Parking in London (Peng et al., 2017).

AIM OF THE BOOK

Due to the significance of the road energy harvesting and speed sensing, we are proposing the development of an edited book (8 chapters) which will serve as a valuable source of cutting-edge information pertaining to smart highways and road energy harvesting and sensing. The edited book will be comprehensive in its scope as all the aspects and technologies of road energy harvesting and sensing will be described.

ORGANIZATION OF THE BOOK

The book is organized into eight chapters. A brief description of each of the chapters follows:

Chapter 1 explores the integration of nanotechnology, energy harvesting, and smart highways into global transportation infrastructure, aiming to create sustainable and efficient systems, and addressing the benefits in traffic management, speed regulation, and dynamic updates. The chapter explores self-healing properties and their potential to reduce maintenance costs.

Chapter 2 presents an innovative strategy for efficient and ecological transportation considering energy harvesting and management technologies into smart roadways designed for electric vehicles. The authors of this chapter have been looking at several renewable energy sources and technologies for maximizing the greener way to power EVs along the road infrastructure.

Chapter 3 demonstrates the design of a piezoelectric speed sensor in smart highways. The piezoelectric phenomenon is briefly explained then utilized to produce a primary sensor. The sensor mathematical model, the signal conditioning circuits, and speed estimation algorithm are explained. The algorithm is tested by simulation and by experiment.

Chapter 4 explores the principles, design considerations, and applications of electrostatic harvesters, with a focus on emerging technologies and trends in the field. The challenges and opportunities in developing electrostatic harvesters for various energy conversion applications will also be discussed, highlighting the potential for these devices to play a key role in the transition to a more sustainable and energy-efficient future.

Chapter 5 explained the major elements of smart highways Infrastructure and their associated numerous technologies. It also covered the smart highways requirements in future in India. The gaps in the current research on and related to smart highways were discussed in this chapter.

Chapter 6 explores the integration of IOT and Machine Learning (ML) algorithms in hazardous gas detection systems, specifically LPG leakage. The integration will create a safer environment, prioritizing safety in connected living and working spaces.

Chapter 7 explores Intelligent Gas Detection Systems: Leveraging IoT and Machine Learning for Early Warning of Hazardous Gas and LPG Leakage.

Chapter 8 presents the simultaneously wireless information and power transfer (SWIPT) enabled cooperative communication for 6th generation (6G) radio networks used in smart roads. These networks integrate many applications and use IoT devices that need large amounts of power to maintain energy threshold level of the networks for long-time. The authors highlight the importance of energy harvesting (EH) as a promising technique to provide sufficient power to small IoT devices and energy constrained user terminals in the 6G radio networks.

Luay Taha
The Pennsylvania State University, Altoona, USA

Sohail Anwar
The Pennsylvania State University, Altoona, USA

REFERENCES

Al-Qadi, I. L., Loulizi, A., Elseifi, M., & Lahouar, S. (2004). The Virginia Smart Road: The impact of pavement instrumentation on understanding pavement performance. *Journal of the Association of Asphalt Paving Technologists, 73*(3), 427-465. https://www.its-platform.eu/wp-content/uploads/ITS-Platform/CorridorDocuments/UrsaMajor/Ursa-Major_DIN%20A4_8S_2018_GB_WEB.pdf

Bello-Orgaz, G., Jung, J. J., & Camacho, D. (2016). Social big data: Recent achievements and new challenges. *Information Fusion, 28*, 45–59. doi:10.1016/j.inffus.2015.08.005 PMID:32288689

Benevolo, C., Dameri, R. P., & D'auria, B. (2016). Smart mobility in smart city: Action taxonomy, ICT intensity and public benefits. In *Empowering organizations: Enabling platforms and artefacts* (pp. 13–28). Springer International Publishing. doi:10.1007/978-3-319-23784-8_2

Birdsall, M. S. (2013). Connected Vehicle: Moving from Research to Deployment Scenarios. Institute of Transportation Engineers. *ITE Journal*, *83*(1), 44.

European Commission. (2014). *COMPANYON Project, technical report: Cooperative Dynamic Formation of Platoons for Safe and Energy-Optimized Goods Transportation*. Seventh Framework Programme.

Georg, J. M., Feiler, J., Diermeyer, F., & Lienkamp, M. (2018, November). Teleoperated driving, a key technology for automated driving? comparison of actual test drives with a head mounted display and conventional monitors. In *2018 21st International Conference on Intelligent Transportation Systems (ITSC)* (pp. 3403-3408). IEEE.

Goulias, D. G. (2000). Development of low cost and innovative smart materials for highways and civil engineering structures. *WIT Transactions on the Built Environment*, 46.

Guerrero-Ibanez, J. A., Zeadally, S., & Contreras-Castillo, J. (2015). Integration challenges of intelligent transportation systems with connected vehicle, cloud computing, and internet of things technologies. *IEEE Wireless Communications*, *22*(6), 122–128. doi:10.1109/MWC.2015.7368833

Hill, D., Agarwal, A., & Tong, N. (2014). *Assessment of piezoelectric materials for roadway energy harvesting*. California Energy Commission. Publication Number: CEC-500-2013-007.

Kokkinopoulos, A., Vokas, G., & Papageorgas, P. (2014). Energy harvesting implementing embedded piezoelectric generators–The potential for the Attiki Odos traffic grid. *Energy Procedia*, *50*, 1070–1085. doi:10.1016/j.egypro.2014.06.126

Liu, C., Du, Y., Ge, Y., Wu, D., Zhao, C., & Li, Y. (2021). New generation of smart highway: Framework and insights. *Journal of Advanced Transportation*, *2021*, 1–12. doi:10.1155/2021/9445070

Narla, S. R. (2013). The evolution of connected vehicle technology: From smart drivers to smart cars to... self-driving cars. *ITE Journal*, *83*(7), 22–26.

Nocera, S., Cavallaro, F., & Irranca Galati, O. (2018). Options for reducing external costs from freight transport along the Brenner corridor. *European Transport Research Review, 10*, 1-18. https://www.seetao.com/details/86814.html

Preface

Peng, G. C. A., Nunes, M. B., & Zheng, L. (2017). Impacts of low citizen awareness and usage in smart city services: The case of London's smart parking system. *Information Systems and e-Business Management, 15*(4), 845–876. doi:10.1007/s10257-016-0333-8

Shi, X. (2020). More than smart pavements: Connected infrastructure paves the way for enhanced winter safety and mobility on highways. *Journal of Infrastructure Preservation and Resilience, 1*(1), 13. doi:10.1186/s43065-020-00014-x

Smith, E. (2022). The future of Intelligent Transportation Systems (ITS): Applying lessons learned from 30 years of Innovation. *Public Roads, 86*(1).

Sontakke, A., Diwakar, A., & Kaur, G. (2019). Intelligent Automatic Traffic Challan on Highways and Payment Through FASTag Card. Indian Journal of Science and Technology, 12(44), 01-06.

Trubia, S., Severino, A., Curto, S., Arena, F., & Pau, G. (2020). Smart roads: An overview of what future mobility will look like. *Infrastructures, 5*(12), 107. doi:10.3390/infrastructures5120107

Vidhya, A. (2023). *Paving the way to a smarter future with smart roads.* https://www.analyticsvidhya.com/blog/2023/05/smart-roads/#:~:text=A.,systems%2C%20digital%20signage%2C%20etc

Wiegand, G. (2019, March). Benefits and Challenges of Smart Highways for the User. IUI Workshops.

Wu, G., & Yu, X. (2012). *Thermal energy harvesting across pavement structure* (No. 12-4492). Academic Press.

Yaqoob, I., Hashem, I. A. T., Mehmood, Y., Gani, A., Mokhtar, S., & Guizani, S. (2017). Enabling communication technologies for smart cities. *IEEE Communications Magazine, 55*(1), 112–120. doi:10.1109/MCOM.2017.1600232CM

Yuan, D., Jiang, W., Sha, A., Xiao, J., Wu, W., & Wang, T. (2023). Technology method and functional characteristics of road thermoelectric generator system based on Seebeck effect. *Applied Energy, 331*, 120459.

Zhu, L., Yu, F. R., Wang, Y., Ning, B., & Tang, T. (2018). Big data analytics in intelligent transportation systems: A survey. *IEEE Transactions on Intelligent Transportation Systems, 20*(1), 383–398. doi:10.1109/TITS.2018.2815678

Ziegler, S., & Crettaz, C. (2013). *IoT6 Project in a Nutshell.* Springer Berlin Heidelberg. doi:10.1007/978-3-642-38082-2_32

Chapter 1
The Current Status and Scope of Smart Highways in India

Aditya Singh
 https://orcid.org/0000-0001-9347-5627
Amrita Vishwa Vidyapeetham, India

ABSTRACT

Due to the advancement in technology, electronic technologies are being implemented on roads as well as on highways. This helps in the improvement of traffic engineering and transportation facility. In this chapter, smart highway will be explained and how they are important in the present scenario. A number of research and scientific publications will be covered to find out the gaps in the current research in regard to the topic. Some smart highway projects in the country India will be covered along with smart highway's present status in the country. Some advantages and challenges will be talked about in this chapter. Then, data from different sources will be collected to perform graphical analyses in order to understand the future of smart highways in the country.

INTRODUCTION

Smart Highways as well as Smart Roads are different from normal highways as well as roads, as electronic technologies could be implemented on them, making them suitable to the changing times. Now, the question comes in the mind that how the system of Smart Highways work? It can be said that such a system is a kind of transport infrastructure which utilizes data analytics in addition to sensors, in order to enhance safety as well as optimize traffic. In the year 1997, it was the first time such a system was executed in the world, which was the country Netherlands, which

DOI: 10.4018/978-1-6684-9214-7.ch001

Copyright © 2024, IGI Global. Copying or distributing in print or electronic forms without written permission of IGI Global is prohibited.

further laid the foundation for more than one thousand systems which are presently active globally. They are utilized to enhance the operations of traffic lights, vehicles which are autonomous as well as connected ones, speed of automobiles, monitoring the road's condition, traffic levels in addition to street lighting. In other words, Smart Highways or Smart Roads have used information as well as communication technologies in order to build a present-day Internet of Things infrastructure, which not only gathers real time traffic but also analyses it. So, utilizing this data, analysts design systems in addition to algorithms in order to improve daily traffic conditions. This helps in understanding the weather conditions, condition of the road surface, and traffic patterns in real time, and accordingly appropriate adjustments are done as needed. Then, Smart Highway infrastructure could also be developed in a way to implement intelligent lighting systems which enables it to adjust the level of brightness automatically according to the time of the day and intensity of the sunlight available. Of course, Smart Highways could also support the novel renewable energy production as well as electric vehicle charging stations. If the system of Smart Highways looked into carefully, then usually it comprises of 3 constituents, which are a data collector, in addition to an info provider, as well as a controller. The info provider is the one accountable for not only managing but also creating the necessary data feeds which are further utilized not only by data collector but also by controller. The traffic data is gathered by the data collector with the assistance of sensors which are present throughout the highway network, whereas the controller utilizes this data to produce prediction in regard to traffic flow pattern as well as traffic congestion.

Figure 1. Smart highway signs at Washington (Wikimedia)

Smart Highway Technologies

Simply put, Smart Highway Technologies are a number of ways through which highways could not only be monitored but also be managed so that safety can be enhanced, energy be conserved, as well as traffic flow could be optimized. It is said that these technologies are utilized globally in numerous forms as well as they are presently expanding quickly into novel applications. It is essential to understand Smart Highway technologies while studying about Smart Highways, which are as follows:

- *Traffic Monitoring:* This system requires the installation of static sensors or cameras along the highway in order to gather data in regard to traffic conditions including direction, speed, and so on. This gathered info is further analysed in order to assist planners to allow adjustments to the flow of the traffic.
- *Real-time Monitoring:* Sensors are installed on the highways, which has the ability to continuously monitor traffic conditions as well as allow adjustments as required. This permits officials to watch the pattern of the traffic in real time, in order to make important alterations as required.
- *Active Transportation:* Design of highways having the characteristics which motivate humans to utilize carpools, bicycles, or other kinds of active transport services, to enhance air quality as well as decrease congestion.
- *Future Road Technology:* It includes various thrilling technologies which are still under development but have the potential to cause a significant impact in the coming years on the management of highways. For instance, lane guidance systems can be found on private roads to aid drivers stay on their respective lanes, etc.
- *Remote Monitoring:* It is possible to monitor highways from a distance, which permit the concerned officials to make alterations without the requirement to visit that site to perform his or her task. This might aid to enhance safety as well as decrease traffic congestion.
- *Autonomously Driving Vehicles:* Automobiles are designed with the goal to be self-driven without the intervention of people, in order to not only enhance traffic flow but also to free up resources.
- *3D Printing:* It is possible to directly print parts of highways and roads with the assistance of digital models, making them flexible to altering conditions as well as customizable. Similar to how 3D Printing could speed up the construction process and reduce manpower as mentioned in the study of Singh (2022).
- *Traffic Signal Management:* This is a smart system which not only allows the monitoring of the flow of traffic at intersections but also to adjust the

timings of the signal consequently, in order to optimize the flow of traffic to decrease congestion as well as aid drivers to evade delays.
- **Road Sign Recognition:** It is a smart system which could help in spotting particular signs along a given highway as well as to send alerts to the concerned drivers when it is required to alter lanes, or an intersection is about come. This aids the drivers in preventing road accidents.
- **Automobile Tracking:** It could track the direction, speed, as well as location of the automobiles while they are driven on the highway. This could aid drivers in avoiding not only congestion but also collisions.
- **Pedestrian Safety:** It could spot pedestrians who are crossing a given highway or road, as well as allows the facility to send alerts to the concerned drivers to make them stop the vehicle safely. It could aid in reducing or stopping fatalities in addition to injuries of the pedestrians.
- **Laser Scanners:** They utilize a light beam in order to map out nearby objects. They are frequently utilized in the planning of highway routes, because of their high accuracy in finding objects as well as in measuring distances. Some of them could even be utilized to detect traffic signs as well as other hindrances on the highway.
- **High-tech Cameras:** They are frequently utilized to shoot photos of highways as well as roads which could be utilized for mapping in addition to planning. For instance, they could be utilized to shoot photos of closed lanes, intersections, as well as other issues on the highway.
- **Traffic Congestion Sensors:** They are devices which utilizes special sensors to find out the time at which there is considerable traffic on the given highway, where the info is later utilized by the highway.
- **Internet of Things Connectivity:** It is possible to connect highways to Internet of Things devices with the assistance of cities, as well as collect weather in addition to traffic data. It enhances energy efficiency, traffic management, as well as safety.
- **Traffic Management Networks:** It utilizes speed cameras to give cautionary signs for dangerous conditions as well as allows sending automated traffic diversion signals which controls traffic, in order to decrease congestion in addition to enhancing safety.
- **Traffic Lights Optimization:** This system utilizes data from smart vehicles or CCTV cameras in order to optimize the traffic lights as well as update travellers on blockage or traffic jams.
- **Solar Powered Roadways:** To pave roads and highways, tempered glass is used to make hexagonal panels where photovoltaic cells are implanted in it. The panels consist of microprocessors, inductive charging capability to

help in the charging of EVs when they are passing by snow melting heating devices, as well as Light Emitting Diodes.
- *Glow in the Dark Roads:* A photo luminescent powder is utilized on glowing markers which is used to paint the current highway surfaces, in order to not only absorb but also store the incoming daylight. However, it is not developed enough to be used practically on a commercial scale.
- *Interactive Lights:* when automobiles approaches, then with the assistance of motion sensors the lights on the highways or roads are activated in order to illuminate a specific section of the highway or road. These lights start becoming dimmer as the automobiles crosses them.
- *Electric Priority Lane for Charging EVs:* To charge Electric Vehicles when passing by on the highway, cables can be implanted on the surface of the highways, which can produce magnetic fields to complete the charging process. The technology of inductive charging is already present for static automobiles, nonetheless in the future wireless technology could be used in the charging of batteries when EVs are moving.
- *Weather Detection:* To detect the conditions of the weather, networks of Artificial Intelligence incorporated sensors are used which affects the safety on the highways. At present, Road Weather Information Systems are utilised, but they have limitations, so in future more development is possible to improve the weather detection facility which could reduce the number of mishaps on the highways due to weather.
- *The Roads That Honk Systems:* In a mission to enhance safety on highways, Leo Burnett India in addition to HP Lubricants joined hands in order to develop this system. National Highway 1 of India was decided to be the place to be their 1st prototype was installed, where it connects the two cities of Jammu and Kashmir union territory of India, which were Srinagar as well as Jammu, as the highway is also known as the extremely dangerous ones globally.

At present, the urban transport authorities in addition to the governments are starting to understand the need of Smart Highways in the world, in regard to provide an important platform for novelty in transportation, and the same can be said for India. Smart Highways have the potential to give governments with unparalleled control as well as visibility over the real time traffic of the automobiles, empowering drivers, as well as powering up the smart vehicles. In fact, a number of metro cities are at present taking responsibility for the Smart Highway project by working on new ways in order to connect Smart Highways with the Automobiles to support automated traffic management in various parts of the world. For instance, in European Continent, it has been a hub of numerous innovations in regard to technology and it was quick to implement smart highway technologies, where a number of projects could be seen in support of it. For example, Ursa Major project was one such projects, with the target to accomplish improve traveller experiences as well as high grade traffic management. To support the project, stakeholders from Netherlands, Germany as well as Italy worked on it, in order to enterprise cross border transport facility with the assistance Intelligent Transport System. Similarly, America is also investing enormous amount of money to implement smart highway technology. The country allocated more than one hundred billion USD in order to repair rundown highways as well as roads to implement intelligent road systems, then sixty-five billion USD for the provision of broadband connectivity to enhance internet access to construct a framework for ITS, in addition to seven and half billion USD for charging stations of EVs. Then in Asian continent, China as well as Japan are making great progress, but India, Vietnam, etc, are also working on it.

Smart Highway Infrastructure Elements

There are some important elements of Smart highway Infrastructure, which are as follows:

- *Sensors:* In order to gather audio as well as visual data on highways, conditions of the weather, as well as the flow of the traffic, cameras in addition to sensors are utilized in large quantities.
- *ITS:* Also known as Intelligent Transport System, is a system which takes the assistance of advanced technologies to utilize real time data in order to help in the guidance of automobiles through a given traffic on the highway, by enhancing safety as well as decreasing traffic congestion.

Figure 2. Intelligent transport system screen at Thailand Highway No: 9 (Wikimedia)

- *Energy Harvesting System:* The paramount quality of Smart Highway is the ability to implement energy harvesting technologies, where mainly sensors as well as cameras decrease the use of external surveillance of the highways. Then, the piezoelectric crystals are implanted beneath the asphalt layer, on the highways, so that the force generated from the wheels of the automobiles could produce electricity, which could be used later on or to power streetlights.
- *Data Analytics as Well As Management Systems:* Since, historical traffic data is utilized to construct Smart Highways, it is required to have a strong facility to analyse data as well as how to manage it properly.

Indian Scenario

India is a major developing country in the world, and it is investing heavily in the recent decade on the development of transport infrastructure as well as incorporating advanced technologies to improve transportation facilities in the country. For instance, EFKON India Pvt. Ltd which is a major company to provide facilities for Smart Highways, like automation of highways in the country, FASTag, ITMS which includes In-Bus Surveillance, tracking of automobiles; ATMS which includes Intelligent Lighting System; IRCS which includes Parking Management and Toll Management; says that it implemented more than one thousand five hundred kilometres of Intelligent Transport Management System throughout the country. According to its claim, it also successfully commissioned more than one thousand infrastructure toll lanes across India as well as more than ten million vehicles passes everyday through their Intelligent Transport Systems in the country. Also, The Eastern Peripheral Expressway is considered as the India's 1st Smart as well as Green Highway, and the inaugural of Smart Roads in Ujjain in Madhya Pradesh state of India in the year 2021 was also a major milestone achieved by the country. Further, India has already adopted Video Incident Detection System, which is a well-organized traffic management system that comprises of real-time video analysis in order to monitor incidents related to traffic and to save lives. It also gives dependable and precise tracking of vehicles on expressways in addition to highways. Further, this system aids in not only making highways safe but also e-Challans generation for the people who violate traffic rules on highways as well as roads. At present, it can detect vehicles which are coming from the opposite direction, gathering of crowds, vehicles which stopped or stalled, deceleration or acceleration of vehicles, triple riding as well as absence of helmets by people driving two wheelers, crossing of pedestrians; status of traffic such as delay, normal, dense, stop as well as go, and so on; status of weather such as fog or smoke, normal conditions; under speeding as well as over speeding of vehicles on the highways, to make sure the flow of traffic remains smooth in addition to the safety of people on the highways. If immediate actions are needed on the highways, then it can also alert the responsible authorities like ambulance, police, road clearing services, and so on. The current limit to cover the maximum number of lanes can be said as 8, and its accuracy to detect violation is about 95%. Currently, it is present on Pune Solapur Expressway, Yamuna Expressway, Tumakuru Smart City Limited, Greater Noida Expressway, Hyderabad ORR, Varanasi Smart City Limited, and Aligarh Smart City Limited, in India. Then, Adaptive Traffic Control Systems are also incorporated in several parts of the country, which is able to detect real time traffic conditions and provides a solution which can automatically adapt the traffic light's timings in order to help in optimizing the traffic flow on the highways and roads. This allows the complete synchronisation of the traffic lights, fixed timing modes, as well as local optimization of vehicles, in order to make the best use of vehicular throughput in addition to decrease of traffic congestion on the highways. So, it is able to not only enhance the effectiveness of the timings of the traffic signals, but also able to decrease the number of stops on a highway lane or a corridor which consequently saves cost on fuels. Currently, it is being utilized at Varanasi Smart City Limited, Tumakuru Smart City Limited as well as Aligarh Smart City Limited. Also, Automatic Traffic Counter and Classifier has been adopted in some parts of the country, which gathers precise, reliable as well as real time info of the flow of vehicles in order to immediately manage the traffic on the highways and roads. It monitors the given real time flow of traffic on a particular section of the highway, and the real time data of traffic which is collected by it comprises of classification of vehicles, counting the passing by vehicles over that lane or section of the highway, average speed of the traffic, occupancy, time headway, volume of traffic, as well as travel direction. The current limit of the classification of vehicles by it is 5 classes, which includes Light Motor Vehicle, Two Wheelers or Motorcycles, Light Commercial Vehicle, Buses or Trucks, as well as others (Machine Equipment Vehicles, Offshore Support Vehicles, Micro Air Vehicle/Micro Aerial Vehicles, and so on). It can cover maximum four lanes of traffic on a highway, as well as able to detect bidirectional vehicles coming from right or left side, also works properly to detect parallel vehicles. Its classification system is on the basis of neural network, and it is applied in infrastructure planning, traffic signal design, free flow tolling or in toll enforcement, in addition to detection of any traffic rules violation by any vehicle on the highway. Its accuracy rate is 95%, with low rate of false classification, and it is at present being used on Greater Noida Expressway, Varanasi Smart City Limited, Yamuna Expressway, as well as Hyderabad ORR. Similarly, Automatic Number Plate Recognition was implemented on Greater Noida Expressway, Tumakuru Smart City Limited, National Highway Authority of India, Yamuna Expressway, Varanasi Smart City Limited, and Aligarh Smart City Limited; Red Light Violation Detection was installed at Varanasi Smart City Limited, Tumakuru Smart City Limited, Aligarh Smart City Limited, and Kerala MVD; in addition

The Current Status and Scope of Smart Highways in India

to Speed Violation Detection was installed on Greater Noida Expressway, Kerala MVD, Yamuna Expressway, Varanasi Smart City Limited, as well as Aligarh Smart City Limited. Then, some projects of Smart Highway are mentioned in the chapter, which further shows India's progress on the path of incorporating Smart Highways in the country, to improve and advance the country's transportation facility. It is expected that with the assistance of Smart Highways, India will be able improve the safety of pedestrians, decrease traffic congestion, enhance E-Tolling, Parking facility, and connectivity up to certain extent.

Figure 3. Electronic toll system at Yamuna Expressway (Wikimedia)

Figure 4. Yamuna Expressway route map (Wikimedia)

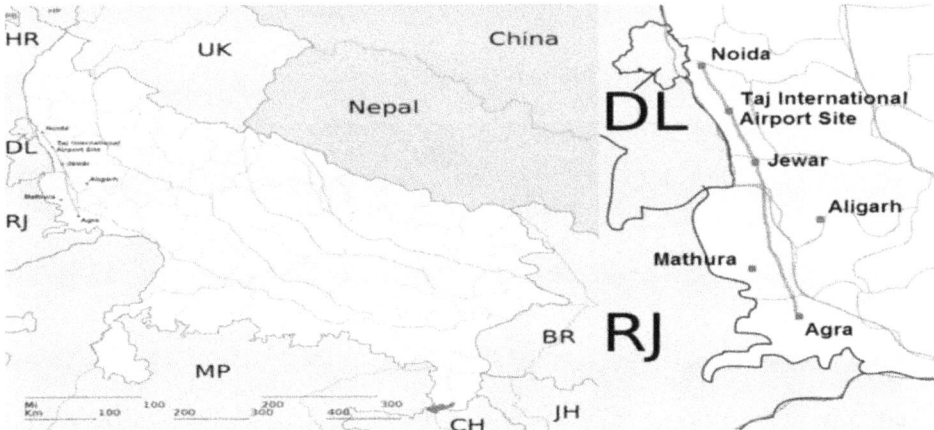

Figure 5. Greater Noida Expressway (Wikimedia)

Figure 6. Noida Expressway (Wikimedia)

Figure 7. Aligarh Smart City limited (Google Earth)

Figure 8. Varanasi Smart City road map (Google Earth)

Figure 9. Tumakuru Smart City Limited (Google Earth)

Figure 10. Pune to Solapur road map (Google Earth)

Figure 11. Kerala MVD (Google Earth)

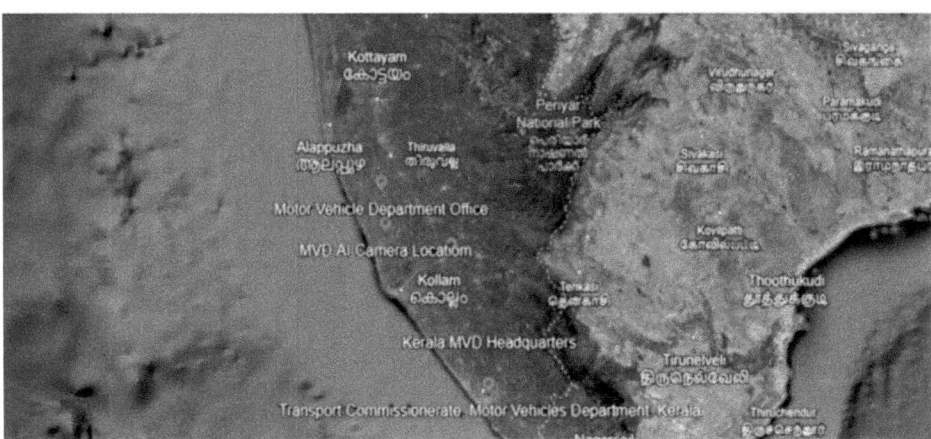

Figure 12. Hyderabad ORR (Google Earth)

Objectives

There are numerous objectives of this chapter, which are:

- To understand Smart Highways in the general sense and briefly understand various Smart Highway Technologies.
- To understand the elements of the Smart Highway Infrastructure and the Smart Highway scenario as well as its need in India.
- To understand the gaps in the research performed on and related to Smart Highways.

- To perform graphical analyses in order to understand the scope of Smart Highways.

MOTIVATION

The author performed the study on Smart Highways, as in the recent decades the Indian Government had been focusing on the transport infrastructure of the country in order to improve the transportation facilities. Particularly, in the last decade a number of transportation engineering projects were going on in India. Then, in the last few years, the Government of India also worked on incorporating advanced technologies like Intelligent Transport Systems, Intelligent Lighting System, E-Tolling, Intelligent Traffic Management Systems, and so on, along with some successful implementation of Smart Highway projects in different parts of the country. Then, Artificial Intelligence is also getting more popular in the country and the 2015 one hundred Smart Cities Mission of India, made the author think how Smart Highways could play a major role in the development of the country. Further, India suffers every year from massive traffic accidents including injuries and fatalities, as well heavy traffic congestions in most of its cities especially the megacities in the country.

What Is the Need of Smart Highways and How Could It Help in the Development of India?

In any form of development, transportation facilities play a vital role, and its importance can't be ignored in the economic growth as well as advancement of a country. According to the report of World Bank as well as Ministry of Road Transport and Highways, India is having more than 480,000 road accidents in a year, making it one of the top countries in the world for highway and road accidents. Almost one third of the people have lost their lives in such mishaps every year. Then, India this year i.e. 2023, became the most populated country in the world, and the number of vehicles on the roads are also increasing in the country day by day. This increases the possibility of accidents on highways and roads even more in the future, which traditional highways and roads won't be able to handle. The megacities and major cities as well as towns in India are already facing problems of heavy congestion, which might increase more in the future (Singh, 2019). So, it's a high time to develop the transportation sector with advanced technologies to overcome such existing traffic and transport problems which are expected to increase further in the future. That's why upgradation of transportation facilities is important from time to time, in order to match with the current times and public need, so to avoid becoming an obstacle on the path of a country's development. However, developing the transportation facilities should be sustainable and have a positive impact on the environment in the long run. When, advanced technologies like Artificial Intelligence, ANN, etc., are being the trend in a number of industries, there is also a need to transform the transportation sector with their assistance. Smart Highways are one of the major ways which can improve the transportation sector significantly and solve a number of existing traffic problems. Singh (2023) mentioned a number of major initiatives and projects by the Indian Government including Smart City Mission, Public Transport, Electric Vehicles, and Green Highway, to promote sustainable planning and development of India. Of course, Smart Highways are compatible with such projects and by decreasing the traffic congestion and reducing the travel time, it will be able to reduce air pollution and carbon footprint considerably. Singh (2023) mentioned a number of major infrastructure projects which are either recently completed or going on in India, including transport infrastructure projects, here it can be said that Smart Highways can be a major game changer in providing advanced transport infrastructure facilities to such large infrastructure projects as well as converting the transport infrastructure projects into Smart Highways. The way AI is important in Smart Construction and promotes sustainable practices (Singh, 2023), similarly it is also important in Smart Highways through ITS. Intelligent Transport System is a major part of Smart Highways, which had been implemented in various parts of the world and also in some parts of India (Singh, 2019). Then, GIS along with ITS can be incorporated to transform the traditional highways further and improve safety, decrease travel time and congestion (Singh, 2018) which will be a step towards Smart Highways. Further, it can be seen how Fuzzy Logic is useful in advancing the traffic system to a new level and especially in the case of high road accidents, too many automobiles, megacities as major cities (Singh, 2022), which will be a part of ITS and subpart of Smart Highways. In the past seven years, Indian Government focused on Digitalization and Digital Transformation, making India enter the Digital Era. Digitalization's benefits and usefulness can be observed in the construction sector (Singh, 2022), but in the similar manner Smart Highways have digital technologies throughout their system to advance transport practices as well as overcome the existing traffic problems along with the reduction of fatalities on the highways. Smart Highways also play a major role in the Digital Era, and it will contribute significantly in India's economic, sustainable and digital development in the future.

LITERATURE REVIEW

Takefuji (2021) studied smart highway and worked on finding out its management policy being an illogical one in the country Japan. Pompigna and Mauro (2022) studied smart roads and they worked on understanding the highway novelties in the modern era which is filled with advanced technologies. Singh et al. (2021) studied Highway 4.0 and worked on highways in order to digitalize them to prepare them for the development of the susceptible road safety with the assistance of ML as well as Internet of Things sensors. Ashwin et al. (2022) studied cattle movement prevention method on the basis of Internet of Things to stop or evade accidents on roads and they worked on its environs aspects particularly for the highways connecting Bhopal and Indore cities present in Madhya Pradesh state in India. Alkhatib et al. (2022) considered the road networks of the cities which are congested ones and they worked on smart traffic scheduling which was intended for such road networks. Angioi et al. (2023) studied road safety in addition to smart on road technologies and they worked on a summarised current or theoretical Smart on Road Technologies systems in order to detect their aims in regard to the way the driving behaviour as well as road safety affects. Yao et al. (2023) studied developed industrial informatics and worked in the direction of sustainable, safe as well as smart roads. They presented an AI algorithm and briefly discussed the various smart sensors which were non embedded ones in order to utilize them to monitor roads. Haider et al. (2023) studied intelligent automobiles and they worked on smart transportation system which was on the basis of CNN intended for them, in order to understand the prevention of cracks on roads. They proposed a post processing algorithm to give choice to a given driver in order to choose the most harmless road among the available options to reach his/her destination. Bianchini et al. (2023) studied the maintenance of smart roads, and they worked on a big data examination method to take advantage of the in-vehicle data which was intended for them. Pascale et al. (2023) studied the assessment of road traffic noise dynamics, and they worked on the process of smart mobility which was intended for them with the assistance of video analysis. Wang et al. (2022) worked on measuring the traffic volume on the basis of a particular smart road stud through their study. Paul et al. (2021) studied smart roads and they worked on wide-ranging survey particularly on vehicular networks aimed at them, in order to concentrate on Internet Protocol based methods. Edward (2023) studied road accidents and they worked on a smart crisis management system aimed at them, on the basis of their proposed altered CNN Particle Swarm Optimization hybrid algorithm through their study. Younes et al. (2023) studied green road intersections and they worked on an algorithm which was a smart effective traffic light scheduling one, intended for them. Wang et al. (2020) worked on smart road stud on the basis of the surveillance of 2 lane traffic through their study. They stated that their developed 2 lane traffic surveillance function has the ability to transmitting as well as gathering traffic data of high resolution along with the ability to guide traffic. Luo and Yang (2021) studied restoration of connectivity through community structure, and they worked in the direction of smart as well as resilient urban road networks by proposing a mixed integer programming model on the basis of community hierarchy to give fast disaster response in the case of urban road networks which are broken or damaged. Zhang et al. (2020) studied thermochromic asphalt binder and they worked on a smart as well as novel road construction material through their research. Salmi and Hokars (2023) studied 3 smart zones present in the country Sweden and they showed them on the basis of sensor data as well as geofencing through their study, in order to accomplish their study in the direction of smart urban traffic zones. Qin et al. (2022) studied smart transportation and they worked to find its demand through their study. They considered the smart transportation era, and they talked about demand management's importance, difficulties as well as prospects in it. Thakur and Han (2023) studied monitoring of smart traffic as well as safety, and they worked on triboelectric Nano-generator which was intended for them. Zhu (2022) studied urban planning as well as smart transportation, and they worked on a model which was a kind of spatial decision making one, on the basis of the IoT in addition to coupling principle. They considered the smart transportation process, and they analysed the coupling in it, in order to enhance the way, the planning of smart transportation could affect the process of decision making in regard to the urban planning space, to accomplish their study. Reddy et al. (2023) studied smart transport system and they worked on a model which consisted of a fog layered intelligent service management, aimed at it, so they could progress in the direction of an innovative green service computing method intended for smart cities. Zhao et al. (2023) studied smart transportation technology and they worked on it to understand the way it encourages green total factor productivity. S and Almutairi (2023) studied smart transport systems and they worked on non-divergent traffic management scheme aimed at them, with the assistance of classification learning. They presented the above-mentioned scheme in order to enhance roadside driving not only for automobiles but also for users through their study. Banafshehvaragh and Rahmani (2023) studied smart vehicles and they worked on detecting anomaly, invasion as well as attack in them. They inspected the various methods of detection used for inter-vehicle networks, Internet of Drones, in-vehicle networks, as well as ground vehicle power stations. Biyik et al. (2021) studied smart mobility and they worked on the way to incorporate it through their review paper. They found that just by constructing new roads would not solve the problems faced in urban transportation globally, which includes ineffective utilization of resources, air pollution, which then obstructs the economic development. Shah et al. (2022) studied smart cars which were based on Internet of Things, and they performed a survey on them in order to understand they're not only challenges but also functionalities. Fernandez-Isabel et al. (2020) studied smart roads and they worked on the way to model multi-agent systems in order to simulate sensor based on them.

RESEARCH GAPS

Numerous published scientific and research papers along with chapters, only the ones published in the recent years were selected. The author found that there was a lack of research of how people are perceiving the existence of Smart Highways and how their day-to-day lifestyle changed, some cases were considered from India, but not enough. There was also lack of research on how Smart Highways could support India's lofty infrastructure development plans especially in the long run.

MAIN FOCUS OF THE CHAPTER

The chapter focused on Smart Highways in the general sense and its numerous technologies which are used in India and around the world. It also highlighted the major elements of such infrastructure and where India stands at present. Also, why Smart Highways are needed in India and the future of Smart Highways in the country. It also covered major issues and problems associated with it and put a limelight on the market of Smart Highways through graphical analyses.

ISSUES, CONTROVERSIES, PROBLEMS

Smart Highways are a major solution to the existing traffic and transport problems not only at present but also in the future. However, it also comes with associated problems as well as issues which hinders its growth, and they can't be ignored. The cost of Smart Highway projects is extremely high, as all the sophisticated components and technologies are pricey. This makes it difficult to convert the traditional highways into Smart Highways on a large scale particularly in the most populated and major developing country India, considering the fact that the country is having 2nd largest road network in the world and 46th ranked in terms of overall quality. Further, technical limitations of each technology, interoperability or amalgamation of incompatible technologies, in addition to data privacy as well as security attracting cyber-attacks and breaches, are the other major problems and issues with Smart Highways. Of course, many people are unfamiliar with such advanced technologies and often people break the traffic rules whenever they see they won't be caught.

SOME MAJOR SMART HIGHWAY PROJECTS IN INDIA

In this section, some major Smart Highway projects are briefly explained which are as follows:

- *Eastern Peripheral Expressway:* This is considered as India's first Smart and Green Highway. It is also known as National Expressway 2 or Kundli-Ghaziabad-Palwal Expressway with a length of 135 km and connects Delhi NCR area (parts of Haryana and Uttar Pradesh), with a width of 6 lanes.
- *Delhi-Meerut Expressway:* Also known as National Expressway 3, with a length of 96 km and connects Delhi with the Delhi NCR areas (particularly the city of Meerut in Uttar Pradesh). Its width is up to 14 lanes, which makes it the widest expressway in the country.
- *NH1:* Also known as the National Highway 1 after the new numbering by the Indian Government. Its total length is 534 km, but its part which connects Jammu and Srinagar cities in the Union Territory of Jammu and Kashmir in India, as mentioned earlier in the chapter, that HP Lubricants as well as Burnett India joined hands to make The Roads That Honk Systems, where SmartLife poles were installed at hairpin bends as well as at sharp curves, making it world's 1st successful execution of anti-collision vehicle management system.
- *Smart Roads in Ujjain:* Two Smart Roads in the first quarter of the year 2021, were inaugurated by TATA Projects Limited in Ujjain present in Madhya Pradesh state in India. Here, for the coming fifty years all utility connections were taken into considerations, in order to avoid digging them for installation of utilities like optical fibre as well as underground electrical lines, sewers, storm water drains, and water supply. They are Road Numbers 2 and 4, with a respective length of 465m and 412m.
- *Dehradun's Smart Road Project:* This Smart Road Project is present in Dehradun in the state of Uttarakhand in India, which is a 1.8 km stretch of Chakrata Road, and part of Dehradun Smart City Project. It extends from Clock Tower of Dehradun to the place named as Kishan Nagar Chowk, which comprises the work of integration of drainage as well as sewage systems, laying of multiple utility ducts underground, in addition to revamping of the road.
- *Davanagere's Smart Roads:* The city Davanagere is present in the Karnataka state and its city corporation (Davanagere City Corporation) planned to enhance ten city roads into Smart Roads, where underground water supply lines, underground utility corridor, smart concrete road with street scraping, LED street lighting, and so on, were considered as the project's constituents.

It is part of Davanagere Smart City project, and it is said by the first quarter of this year i.e. 2023, 6 major Smart Roads were completed.

- **Tumakuru's Smart Road Project:** The project location is Tumakuru city close to Bengaluru city in the Karnataka state in India. The total length covered by it is 14.53 Km with Right of Way of more than 12 metres as well as covered eighteen prime roads in the ABD area. Utility corridor, storm water drain, intelligent signalling, and so on will be part of it and they will support the Tumakuru Smart City project.

Figure 13. Eastern Peripheral Expressway (Wikimedia)

Figure 14. Delhi-Meerut Expressway (Wikimedia)

Figure 15. Ujjain Smart City Limited (Google Earth)

Figure 16. Dehradun map (Google Earth)

Figure 17. Davanagere Smart City Limited (Google Earth)

Figure 18. Tumakuru Smart City Limited (Google Earth)

METHODOLOGY

In this section, some major definite issues which were observed during the course of Smart Highway projects which were mentioned in the chapter earlier, are discussed:

- The target for construction of Eastern Peripheral Expressway was of 400 days which was almost impossible to achieve, as it was unable to include technical problems as well as land acquisition issues faced during the project.
- Safety of commuters, Right of Way being restricted, environmental challenges as well as some the construction of unsatisfactory roadside facilities provided by some concerned contractors to save cost, were the major hurdles in the Delhi-Meerut Expressway project.
- The technological challenges, lack of practical prior experience and terrain difficulties were the major obstacles in the construction of The Road That Honk System on NH1 to connect Srinagar and Jammu cities.
- Mostly, technical challenges were the major issues while constructing the Smart Roads in Ujjain.
- Inconvenience to the residents and daily commuters, some poor implementations, technical challenges in addition to delays were the major problems in the construction of Smart Road Project in Dehradun.
- Planning and implementation of the Smart Roads while considering the existing environmental issues, traffic congestion along with land acquisition and removal of encroachers were some major issues faced during the project.

- Lack of proper planning as well as coordination in the implementation of the project, along with frequent complaints or dissatisfaction of the locals were some of the major problems faced in the Tumakuru's Smart Road project.

RESULTS AND DISCUSSION

In this section the data was collected from Markets and Markets, Data Bridge, and many other sources to perform graphical analysis with the aim to support the study.

Table 1. Comparison of smart city projects in China and India

Smart City Projects in India and China	Number
Smart City Projects in India	100
Smart City Pilot Projects in China	900

Source: MagicBricks, Statista

Table 2. Segment wise market share of smart city projects in China

Segment Wise Market Share of Smart City Projects in China	In (%)
Smart Mobility	4.7
Smart Logistics	28.9
Others	66.4

Source: Statista

Table 3. Share of transportation sector in GHG in USA

Share of Transportation Sector in Emission of Greenhouse Gases in USA	In (%)
Transportation	28
Others	72

Source: USEPA, Mordor Intelligence

Table 4. Growth of automobiles utilizing ETCS on Japan's expressways

Share of Automobiles Utilizing Electronic Toll Collection System on Japan's Expressways	In (%)
Automobiles Utilizing Electronic Toll Collection Systems in Year 2017	90.3
Automobiles Utilizing Electronic Toll Collection Systems in Year 2022	93.8

Source: MLIT (Japan), Mordor Intelligence

Table 5. Growth of automobiles utilizing ETCS 2.0 on Japan's expressways

Share of Automobiles Utilizing Electronic Toll Collection Systems 2.0 on Japan's Expressways	In (%)
Automobiles Utilizing Electronic Toll Collection Systems 2.0 in Year 2017	12.4
Automobiles Utilizing Electronic Toll Collection Systems 2.0 in Year 2022	28.1

Source: MLIT (Japan), Mordor Intelligence

Table 6. Growth of automobiles

Rise in the Number of Automobiles	In Numbers
Vehicles Sold in Year 2010 by IOMVM	75,000,000
Vehicles Sold in Year 2018 by IOMVM	95,000,000

Source: International Organization of Motor Vehicle Manufacturer, Data Bridge

Table 7. Road accidents per year

Road Accidents Per Year	In Numbers
Minimum Non-Fatal Accidents	20,000,000
Maximum Non-Fatal Accidents	50,000,000
Road Fatalities	1,350,000

Source: WHO

Table 8. Component wise market share of smart highway technologies

Component Wise Market Share of Smart Highway Technologies	In (%)
Hardware	50
Others	50

Source: IMARC Group

Table 9. Expected market growth of the leading continents or regions

Expected Market Growth of Continents or Regions	In (%)
North American Continent (2023-2031)	14.7
Asia Pacific Region (2023-2031)	21.3

Source: Straits Research

Table 10. Revenue share in smart highway market

Revenue Share in Smart Highway Market	In (%)
North American Continent (2022)	45.8
Others (2022)	54.2

Source: Market Research Future

Table 11: Growth of Indian smart highway market

Indian Smart Highway Market Growth	In USD
Market Value in 2023	2,660,000,000
Expected Market Value in 2029	7,880,000,000

Source: Global Incorporation, Inc.

Table 12. Growth of national highways in India

National Highway Growth in India	In (1000 Km)
Length of NHs in Year 2013-2014	91.287
Length of NHs in Year 2022-2023	145.24

Source: IBEF

Table 13. Growth of smart highway market in the world

Smart Highway Market Growth	In USD
Value in 2014 by MM	12,564,400,000
Value in 2018 by AMR	23,670,000,000
Expected Value in 2019 by MM	27,992,000,000
Value in 2022 by DB	38,910,030,000
Value in 2022 by GMR	45,140,000,000
Value in 2023 by MI	46,560,000,000
Expected Value in 2026 by AMR	92,380,000,000
Expected Value in 2028 by MI	115,130,000,000
Expected Value in 2029 by DB	145,268,700,000
Expected Value in 2031 by GMR	191,240,000,000

Source: Markets and Markets, Allied Market Research, Data Bridge, Growth Market Reports and Mordor Intelligence

Figure 19. Comparison of smart city projects in China and India (MagicBricks, Statista)

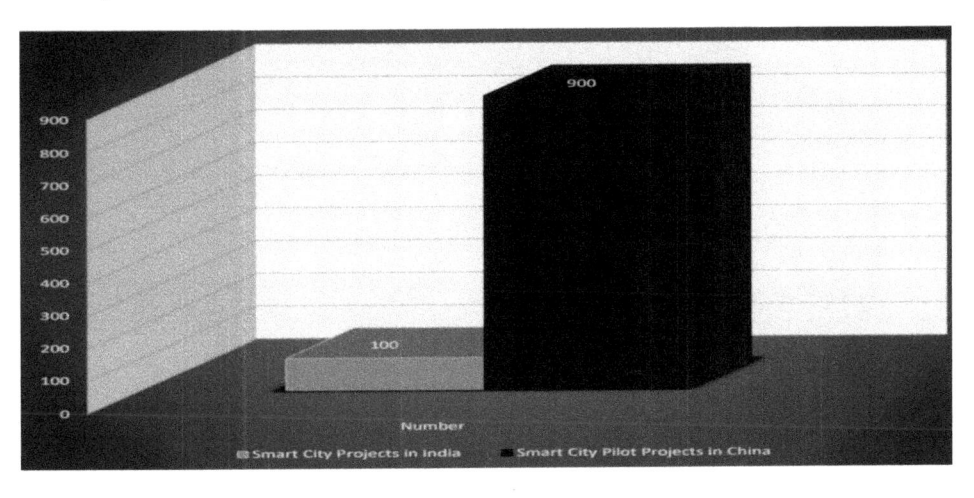

Figure 20. Segment wise market share of smart city projects in China (Statista)

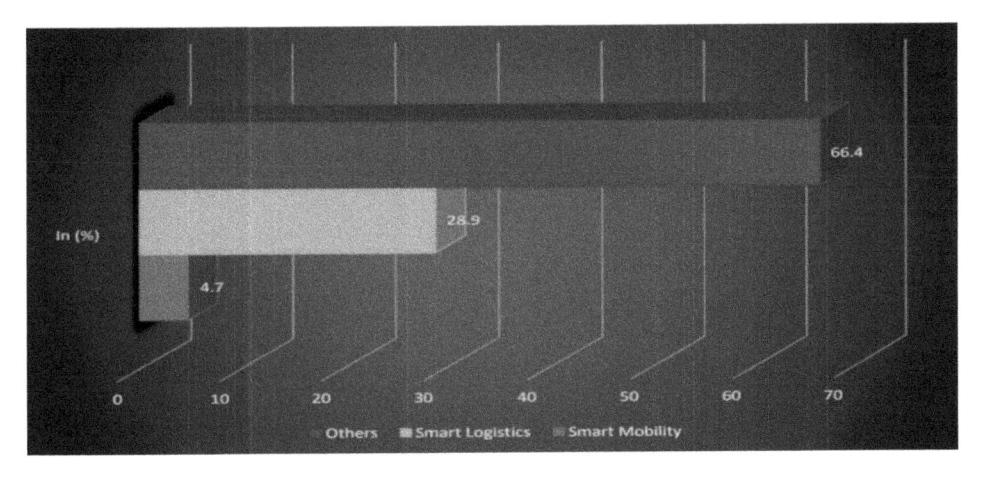

Figure 21. Share of transportation sector in GHG in USA (USEPA, Mordor Intelligence)

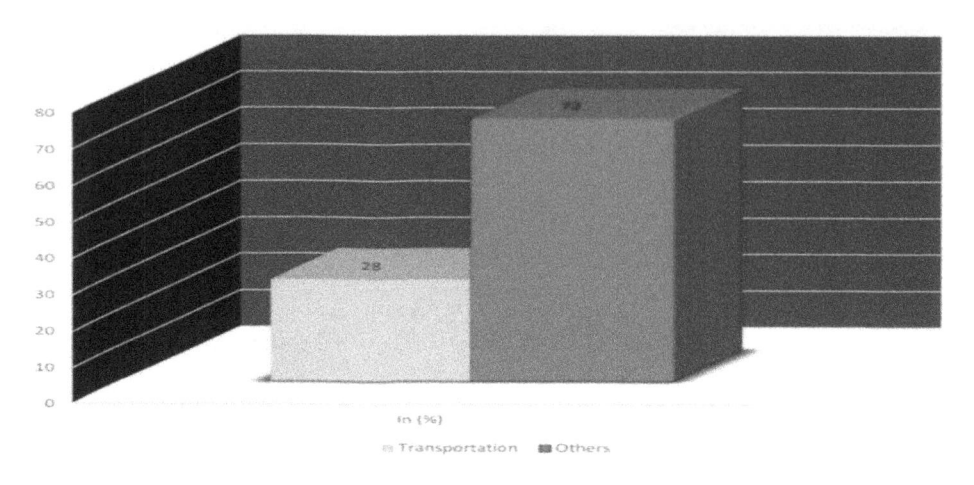

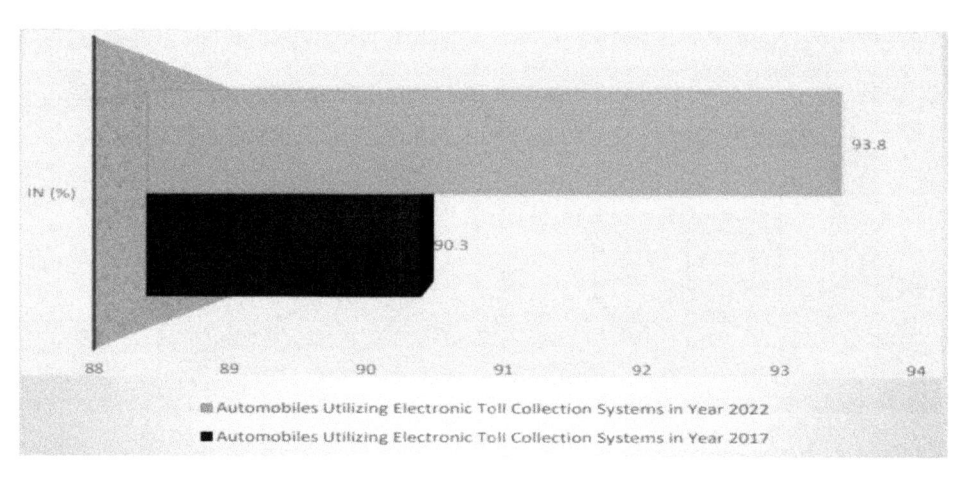

Figure 22. Growth of automobiles utilizing ETCS on Japan's expressways (MLIT (Japan), Mordor Intelligence)

Figure 23. Growth of automobiles utilizing ETCS 2.0 on Japan's expressways (MLIT (Japan), Mordor Intelligence)

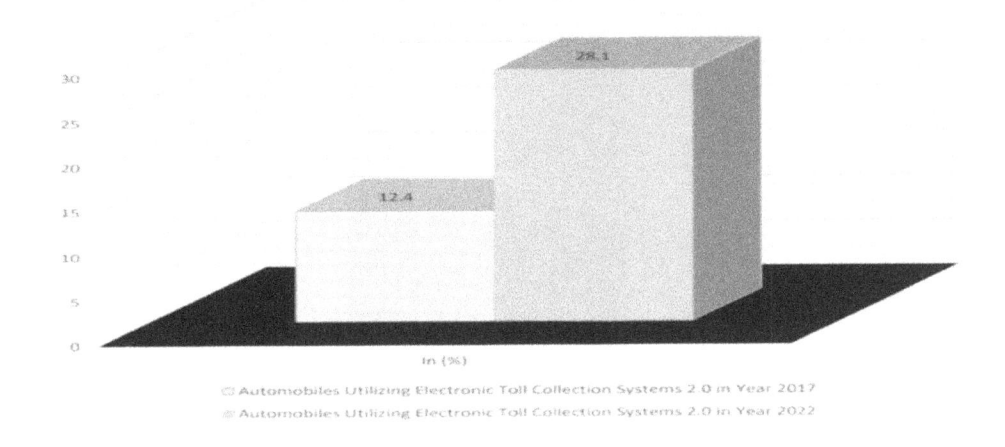

Figure 24. Growth of automobiles (International Organization of Motor Vehicle Manufacturer, Data Bridge)

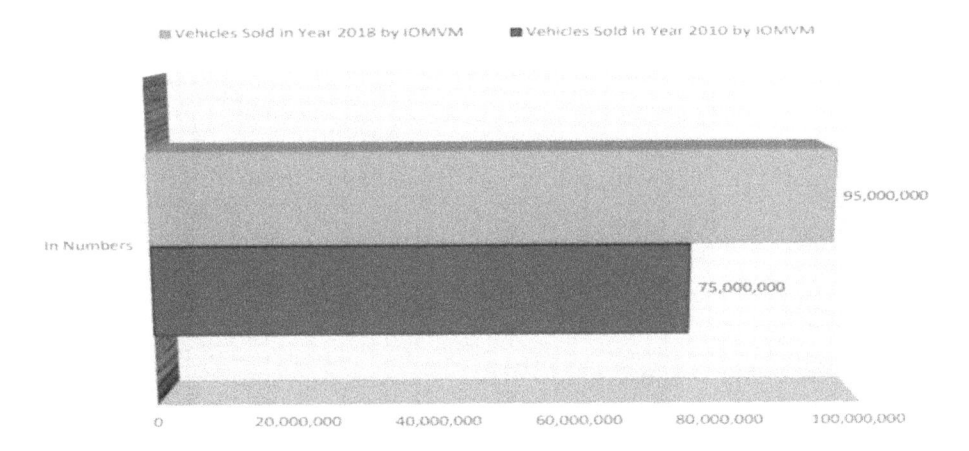

Figure 25. Road accidents per year (WHO)

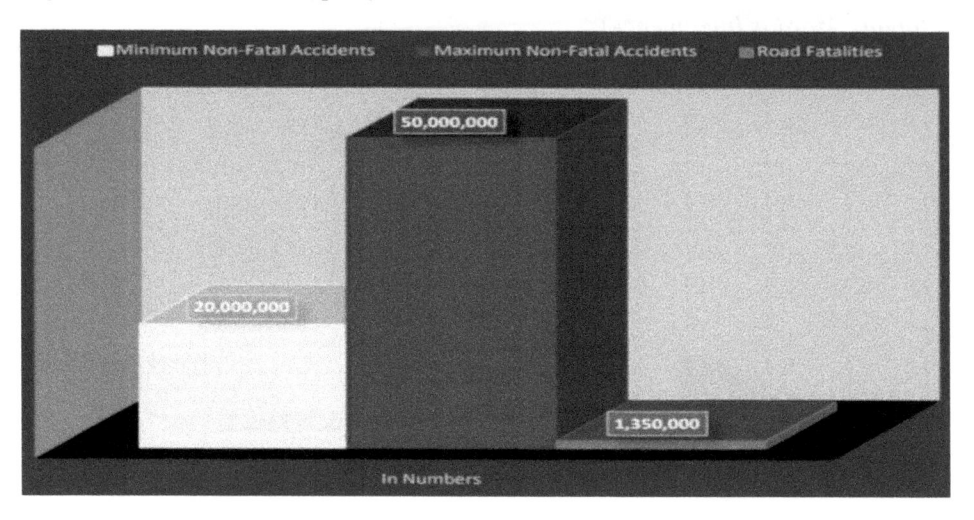

Figure 26. Component wise market share of smart highway technologies (IMARC Group)

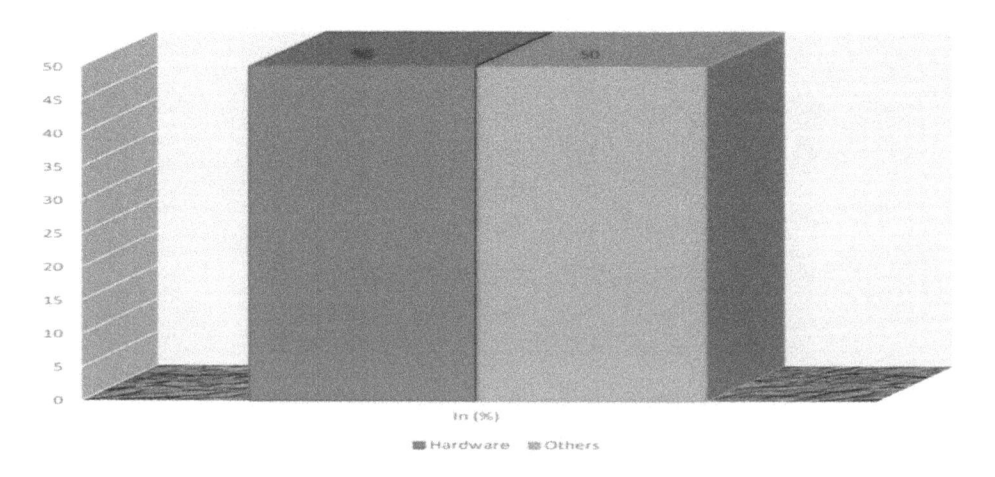

Figure 27. Expected market growth of the leading continents or regions (Straits Research)

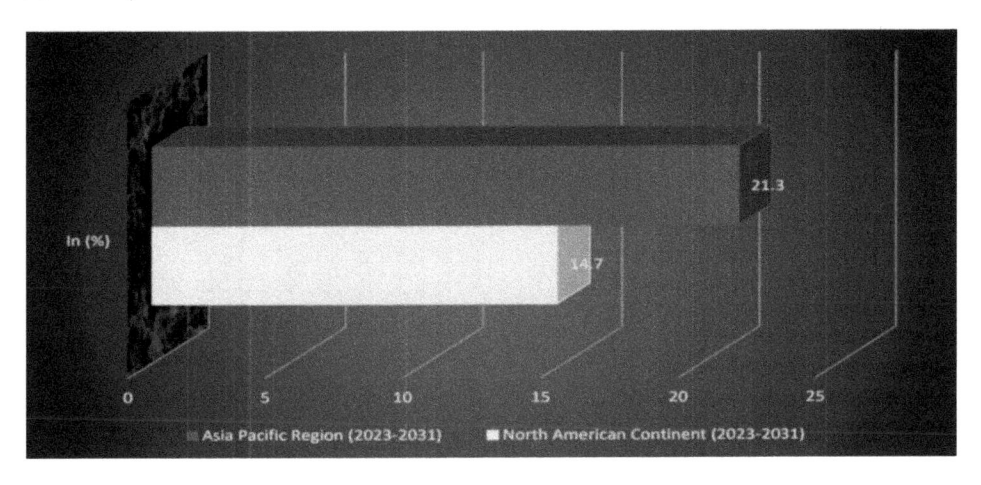

Figure 28. Revenue share in smart highway market (Market Research Future)

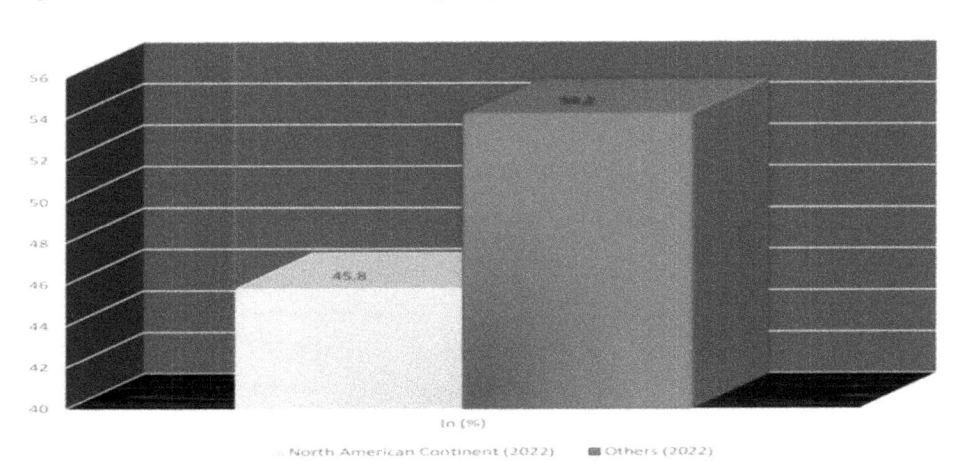

Figure 29. Growth of Indian smart highway market (Global Incorporation, Inc.)

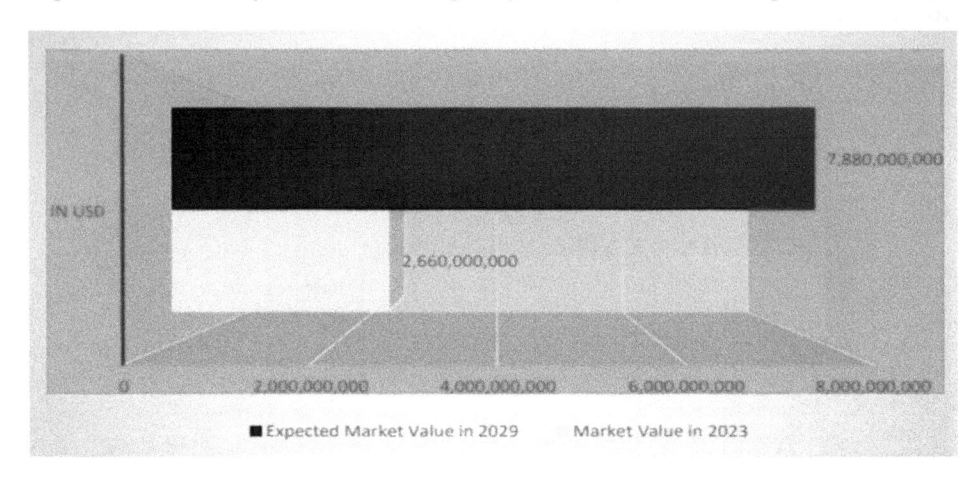

Figure 30. Growth of national highways in India (IBEF)

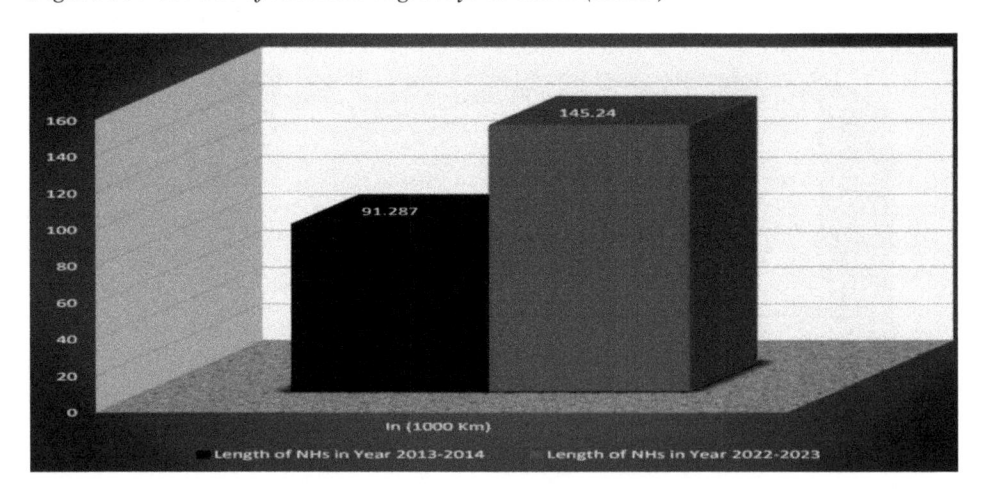

Figure 31. Growth of smart highway market in the world (Markets and Markets, Allied Market Research, Data Bridge, Growth Market Reports and Mordor Intelligence)

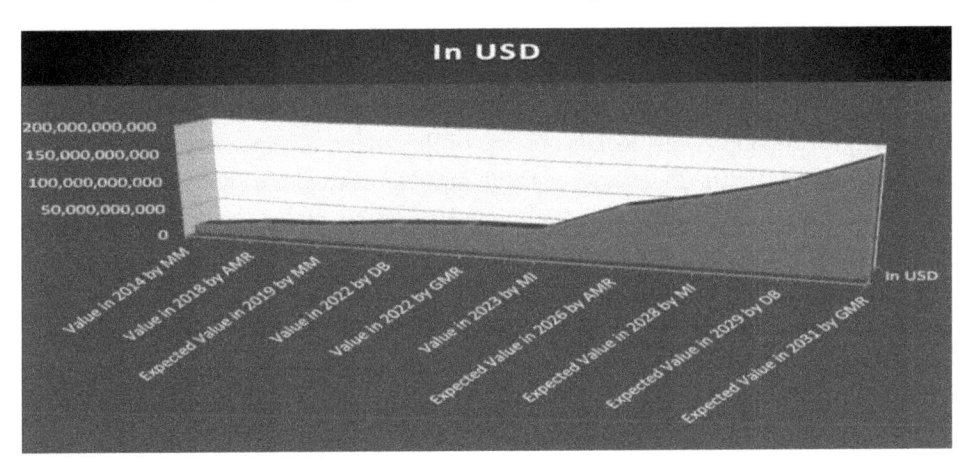

Major Advantages and Challenges of Smart Highways

In this subsection, some major advantages as well as challenges of Smart Highways based on the above figures are as follows:

- According to Figure 19, India had been investing heavily in the development of 100 Smart Cities whereas China had been developing over 900 Smart Cities Pilot projects. Since, Smart Cities also need Smart Highways and Smart Roads, so it is safe to say that the demand and scope of Smart Highways will be significant in these two major developing countries of the world not only now but also in the future.
- Based on Figure 20, Smart Logistics and Smart Mobility have 33.6% or slightly above one third of the market share in the Smart City projects in China, but without Smart Highways and Smart Roads, they will be incomplete. This also indicates the importance of Smart Highways and Smart Roads in the Smart City projects.
- According to Figure 21, the Transportation Sector is contributing 28% in the emission of Greenhouse Gases, which shows the need of more improvement in the Transportation Sector in a sustainable way. This could be achieved by incorporating more Smart Roads and Smart Highways to overcome traffic related problems as well as reduce the air pollution caused due to traffic jams.
- Based on Figure 22 and Figure 23, more vehicles started using ECTS as well as ECTS 2.0 from the year 2017 to the year 2022 on the Expressways in

The Current Status and Scope of Smart Highways in India

Japan. This shows the growth of Japan towards Smart Highways in the recent years.
- According to Figure 24, the growth of vehicles increased considerably from the year 2010 to the year 2018. It is safe to say that the vehicles in the future will increase further which will increase traffic related problems. Smart Highways and Smart Roads are one of the major ways to solve this problem.
- Based on Figure 25, every year high number of road accidents including both fatal and non-fatal ones, which occur more on highways than roads in the world. These fatal and non-fatal road accidents could be significantly reduced with the implementation of Smart Highways and Smart Roads.
- According to Figure 26, the hardware has a 50% share in the Smart Highway market, whereas others including software and services have the remaining half of the Smart Highway market. This shows that the hardware is having more demand, and it will be wise to invest more on the hardware component of Smart Highway in the future, which will be profitable.
- Based on Figure 27, Asia Pacific Region is expected to have even higher growth in Smart Highways market compared to North American Continent in the coming years. This also means it will be wise and profitable for India to invest more on Smart Highways in the coming years.
- According to Figure 28, North American continent was having over 45% share in the Smart Highways, which shows the dominance of North American continent in the world in regard to Smart Highways. It is safe to say that developing countries like India will also need Smart Highways to match with the development of advanced countries in the future.
- Based on Figure 29, Indian Smart Highway market is expected to increase up to almost 3 times in the next six years, which shows positive trend. This indicates that the demand of Smart Highways will also increase in the future in India.
- According to Figure 30, the length of National Highways in India increased slightly above 1.59 times in the span of 9 years. This only shows the significant growth of National Highways in India, but if State Highways, other major roads and rural roads, are considered, then it is clearly noticeable India's incline towards improving its Transportation Sector. This also shows the wide scope of Smart Highways in India in the future, as Smart Highways will help in the advancement of the country's Transportation Sector further.
- Based on Figure 31, it can be seen that from the year 2014 till the year 2023, the Smart Highways market grew considerably in the world and by the year 2031, it is expected that the Smart Highways market will rise even more rapidly. This means that the demand of Smart Highways will rise with time and investing on them will be profitable in the future.

LIMITATIONS

This chapter mentioned a number of Smart Highway projects which were either completed successfully or going on in India. However, it only briefly covered such Smart Highway projects, and the book chapter was not able to go in much depth in any particular Smart Highway project in India. Then, Smart Highway projects in other countries were not covered in the chapter.

SOLUTIONS AND RECOMMENDATIONS

Of course, currently the high cost of implementing such advanced digital technologies on highways to make them Smart Highways, can't be helped with, but as more technological advancements take place, then the cost of current technologies will automatically lower. Similarly, the technological limitations of every technology could only be solved with more development of sophisticated technologies with time. In regard to interoperability or amalgamation of incompatible technologies, that could be discerned by testing and checking the compatibility of devices with each other. For data privacy and security, this problem will rise more in the future, so the Government needs to implement strong policies which would prevent cyber-attacks and breaches even in the future. People also needs to be more aware and gets accustomed with Smart Highways and not to break traffic rules intentionally, which could be achieved with the passage of time.

FUTURE RESEARCH DIRECTIONS

Since, a number of major Smart Highway projects were covered in the chapter briefly, so in the future research could be done in detail in a specific Smart Highway project in India, to better understand their respective utility in the areas where they are located and the possible change they brought with their existence. Similarly, Smart Highway projects in foreign countries could also be considered as possible cases in the future research.

CONCLUSION

The chapter explained about Smart Highways as well as its associated numerous technologies. It also covered the major elements of Smart Highway Infrastructure and the need as well as Smart Highways future in India. The gaps in the current research on and related to Smart Highways were talked about in this chapter. Some major Smart Highway projects in India were briefly highlighted and with the help of graphical analyses, the scope of Smart Highways was understood in the book chapter. Then, the cost of Smart Highway projects is extremely high, as all the sophisticated components and technologies are pricey. This makes it difficult to convert the traditional highways into Smart Highways on a large scale particularly in the most populated and major developing country India, considering the fact that the country is having 2nd largest road network in the world and 46th ranked in terms of overall quality. Further, technical limitations of each technology, interoperability or amalgamation of incompatible technologies, in addition to data privacy as well as security attracting cyber-attacks and breaches, are the other major problems and issues with Smart Highways. Further, every year high number of road accidents including both fatal and non-fatal ones, which occur more on highways than roads in the world. These fatal and non-fatal road accidents could be significantly reduced with the implementation of Smart Highways and Smart Roads. Indian Smart Highway market is expected to increase up to almost 3 times in the next six years, which shows positive trend. This indicates that the demand of Smart Highways will also increase in the future in India.

REFERENCES

Alkhatib, A. A., Maria, K. A., AlZu'bi, S., & Maria, E. A. (2022). Smart Traffic Scheduling for Crowded Cities Road Networks. *Egyptian Informatics Journal, 23*(4), 163–176. doi:10.1016/j.eij.2022.10.002

Allied Market Research. (n.d.). https://www.alliedmarketresearch.com/smart-highway-market

Analytics Vidhya. (n.d.). https://www.analyticsvidhya.com/blog/2023/05/smart-roads/#:~:text=A.,systems%2C%20digital%20signage%2C%20etc

Angioi, F., Portera, A., Bassani, M., de Ona, J., & Stasi, L. L. D. (2023). Smart on-Road Technologies and Road Safety: A short overview. *Transportation Research Procedia, 71*, 395–402. doi:10.1016/j.trpro.2023.11.100

Ashwin, M., Algahtani, A. S., Mubarakali, A., & Parthasarathy, P. (2022). Environmental aspects – IoT based cattle movement prevention to avoid road accident for Bhopal-Indore highways in India. *Sustainable Energy Technologies and Assessments, 50*, 101847. doi:10.1016/j.seta.2021.101847

Banafshehvaragh, S. T., & Rahmani, A. M. (2023). Intrusion, anomaly, and attack detection in smart vehicles. *Microprocessors and Microsystems*, *96*, 104726. https://bangaloremirror.indiatimes.com/bangalore/others/tumakuru-road-could-be-citys-future/articleshow/95613044.cms

Bhowmik, S., Singh, A., & Misengo, C. (2019). A Case Study on Intelligent Transport System Using Traffic Lights. *Our Heritage*, *67*(7), 96–110.

Bianchini, D., De Antonellis, V., & Garda, M. (2023). A big data exploration approach to exploit in-vehicle data for smart road maintenance. *Future Generation Computer Systems*, *149*, 701–716. doi:10.1016/j.future.2023.08.004

Biyik, C., Abareshi, A., Paz, A., Ruiz, R. A., Battarra, R., Rogers, C. D. F., & Lizarraga, C. (2021). Smart Mobility Adoption: A Review of the Literature. *Journal of Open Innovation*, *7*(2), 146. doi:10.3390/joitmc7020146

Construction World. (n.d.-a). https://www.constructionworld.in/gold/urban-infrastructure/smart-cities-projects/this-smart-city-has-completed-six-major-smart-roads/39322

Construction World. (n.d.-b). https://www.constructionworld.in/urban-infrastructure/smart-cities-projects/this-city-has-taken-up-the-development-of-smart-roads-/37484

Construction World. (n.d.-c). https://www.constructionworld.in/transport-infrastructure/highways-and-roads-infrastructure/construction-challenges-of-delhi-meerut-expressway-package-3/22477

ConstructionPlacements. (n.d.). https://www.constructionplacements.com/smart-road-technology/#google_vignette

Data Bridge. (n.d.). https://www.databridgemarketresearch.com/reports/global-smart-highway-market

Edward, V. C. P. (2023). Smart crisis management system for road accidents based on Modified Convolutional Neural Networks–Particle Swarm Optimization hybrid algorithm. *Advances in Computers*, *131*. doi:10.1016/bs.adcom.2023.07.002

Fernandez-Isabel, A., Fuentes-Fernandez, R., & de Diego, I. M. (2020). Modeling multi-agent systems to simulate sensor-based Smart Roads. *Simulation Modelling Practice and Theory*, *99*, 101994. https://www.giiresearch.com/report/tsci1371381-india-smart-highway-market-by-technology-by.html

Growth Market Reports. (n.d.). https://growthmarketreports.com/report/smart-highway-market-global-industry-analysis

Haider, M., Peyal, M. K., Huang, T., & Xiang, W. (2023). Road crack avoidance: a convolutional neural network-based smart transportation system for intelligent vehicles. *Journal of Intelligent Transportation Systems*. https://www.iasparliament.com/current-affairs/smart-highways-are-the-road-to-the-future

IMARC Group. (n.d.). https://www.imarcgroup.com/smart-highway-market

IndiaE. F. C. O. N. (n.d.). https://efkonindia.com/products-anpr.php?PRODUCTS&ANPR

India Briefing News. (n.d.). https://www.india-briefing.com/news/eastern-peripheral-expressway-india-infrastructure-challenges-investors-16825.html/#:~:text=Moreover%2C%20the%20EPE%2C%20touted%20as,challenges%20in%20India%27s%20infrastructure%20sector

IndiaToday. (n.d.). https://www.indiatoday.in/auto/auto-news/story/hp-lubricants-and-leo-burnett-india-create-innovative-intelligent-road-systems-974118-2017-04-28

Luo, Z., & Yang, B. (2021). Towards resilient and smart urban road networks: Connectivity restoration via community structure. *Sustainable Cities and Society*, 75, 103344. https://www.magicbricks.com/blog/smart-cities-in-india/132271.html

Market Research Future. (n.d.). https://www.marketresearchfuture.com/reports/smart-roads-market-1000

Markets and Markets. (n.d.). https://www.marketsandmarkets.com/Market-Reports/smart-highway-market-90651023.html

Mint. (n.d.-a). https://www.livemint.com/news/india/delhimeerut-expressway-gadkari-unhappy-with-poor-road-side-amenities-11619230500343.html

Mint. (n.d.-b). https://www.livemint.com/news/india/delhi-to-meerut-in-45-minutes-7-facts-to-know-about-this-smart-highway-11617338840541.html

Mint. (n.d.-c). https://www.livemint.com/news/india/nitin-gadkari-launches-intelligent-transport-system-on-eastern-peripheral-expressway-11640306946337.html

Mordor Intelligence. (n.d.). https://www.mordorintelligence.com/industry-reports/smart-highway-market

MoRTH Annual Report for the Year 2022-23. (n.d.). https://morth.nic.in/sites/default/files/MoRTH%20Annual%20Report%20for%20the%20Year%202022-23%20in%20English.pdf

News18. (n.d.). https://www.news18.com/news/lifestyle/national-safety-day-countries-with-highest-number-of-road-accidents-and-where-does-india-stand-4833212.html

NewswireP. R. (n.d.). https://www.prnewswire.com/in/news-releases/smart-roads-inaugurated-in-ujjain-833646277.html#:~:text=MLA%20recently%20inaugurated%20these%20Smart,Mahadev%20Temple%20to%20Jain%20Mandir

Pascale, A., Macedo, E., Guarnaccia, C., & Coelho, M. C. (2023). Smart mobility procedure for road traffic noise dynamic estimation by video analysis. *Applied Acoustics*, *208*, 109381. doi:10.1016/j.apacoust.2023.109381

Paul, J., Chris, Y. S., Tom, T. O., Cespedes, S., Benamar, N., Wetterwald, M., & Harri, J. (2021). A comprehensive survey on vehicular networks for smart roads: A focus on IP-based approaches. *Vehicular Communications*, *29*, 100334. doi:10.1016/j.vehcom.2021.100334

Pompigna, A., & Mauro, R. (2022). Smart roads: A state of the art of highways innovations in the Smart Age. *Engineering Science and Technology, an International Journal*, *25*, 100986. https://www.prescouter.com/2021/06/4-smart-road-technologies-shaping-the-future-of-transportation/

Qin, X., Ke, J., Wang, X., Tang, Y., & Yang, H. (2022). Demand management for smart transportation: A review. *Multimodal Transportation*, *1*(4), 100038. doi:10.1016/j.multra.2022.100038

Reddy, H. K., Goswami, R. S., & Roy, D. S. (2023). A futuristic green service computing approach for smart city: A fog layered intelligent service management model for smart transport system. *Computer Communications*, *212*, 151–160. doi:10.1016/j.comcom.2023.08.001

S, M., & Almutairi, S. (2023). Non-divergent traffic management scheme using classification learning for smart transportation systems. *Computers and Electrical Engineering*, *106*, 108581.

Salmi, A.-K., & Hokars, F. (2023). Smart urban traffic zones – demonstration of three smart zones in Sweden based on geofencing and sensor data. *Transportation Research Procedia*, *72*, 1459–1466. doi:10.1016/j.trpro.2023.11.611

Shah, K., Sheth, C., & Doshi, N. (2022). A Survey on IoT-Based Smart Cars, their Functionalities and Challenges. *Procedia Computer Science*, *210*, 295–300. doi:10.1016/j.procs.2022.10.153

Simplilearn. (n.d.). https://www.simplilearn.com/smart-road-technologies-article#:~:text=It%20can%20help%20planners%20determine,produce%20a%20lot%20of%20pollution

Singh, A. (2018). Use of Geographical Information System in Intelligent Transport. *International Journal of Research and Analytical Reviews*, *5*(4), 882–890.

Singh, A. (2019). Congestion pricing in developing countries -a case study. *International Journal of Research and Analytical Reviews*, *6*(1), 692–701.

Singh, A. (2019). Impact of work culture on traffic. *International Journal of Research and Analytical Reviews*, *6*(1), 702–711.

Singh, A. (2022). Importance of Fuzzy Logic in Traffic and Transportation Engineering. In P. Vasant, I. Zelinka, & G. W. Weber (Eds.), *Intelligent Computing & Optimization. ICO 2021. Lecture Notes in Networks and Systems* (Vol. 371). Springer. doi:10.1007/978-3-030-93247-3_10

Singh, A. (2023a). Mechatronics Engineering in the Modern World and Its Use in the Construction Industry. In M. Mellal (Ed.), *Trends, Paradigms, and Advances in Mechatronics Engineering* (pp. 64–92). IGI Global. doi:10.4018/978-1-6684-5887-7.ch004

Singh, A. (2023b). The Current Scenario of Sustainable Planning and Development of the Indian Built Environment. In N. Cobîrzan, R. Muntean, & R. Felseghi (Eds.), *Circular Economy Implementation for Sustainability in the Built Environment* (pp. 106–138). IGI Global. doi:10.4018/978-1-6684-8238-4.ch005

Singh, A. (2023c). The Current Trends and Need of Infrastructure Projects in the Country India. In C. Popescu, P. Yu, & Y. Wei (Eds.), *Achieving the Sustainable Development Goals Through Infrastructure Development* (pp. 235–261). IGI Global. doi:10.4018/979-8-3693-0794-6.ch010

Singh, A. (2023d). The Present and Future Role of Artificial Intelligence in the Promotion of Sustainable Practices in the Construction Sector. In C. Kahraman, I. U. Sari, B. Oztaysi, S. Cebi, S. Cevik Onar, & A. Ç. Tolga (Eds.), *Intelligent and Fuzzy Systems. INFUS 2023. Lecture Notes in Networks and Systems* (Vol. 759). Springer. doi:10.1007/978-3-031-39777-6_2

Singh, A. (2023e). The Recent Trend of Artificial Neural Network in the Field of Civil Engineering. In *Intelligent Computing and Optimization. ICO 2023. Lecture Notes in Networks and Systems* (Vol. 855). Springer. doi:10.1007/978-3-031-50158-6_32

Singh, A. (2023f). The Significance of Digitalization of the Construction Sector. In P. Vasant, G. W. Weber, J. A. Marmolejo-Saucedo, E. Munapo, & J. J. Thomas (Eds.), *Intelligent Computing & Optimization. ICO 2022. Lecture Notes in Networks and Systems* (Vol. 569). Springer. doi:10.1007/978-3-031-19958-5_100

Singh, R., Sharma, R., Akram, S. V., Gehlot, A., Buddhi, D., Malik, P. K., & Arya, R. (2021). Highway 4.0: Digitalization of highways for vulnerable road safety development with intelligent IoT sensors and machine learning. *Safety Science*, *143*, 105407. https://smartnet.niua.org/content/2b9bf34e-b033-44b8-90a6-3004307099d5

Spread Design and Innovation. (n.d.). https://spread.ooo/portfolio/launching-indias-1st-smart-expressway/

Statista. (n.d.-a). https://www.statista.com/statistics/1135346/china-smart-city-market-share-by-segment/

Statista. (n.d.-b). https://www.statista.com/topics/5794/smart-city-in-china/#topicOverview

Stormwater Solutions. (n.d.). https://www.stormwater.com/bmps/article/21217281/smart-roads-inaugurated-in-india

Straits Research. (n.d.). https://straitsresearch.com/press-release/global-smart-highway-market

Takefuji, Y. (2021). Illogical smart highway management policy in Japan. *Transportation Engineering*, *3*, 100051. https://infolks.info/blog/smart-roads-ai-enabling-intelligent-highways-for-the-future/

Thakur, V. N., & In Han, J. (2023). Triboelectric nanogenerator for smart traffic monitoring and safety. *Journal of Industrial and Engineering Chemistry*, *124*, 89–101. doi:10.1016/j.jiec.2023.04.028

The Hindu BusinessLine. (n.d.). https://www.thehindubusinessline.com/opinion/smart-highways-are-the-road-to-the-future/article64594234.ece

TheGlobalEconomy. (n.d.). https://www.theglobaleconomy.com/rankings/roads_quality/

Times of India. (n.d.-a). https://timesofindia.indiatimes.com/city/dehradun/smart-city-construction-near-bindal-tiraha-giving-commuters-tough-time/articleshow/104056019.cms

Times of India. (n.d.-b). https://timesofindia.indiatimes.com/city/dehradun/smart-road-project-to-miss-deadline-stretches-to-stay-dug-up-till-year-end-in-dehradun/articleshow/99543664.cms

Tumakuru-SCM-Report-Dec-2021-2.pdf. (n.d.). https://www.cenfa.org/wp-content/uploads/2022/01/Tumakuru-SCM-Report-Dec-2021-2.pdf

Tumakuru Smart City Limited Brochure. (n.d.). https://issuu.com/smartcitytumakuru/docs/dubai_expo21_final_1_

Vrio. (n.d.). https://vrioeurope.com/en/smart-road-technology-digital-highways-of-the-future/

Wang, H., Quan, W., & Ochieng, W. Y. (2020). Smart road stud based two-lane traffic surveillance. *Journal of Intelligent Transport Systems*, *24*(5), 480–493. doi: 10.1080/15472450.2019.1610405

Wang, H., Sun, Y., Quan, W., Ma, X., & Ochieng, W. Y. (2022). Traffic volume measurement based on a single smart road stud. *Measurement*, *187*, 110150. doi:10.1016/j.measurement.2021.110150

WHO. (n.d.-a). https://www.who.int/news-room/fact-sheets/detail/road-traffic-injuries#:~:text=Overview,with%20many%20incurring%20a%20disability

WHO. (n.d.-b). https://www.un.org/en/observances/road-traffic-victims-day#:~:text=The%20Global%20status%20report%20on,people%20aged%205%2D29%20years

Wikimedia Commons. (n.d.-a). https://commons.wikimedia.org/wiki/File:Delhi%E2%80%93Meerut_Expressway.jpg

Wikimedia Commons. (n.d.-b). https://commons.wikimedia.org/wiki/File:Delhi-Meerut-Express-Highway-India.png

Wikimedia Commons. (n.d.-c). https://commons.wikimedia.org/wiki/File:Greaternoida_expressway_Q2.jpg

Wikimedia Commons. (n.d.-d). https://commons.wikimedia.org/wiki/File:India_ne2.png

Wikimedia Commons. (n.d.-e). https://commons.wikimedia.org/wiki/File:Noida_expressway.jpg

Wikimedia Commons. (n.d.-f). https://commons.wikimedia.org/wiki/File:Smart_highway_signs_on_Interstate_90_in_Mercer_Island,_Washington.jpg

Wikimedia Commons. (n.d.-g). https://commons.wikimedia.org/wiki/File:The_Intelligent_Transport_System_screen_on_Thailand_Highway_9.jpg

Wikimedia Commons. (n.d.-h). https://commons.wikimedia.org/wiki/File:Yamuna_Expressway_Detailed_Map.png

Wikimedia Commons. (n.d.-i). https://commons.wikimedia.org/wiki/File:Yamuna_Expressway_Map_Large.png

Wikimedia Commons. (n.d.-j). https://commons.wikimedia.org/wiki/File:Yamuna_Expressway_Toll_Plaza_Delhi_Agra_India.jpg

Wikipedia. (n.d.-a). https://en.wikipedia.org/wiki/Eastern_Peripheral_Expressway

Wikipedia. (n.d.-b). https://en.wikipedia.org/wiki/Roads_in_India#:~:text=Roads%20in%20India%20are%20an,world%2C%20after%20the%20United%20States

Wikipedia. https://en.wikipedia.org/wiki/Smart_highway

Yao, H., Xu, Z., Hou, Y., Dong, Q., Liu, P., Ye, Z., Pei, X., Oeser, M., Wang, L., & Wang, D. (2023). Advanced industrial informatics towards smart, safe and sustainable roads: A state of the art. *Journal of Traffic and Transportation Engineering*, *10*(2), 143–158. doi:10.1016/j.jtte.2023.02.001

Younes, M. B., Boukerche, A., & De Rango, F. (2023). SmartLight: A smart efficient traffic light scheduling algorithm for green road intersections. *Ad Hoc Networks*, *140*, 103061. doi:10.1016/j.adhoc.2022.103061

Zhang, H., Chen, Z., Zhu, C., & Wei, C. (2020). *An innovative and smart road construction material: thermochromic asphalt binder*. New Materials in Civil Engineering.

Zhao, C., Jia, R., & Dong, K. (2023). How does smart transportation technology promote green total factor productivity? The case of China. *Research in Transportation Economics*, *101*, 101353. doi:10.1016/j.retrec.2023.101353

Zhu, W. (2022). A spatial decision-making model of smart transportation and urban planning based on coupling principle and Internet of Things. *Computers & Electrical Engineering*, *102*, 108222. doi:10.1016/j.compeleceng.2022.108222

Chapter 2
Renewable Energy-Integrated Electric Vehicle Charging Infrastructure Across Cold Region Roads

Dharmbir Prasad
https://orcid.org/0000-0002-9010-9717
Asansol Engineering College, India

Rudra Pratap Singh
https://orcid.org/0000-0001-7352-855X
Asansol Engineering College, India

Tanuja Tiwary
Asansol Engineering College, India

Ranadip Roy
https://orcid.org/0000-0003-2111-2581
Sanaka Educational Trust's Group of Institutions, India

Md. Irfan Khan
IAC Electricals Pvt. Ltd., India

ABSTRACT

The growing popularity of electric vehicles and the need for environmentally friendly transportation have made it necessary to build smart roadways with charging station equipped by energy harvesting and management systems. In order to collect and transform ambient energy into electrical power, the suggested system makes use of a variety of energy harvesting technologies embedded in the road surface, including

DOI: 10.4018/978-1-6684-9214-7.ch002

solar panels, wind energy, etc. A complex energy management system integrated into the smart road infrastructure effectively stores, distributes, and transfers the captured energy in real time to electric vehicle charging stations or to the vehicles themselves. The suggested system comprises of essential elements such as intelligent inverters, decentralized energy storage units, and a central management system that manages the entire process. In the proposed renewable energy integrated charging system having energy production of 3,00,233 kWh/year with energy storage of 161 kWh.

List of Abbreviations

BEHF-SRM	Bidirectional energy harvesting floor with sustained-release regulation mechanism
BESS-VSG	Battery energy storage system - virtual synchronous generator
BEV	Battery electric vehicle
BIPV	Building integrated PV
BMW	Berlin Motor Work
BRTDP	Batch reinforcement transfer decision process
CAGR	Compound annual growth rate
CGP	Control generating parameter
CS	Charging station
DWC	Dynamic wireless charging
ESS	Energy storage system
EV	Electric vehicle
EVCS	Electric vehicle charging station
EVCDS	Electric vehicle charging/discharging
EVSC	Electric vehicle supply chain
FAME	Faster Adoption and Manufacturing of (Hybrid &) Electric Vehicles in India
GEN	Diesel generator
GHI	Global horizontal irradiance
GOI	Government of India
G2V	Grid to vehicle
IDA-PEH	Improved design algorithm for photovoltaic energy harvesting
IRR	Internal rate of return
LCOE	Levelized cost of electricity
MMOSSA	Multi objective shuffled salp swarm algorithm
NPC	Net present cost
PHEV	Photovoltaic energy vehicle

PV	Photovoltaic
PVEV	Photovoltaic electric vehicle
REHS	Renewable energy harvesting system
ROI	Return on investment
TAC	Total annualized cost
TNPC	Total Net Present Cost
WPT	Wireless power transfer
WT	Wind turbine
UAV	Unmanned aerial vehicle
VRF	Vanadium redox flow
VSI	Voltage stability indicator
V2G	Vehicle to grid

INTRODUCTION

In the midst of accelerated urbanization, escalating environmental apprehensions and a global shift towards sustainable technologies, the automotive industry is experiencing a paradigm shift with the ascendancy of EVs. As EV adoption gains momentum, the imperative for an efficient and sustainable charging infrastructure becomes more pronounced. Against this backdrop, the concept of smart roads emerges as an innovative solution, integrating energy harvesting and management systems to forge an intelligent and self-sufficient ecosystem tailored for electric vehicles. This integration transcends mere energy provisioning; it aspires to establish a dynamic and adaptive infrastructure adept at responding to real-time demands and fluctuating environmental conditions.Conventional approaches to electric vehicle charging predominantly hinge on centralized charging stations tethered to the grid, presenting challenges associated with grid capacity, energy distribution and environmental repercussions. The infusion of energy harvesting technologies into road infrastructure confronts these challenges head-on, redefining roads as active contributors to the energy supply chain. This paradigmatic shift not only alleviates pressure on conventional power grids but also aligns with the overarching objective of cultivating a cleaner and more sustainable transportation milieu.

> *(i).Driving Factors for EV Market:* India's market for EV charging was estimated to be worth USD 449.06 million in 2022 and is expected to grow at a CAGR of 37.7% from 2023 to 2030, reaching USD 5695.6 million as shown in Figure 1 (Next Move Strategy Consulting, 2023). The speed at which they supply electricity to the car's battery is what distinguishes electric vehicle chargers.

The market for charging stations for electric vehicles in India is somewhat fragmented viz., ABB Ltd., ExicomTelesystems Ltd., Mass-tech Controls Pvt Ltd., Charzer Tech Pvt Ltd. and Tata Power Company Limited are a few of the major participants. Following are the key driving factors:

- India's EV charging infrastructure is expected to experience a peak-level expansion in the upcoming years as a result of national regulations and initiatives. With its e-mobility programs, the Indian government is set to achieve a high rate of decarbonization in the transportation sector.
- As of February 2022, there were approximately 1640 active EV charging stations in India, of which 940 were located in nine megacities, including Ahmedabad, Pune, Delhi, Bangalore, and so on. The surge can be attributed to the elevated need for electric automobiles. In India, sales of BEVs were expected to be 12,000 in 2021. The Indian government's and PSUs' efforts to speed up the market are what's caused the acceleration during the past five years. In response to the changing environment, the government has even more EV charging station projects planned.
- For instance, the Phase-II of the FAME scheme, which aims to install 2877 EV charging stations in 68 cities and 1576 stations across highways and more, will be implemented with funding allocated by the central government in December 2021.
- The national government has also requested that public sector businesses establish new EV charging stations across the nation. The Ministry of Power declared in February 2022 that 22,000 EV charging stations would be installed nationwide on national roads and in developed cities by PSUs, IOCL, BPCL and HPCL. The rules and specifications that must be adhered to for the EV charging infrastructure have already been released by the government.

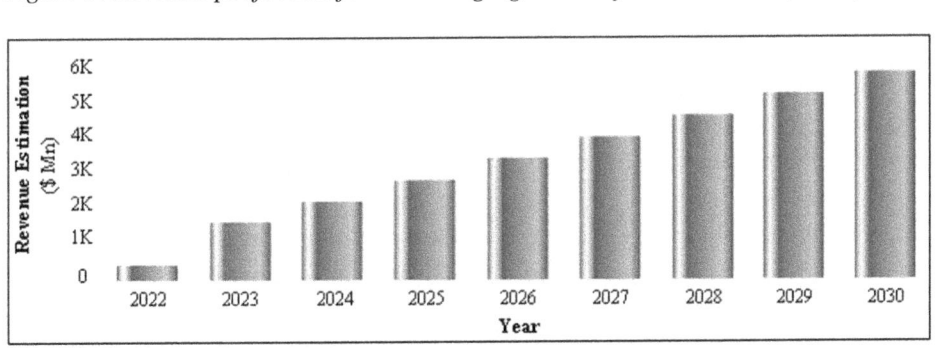

Figure 1. Revenue projection for EV charging market for 2022-2030 ($ Mn)

It is anticipated that these advancements will have the significant impact on the EV charging infrastructure market in the near future. In this context, the present study focussed on renewable energy supply based EV charging station at Leh, Ladakh (India). Its location mapping is shown in Figure 2 and energy access significance is illustrated as followings:

Figure 2. Location mapping of the proposed green energy powered EV charging station

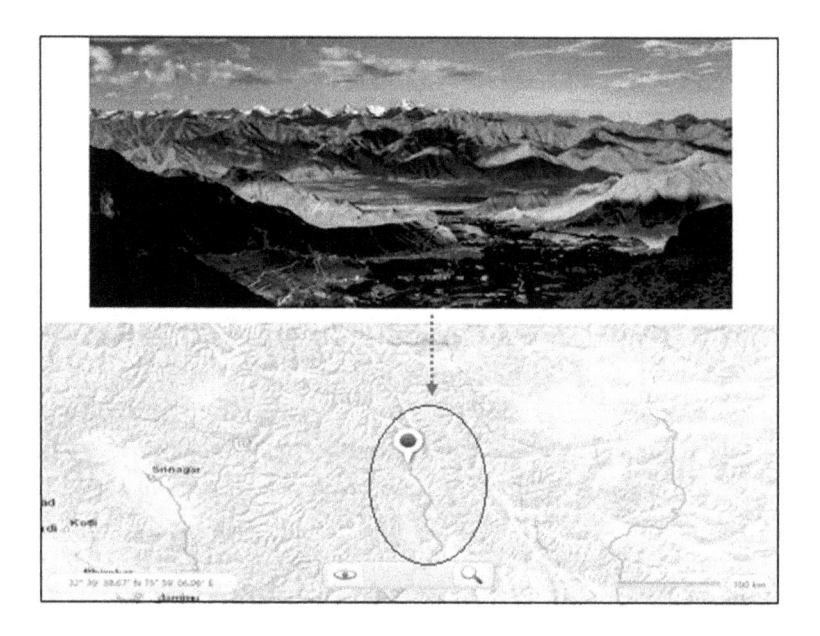

The historical capital of the Kingdom of Ladakh was Leh, which is situated in the Leh district. The city is located on the bank of the Indus River. Built in the same design and about the same period as the Potala Palace in Tibet, Leh Palace, the seat of the monarchy, was once the home of the Ladakh royal dynasty. The mountains dominate the landscape around the Leh, as it is at an altitude of 3,500m. Peaks such as Nanga Sago can reach well above 5,500m. Although the roads from Srinagar and Manali are often blocked by snow in winter, the local roads in the Indus Valley usually remain open due to the low levels of snowfall. Leh-Manali Highway, also known as National Highway 3, is the second land access to Ladakh. Because Leh is in a cold desert climate zone, it experiences lengthy, bitter winters with temperatures below freezing from late November to early March. Because of its high altitude, the city occasionally experiences snowfall in the winter, which is unusually chilly for India. The subsequent months often have pleasant, warm daytime temperatures.

(ii).EV Charging Infrastructure Expansion along Cities and Highways: The absence of incentives and worries about the expensive installation costs of EV chargers could pose challenges to the expansion of the EV charging industry. The considerable upfront cost of level 3 and ultra-fast chargers is one of the biggest obstacles to the industry's growth. Although it can take up to 16 hours to fully charge level 1 and level 2 chargers, consumers are used to refuelling their conventional fossil fuel automobiles in as little as 5 to 7 minutes (Next Move Strategy Consulting, 2023). The government's increasing incentives to EV manufacturers to install multiple car charging stations on highways, expressways, and in cities are anticipated to further fuel the market's expansion for EV chargers. For example, the GoI had started promoting the use and production of electric cars in the nation (Next Move Strategy Consulting, 2023). The government also intended to work with both public and private organizations to improve the infrastructure for public charging. In response to consumer demand, numerous private organizations have offered to install EV charging stations and create a user-friendly charging network grid.

An essential overland route between Ladakh and the Kashmir valley is the 434 km National Highway 1. Traffic usually flows through it from April/May to October/November. The trip involves climbing the challenging Zoji-la Pass in the Great Himalayan Wall, which is located at a height of 3,505 m. Regular bus services between Srinagar and Leh are run by the Jammu and Kashmir State Road Transport Corporation, including an overnight stop at Kargil. For this trip, taxis in Srinagar are also accessible, including automobiles and jeeps. Earlier the main source of electricity was diesel generators. Ladakh is known as the 'roof of the world' and it has abundant sunlight and clear air making it suitable for solar energy technologies. Electrification of Ladakh with solar energy is the main concern. There is a large amount of barren land available to the solar power developer. On average in Ladakh there are 325 sunny days in a year. The daily average solar energy available in per square meter in Leh district is 6-12 kWh. Leh receives only an average of 35 mm of yearly rainfall, despite the region's generally arid climate. However, in 2010, flash floods occurred, posing a serious environmental threat to the city and tragically taking over 100 lives. This instance highlights how susceptible desert regions are too harsh weather, especially in places where precipitation is often infrequent.

Rest of the chapter is organised as follows: starting with research gap and contribution, then mathematical modelling; further, modelling of RE powered EV charging station, energy harvesting potential access. Next, continued results and discussion, while, at the end the study has been concluded.

PROBLEM STATEMENT

The current study on EV charging station systems uses renewable energy to meet load demand; however, when the total power generated by both systems is greater than zero, the extra power is used to charge the batteries to their full capacity. Any remaining power can be used for deferrable loads even after it has been used. On the other hand, battery backup will assist in meeting demand until it is entirely met if solar and wind resources are insufficient to supply it. One of the primary goals of this study is to ensure that there is never an unmet load requirement by carefully selecting the sizes of the various components.

RESEARCH GAP AND CONTRIBUTION

Electric vehicles have become more commonplace in recent years due to the increased focus on lowering greenhouse gas emissions and using sustainable energy sources. The way we commute is changing as a result of electric cars' cheaper operating costs, less maintenance needs and better environmental effect than those of internal combustion engine vehicles. Above all, EVs signify advancements that go beyond the automobile sector, such as energy storage technologies and the incorporation of renewable energy into smart grid systems (The Times of India, 2023). Following are the recent research centred on electric mobility:

(i). Charging Station for e-Mobility: The growing demand for charging stations to accommodate EVs is a result of their increasing use. By acting as a natural aggregator of EVs, charging stations allow EVs to take part in power system management by utilizing renewable energy sources and energy storage technologies. Following are the significant EV charging station oriented researches conducted:

Locations that offer quick charging for batteries will become more in demand as electric vehicles become more common. Using energy storage devices, namely VRF batteries, appears to be a viable way to lower the installed power while implementing a peak shaving plan (Cunha et al., 2016). Venugopal et al. examined the functional integration of several future highway components and examines a system-level analysis of a real-highway installation in the Netherlands (Venugopal et al., 2018). The charging schedule issue for the office parking lot that is powered by the grid and photovoltaic system was discussed in (Zhang & Cai, 2018). A study of (Ghosh & Aggarwal, 2018) suggested a menu-based pricing strategy for V2G services, in which the charging station chooses fees for new customers depending on the amount of battery usage, the deadline, and various charging quantities.

The transportation and infrastructure sectors are growing more dynamic and laying the groundwork for a smarter future innovation from both businesses and academic organizations. From energy production to noise barriers, solar technologies are integral to the whole development framework (Solar Power World, 2019).Because renewable energy is available and grid power quality effects are minimal, stand-alone renewable energy powered charging stations are replacing grid-connected charging stations for electric cars (Joseph & Devaraj, 2019). A UK initiative named smart infrastructure aims to add roads that use the energy from passing automobiles and trucks to local road networks. The goal of developing new road surfaces is to capture energy from moving vehicles (The Engineer, 2019).The UK government's interest has been drawn to electrified roads, which automatically charge automobiles as electric vehicles become more prevalent. In order to assist counter future power supply difficulties, intelligent roadways might also capture sunlight and create electricity, creating illumination at night and melting snow and ice during the winter (iNews, 2020).

The microgrid was powered by tiny turbines, wind and sun energy and was linked to the upstream grid. Throughout the planning horizon, the plan's goal was to reduce both the operating and investment costs associated with all capacity resources (Mehrjerdi, 2020).For the PV aided EVCS, a finite-horizon model and the accompanying solution technique, the BRTDP algorithm, are developed in order to capitalize on the benefits of incorporating renewable energy sources into EVCS (Wu et al., 2020). Reducing microgrid's reliance on the electricity grid requires ESSs. This quantity of charging stations was selected since it needs fewer infrastructures(AbuElrub et al., 2020).The study evaluated and provided a revolutionary idea for road piezoelectric energy collecting. In order to overcome the constraints of low electrical power and low durability, an IDA-PEH was designed and built with a deformation-guiding auxiliary structure at extremely low road displacement conditions of 1 mm (Jeon et al., 2021).

Electric cars can lower the cost of charging and road infrastructure by using part of the energy produced by the road's surface. The vast expanse of roads in most cities makes them costly to maintain while taking into account the expense of surface upkeep, light posts, signage, and charging stations for electric vehicles (TheNextWeb, 2022).This spring, the city of Lenexa, Kansas, decided to proceed with a test pilot of smart pavements. This was a positive development after almost ten years of work for Integrated Roadways - not just for the Tim Sylvester-founded company, but also for attempts to improve road technology in an industry that has been preoccupied with smarter cars (Engineering New-Record, 2022).CarBuzz reported that BMW is developing a novel suspension system that can capture energy from an EV's wheel movement as it absorbs bump stress, according to a recent patent submitted to the German patent office (Electrek, 2022).

In the EV charging ecosystem, the most widespread ecosystems globally are highly vulnerable to cyber attacks due to serious flaws in its most basic components, EVCS and their management systems (Nasr et al., 2022). A thorough techno-economic analysis of an innovative hybrid renewable energy-based standalone rapid electric car charging station is conducted in (Al Wahedi & Bicer, 2022). A study evaluated range and energy consumption for several vehicle segments with varied drive cycles, analyzing the performance of PVEV on battery and fuel cell power trains (Sagaria et al., 2022). A clustered charging station for EVCDS is intended to be located in business parking lots. An intelligent power management plan must be put into place at charging stations with different power sources and consumers (Ozkan & Erol-Kantarci, 2022). In light of the growing popularity of EVs, academics have recently placed a great deal of emphasis on EVSC. Several EVSC elements were covered in the research of (Sadeghian et al., 2022).

Moreover, Yap et al. offered a thorough understanding of the most recent advancements in solar energy for BEV CS. The acceleration of solar energy in EV CS is also aided by developments in solar energy, such as flexible solar, rooftop solar and increased efficiency (Yap et al., 2022). In next study, authors resolved optimization issue by considering the operators of EVs and UAVs and applying weights to service quality, profit and cost (Qin et al., 2022). Another study outlined potential future research issues and evaluates the most recent research on transportation systems management taking DWC EVs into consideration (Li et al., 2023). In Canberra, Australia, the energy, economic and environmental performances of BIPV with EV charging systems are assessed through modelling with the PV system (Khan et al., 2023). In light of the significantly varying usage intensity displayed across EVs, an 11-month continuous consumption data set from 3,777 BEVs and 5,973 PHEVs sampled in Shanghai is analyzed in (Li et al., 2023).

The results of the optimization in (Bilal et al., 2023), showed that, in certain locations, grid-based, solar photovoltaic, and wind-powered electric car charging stations are the best means of meeting the demand for EVs. In a single-load scenario, the magnetic coupler can transfer energy with an acceptable output voltage fluctuation and up to 120 W of output power or 85% WPT efficiency(Lyu et al., 2023). Within the framework of a non-cooperative game, a novel distributed charging and discharging approach integrating load limits and the time anxiety idea was developed in (Kang et al., 2023) research. The research of Hasan et al. offered a thorough grid-connected EVCS concept for Chattogram, Bangladesh's CGP International Airport (Hasan et al., 2023). A MMOSSA approach for allocating RES, BESS-VSG and EVCS on microgrids in consideration of carbon emissions, frequency variation, TNPC, LCOE, network energy loss, and VSI is presented in (Abid et al., 2024). In order to support the future development of the market for electricity-assisted services, (Lin et al., 2024) study suggests a workable solution to the issues related to aggregator dispatching EV charging capacity.

(ii).Non-conventional Approach for Energy Harvesting: With the rapid industrial development and ongoing scientific advancements in human society, energy consumption is rising and environmental issues are becoming more and more significant. The Earth's supplies of conventional fossil fuels are finite, and using them extensively will worsen already-existing environmental issues like global warming and ozone depletion. Emissions will also rise significantly as a result of urbanization and industrialization, but they can be reduced by 1.2% for every 1% increase in the usage of renewable energy. Thus, ambient energy sources like photovoltaic power, wind power, hydroelectric power, geothermal plants, etc. that are renewable and do not harm the environment with pollutants are gaining more and more attention. Following are recent energy harvesting studies for multiple applications:

California's highways almost buzz with road vibrations activity as an energy source. Still, the state has begun testing the concept to determine whether it might be implemented on a large scale (Government Technology, 2016). The goal of another study was to create a piezoelectric road energy harvester that may be utilized in smart roads to power wireless sensor networks (Yang et al., 2017). In (Wang et al., 2017) research, they examined a hybrid architecture that combines solar energy harvesting with wireless charging benefits. Hybrid cars, on the other hand, have an internal combustion engine that runs on regular fuels to recharge the batteries and increase the vehicle's range. Due to the batteries' low capacity, battery-powered vehicles are only suitable for brief travel distances (AltEnergyMag, 2018). In (Zou et al., 2019) study, mechanical modulation can be used for other electromechanical conversion methods, such as electrostatic energy harvesting, turboelectric nano generator and others, even though piezoelectric and electromagnetic energy harvesting make up the majority of the design examples covered in this review.

In (Kim et al., 2019) study, authors presented an energy harvester for self-tuning stochastic resonance that can power smart tires. Centrifugal force is used in a rotating tire to build a passively tuned system. The majority of energy harvesting methods is put into practice by embedding the energy producers into the structures of vehicles or roads, which harms the structures as a whole and eventually ruins the energy collecting system (Pei et al., 2021). For near-zero-energy toll stations on expressways, this work presents the design, simulation, experimentation and estimation of a unique REHS based on spatial double V-shaped mechanisms (Sun et al., 2021). In further study, the suggested energy collecting device has a lot of potential to be used as a power source and a weigh-in-motion and vehicle speed sensor on the future smart highway (Chen et al., 2021).Reducing emissions in the transportation sector and increasing the total energy efficiency of vehicles are important for sustainable development (Bai & Liu, 2021).

Khan et al. suggested two main renewable energy sources that Saudi Arabia is expected to build are solar and wind energy sources to improve the sustainability of their building sectors, encourage energy diversity and see faster growth in the prosumers-integrated energy market (Khan et al., 2023).In (Li et al., 2023), study examined usage of wireless energy transfer technology in electric cars, with a focus on its use in the motors that power new energy vehicles' charging and driving systems. Further, a study presented a BEHF-SRM approach, which can harvest more energy and improve the quality factor of output power for immediate utilization (Zou et al., 2024).

MATHEMATICAL MODELING

Electrical power generated from PV module is affected by the solar irradiance and PV panel cell temperature. In the present study, power is calculated by (1) (AbuElrub et al., 2020),

$$P_{Direct\_Current} = P_{STC} \times \frac{G_A}{G_{STC}} \times \left[1 + (T_C - T_{STC})C_T\right] \qquad (1)$$

where, P_{STC}: power output of the PV module under STC conditions (W), G_A: in-plane radiance on the surface of the PV module (W/m²), G_{STC}: STC irradiance ((W/m²)), T_C: PV cell operating temperature (°C), T_{STC}: STC temperature (°C), C_T: temperature power coefficient.

The output power of the wind turbine under normal condition is given by (2) (Bilal et al., 2023),

$$P_{Wind\_Turbine}^{N} = P_{WT}(V) \times \frac{\rho}{1.225} \qquad (2)$$

where, $P_{Wind\_Turbine}^{N}$: power rating of each wind turbine for the lifespan of the project in years (W), P_{WT}: power rating of each turbine (W), ρ: air density (kg/m3).

When there is an excess power supply from SPV and WT during the day, the battery units are used to store the extra energy and provide it to the load when SPV and WT performs inadequately to meet the required power requirements. The total energy stored in a battery is given by (3) (Bilal et al., 2023),

$$E_{battery} = E_{battery}^{ini} + \int_0^T V_{battery} I_{battery} dt \qquad (3)$$

where, $E_{battery}$: battery energy storage, $E_{battery}^{ini}$: initial energy of the battery energy storage, $V_{battery}$: voltage of the battery, $I_{battery}$: current of the battery.

The battery's state of charge is expressed by the equation (4) (Bilal et al., 2023),

$$SoC_{battery} = \frac{E_{battery}}{E_{battery}^{maximum}} \times 100 \qquad (4)$$

where, $SoC_{battery}$: state of charge of the batteries, $E_{battery}$: total energy stored in the battery, $E_{battery}^{maximum}$: maximum energy stored in the battery.

The inverter's primary function is to control the system's power flow by converting DC electricity from the battery and SPV to AC power and vice versa. The input power of the bi-directional input power is given by the equation (5) (Bilal et al., 2023),

$$P_{ip} = \frac{P_{op}}{\eta_{bidinverter}} \qquad (5)$$

where, P_{ip}: dc input power, P_{op}: ac output power, $\eta_{bidinverter}$: efficiency of the inverter.

The whole cost of each component in the energy system is shown by the total net present cost, or TNPC. The mathematical formula for the TNPC is given (6) (Bilal et al., 2023),

$$TNPC = \frac{C_t^{ann}}{CRF(i,N)} \qquad (6)$$

where, *TNPC*: total net present cost, C_t^{ann}: total annualized cost ($), *i*: actual discount rate (%), *N*: lifespan of the project expressed (year), *CRF*: capital recovery factor.

An integrated energy system's economics are assessed using the LCOE. Equation (7) (Bilal et al., 2023), determines the LCOE as the ratio of the yearly cost of system components to the total amount of energy produced to meet demand,

$$LCOE = \frac{C_t^{ann}}{E_P} \tag{7}$$

where, C_t^{ann}: total annualized cost (\$), E_p: total energy production (year), and *LCOE*: levelized cost of energy.

MODELING OF RE POWERED EV CHARGING STATION

The modern cutting-edge technologies empower the continual monitoring and optimization of energy production and consumption, ushering in an era of intelligent and data-driven energy management. Additionally, the proposed system contemplates the assimilation of decentralized energy storage units, ensuring a dependable and resilient power supply even in scenarios devoid of direct sunlight or wind speed availability. In the present analysis, HOMER Pro simulates the operation of a hybrid or distributed energy system for an entire operating year, evaluating and optimizing the electrical system design, load profiles, components, fuel costs and environmental variables. The simulation produces key information on technical performance, risk-mitigation, and projected cost-savings to inform system design and optimization. The system modeling is illustrated in Figure 3.

Figure 3. Schematic of the integrated green energy system for charging infrastructure

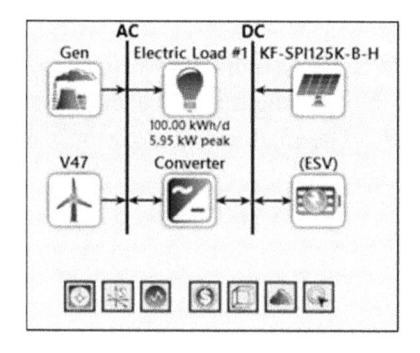

ENERGY HARVESTING POTENTIAL ACCESS

An enormous amount of the hazardous gas emissions is caused by conventional cars. Compared to conventional cars, EVs have a lot of benefits, such as less initial costs, less environmental effect and increased efficiency. However, a large number of EVs charging at once could negatively impact the traditional grid. As a result, some microgrids powered by renewable energy are now integrated with EV charging stations. In the case of the study location, it's surrounded by Jammu and Kashmir on the West and Himachal Pradesh on the South. The borders to the north and east are in dispute with Pakistan and China. This adds another layer of consideration because the location has specific geopolitical factors that need attention. A solar photovoltaic power plant needs two factors a lot of light and heat and Ladakh has both. Ladakh has highest amount of sunlight that falls in the panels (coordinates: latitude: 34.1526°N, longitude: 77.5771°E). There is a large amount of barren land available to the solar power developer. On average in Ladakh there are 325 sunny days in a year. The daily average solar energy available in per square meter in Leh district is 6-12 kWh.

Table 1. Meteorological resources of study site

Month	Clearness Index	Daily Radiation (kWh/m²/day)	Average Wind Speed (m/s)	Daily Temperature (°C)
Jan	0.547	2.860	2.990	-16.990
Feb	0.539	3.540	3.280	-15.680
Mar	0.531	4.450	3.410	-10.680
Apr	0.540	5.430	3.300	-4.820
May	0.556	6.190	3.130	-0.440
Jun	0.600	6.910	2.840	4.220
Jul	0.596	6.720	2.680	8.170
Aug	0.594	6.140	2.470	7.800
Sep	0.642	5.660	2.520	2.940
Oct	0.685	4.790	2.460	-4.050
Nov	0.660	3.610	2.580	-9.460
Dec	0.573	2.760	2.870	-14.030

The rising popularity of EV technology and the declining cost of PV module and wind turbine systems have led to the recent start of commercial manufacturing of solar-powered battery charging stations for electric vehicles. The growing awareness of environmental issues and technological advancements are both contributing factors to the popularity of electric vehicles. The choice of location to set up a GEN/PV/WT battery hybrid system in India is important to harvest green energy at its maximum potential. The best place depends on sunlight and wind the place is getting because these are the sources of energy. It's also important to look at where need of electricity is most important. So, finding the right energy spot helps to generate power efficiently and uses nature's resources well to meet the electricity needs of the people. This assessment is crucial because it involves looking at places where the sun shines a lot, the wind blows well, and there is a demand for power.From the table it is observed that solar GHI solar radiation ranges from 2.760 - 6.910 kWh/m²/day with a scaled annual average value of 4.92 kWh/m²/day. This data is maximum for the month of June is6.910 kWh/m²/day and minimum for the month of December is2.760 kWh/m²/day.

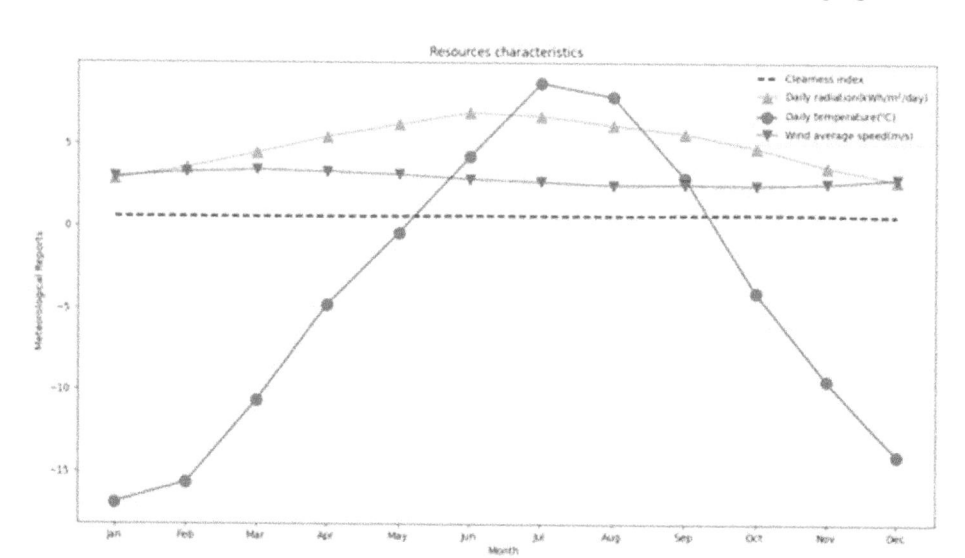

Figure 4. Meteorological characteristics for green energy powered EV charging station

Further, in the present analysis, Table 1 and Fig. 4 present monthly average wind speed data for Leh, Ladakh. From the graph it is observed that the monthly average wind speed data varies for the whole year. However, from the tabularresults, it is observed that the wind speed is maximum for March 3.410m/s and minimum for October 2.460m/s. The annual average wind speed is 2.88 m/s, while, the graph it is observed that the monthly average temperature data varies for the whole year. The temperature is maximum for July is 8.170°C and minimum for January is -16.990, while average temperature is -4.42°C.

The energy produce from the renewable energy sources will be used to power the EV charging station in Leh, Ladakh. The charging stations could be placed in strategic areas using the important halt areas for travelers along the route using Urvashi's retreat, Ride Inn Cafe and resort and the Ambika HP fuel outlet in Manali.

The Unicorn Hotel and The Royal Enfield Showroom in Khangsar and Hotel Ibex and Padma Lodge in Jispa. Along with Hemis Monastery and the Lato Guesthouse in Lato. The HP fuel outlets of Buddha Filling center and Ladakh Autonomous Hill Development Council in Leh have also received the infrastructure, with the Abduz and Grand Dragon hotel in Leh. And finally, the points chosen in Kargil are Hotel Kargil, Nubra Organic Retreat and Cafe-wala Chai. So, that the travelers can charge their electric vehicles overnight at these facilities(Encyclopaedia Britannica, 2021). The batteries connected stores extra energy that can be used for later purposes. This system produces a sufficient steady and self sufficient green energy supply of electricity.

RESULTS AND DISCUSSION

Numerous nations have put in place a range of agreements to preserve environmental sustainability on a worldwide scale as a result of the numerous ways that global economies damage the environment. Ladakh has an abundance of renewable energy sources, thus PV modules and WT systems are undoubtedly among the primary components. This study is an analytical investigation of the green energy supply to EV charging stations. This section covers energy consumption and power generation in the succeeding sub-sections:

(i). Energy Consumption: The present research reveals regional heterogeneity; that is, energy intensity targets in resource-rich and non-resource-rich areas exhibit positive and negative impacts, respectively, on renewable energy consumption, resulting in significant energy conservation and emission reduction. Energy intensity targets in local and neighboring provinces significantly promote the consumption of renewable energy.

Table 2. Energy production summary at the proposed cold region EV charging station

Production	kWh/Year (%)
Solar module (KEHUA France KF-SPI125K-B-H with Generic PV)	17,267 (5.75%)
Diesel generator (Autosize Genset)	214 (0.0712%)
Wind turbine (Vestas V47 660kW)	282,738 (94.2%)
Total	300,218 (100%)

Table 2-3 present the energy production, consumption for Leh, Ladakh and Figure 5 represent electrical power demand summary for Leh, Ladakh. This characteristics curve illustrates the percentage of energy demand fulfilled by various renewable energy sources such solar, wind and generator. The energy produced by the wind turbine is highest. The energy produced is maximum for the month of March and minimum for the month of October.

Table 3. Energy consumption optimization resultsforEV charging station

Consumption	kWh/year (%)	Consumption	kWh/year (%)
AC primary load	36,500 (100%)	Unmet electric load	0 (0%)
DC primary load	0 (0%)	Capacity storage	0 (0%)
Deferrable load	0 (0%)	Renewable fraction	99.4 (%)
Excess electricity	260,304 (76.0%)	-	-
Total	36,500 (100%)	Maximum renewable penetration	24,343(%)

Figure 5. Characteristics for load demand daily, seasonal and yearly profile

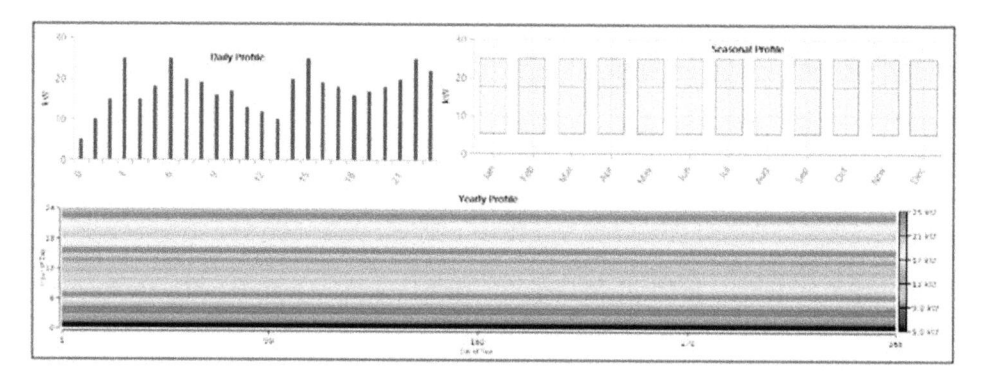

(ii).Optimization of Green Energy Generation: One of the most worldwide issues now is climate change and greenhouse gas emissions, primarily composed of CO_2, are thought to be the main reasons. Green projects aimed at producing renewable energy must be continuously financed in order to assuring energy demand and supply continuity as given in Table 4 and Figure 6. In this context, green financing plays a critical role. The following ways for energy harvesting affect the production and consumption of renewable energy:

Table 4. Renewable energy generation and penetration for EV charging station

Parameters	Value
Nominal renewable capacity divided by total nominal capacity (%)	99.0
Usable renewable capacity divideed by total capacity (%)	99.0
Total renewable production divided by load (%)	822
Total renewable production divided by generation (%)	99.9
One minus total nonrenewable production divided by load (%)	99.4
Renewable output divided by load (%)	24,343
Renewable output divided by total generation (%)	100
One minus nin renewable output divided by total load (%)	100

Figure 6. Characteristics for renewable penetration over a year for EV charging station

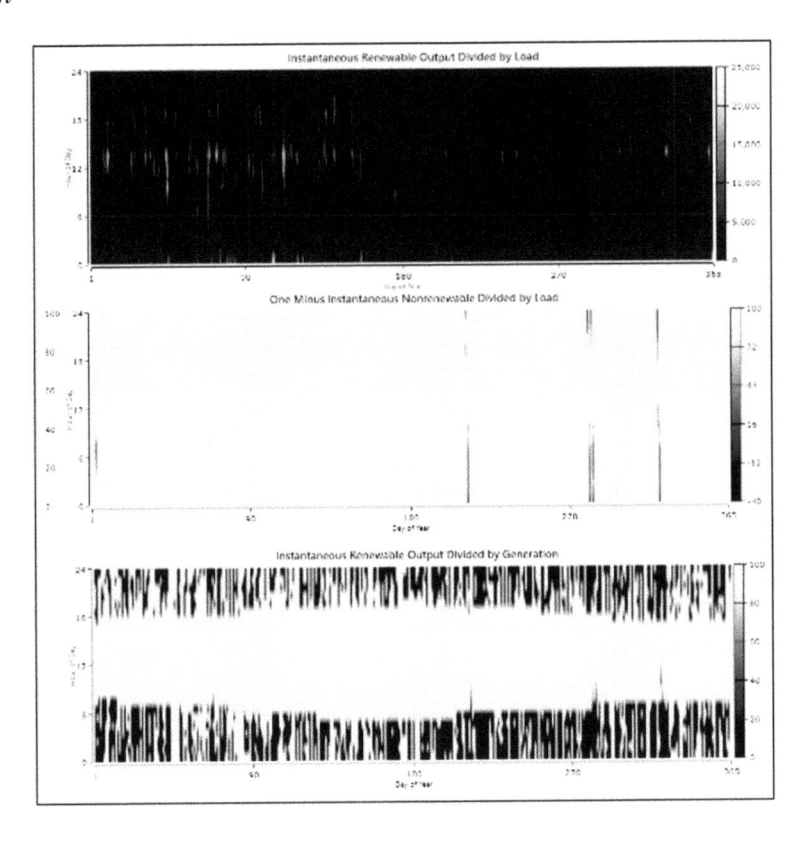

(a). Solar Power Generation: PV modules use solar radiation to create DC electricity in direct proportion to the amount of solar radiation present at any given moment at a certain location. Even though these early solar cells' efficiencies were not as high as those of conventional power plants at the time, significant improvements in solar cell technology have mostly lain dormant until recently. In this study, the solar flat module having KEHUA France KF-SPI125K-B-H with rated capacity, derating factor of 10 kW and 88%, respectively. Its dimension is followings: weight: 85 kg, L: 700mm; I: 310mm; H: 870mm. The capital, replacement and O&M cost is $7000, $7000 and $140, respectively. Table 5 and Figure 7 provide the output power information about the solar PV panels used for generating electricity. The PV panels generate 17,267kWh/year over a year. The levelized cost is 0.0634 $/kWh, which shows the cost effectiveness of the system. The characteristics curve of solar PV power output. The curve is plotted against day of year (x-axis) and hour of day (y-axis). The figure shows the PV power output daily profile for Leh, Ladakh.

Table 5. Solar module configuration and power generation results for EV charging station

Operational Parameters	Results
Rated power capacity (kW)	10.0
Mean power output (kW)	1.97
Mean power output (kWh/day)	47.3
Capacity factor (%)	19.7
Total energy production (kWh/year)	17,267
Minimum power output (kW)	0
Maximum power output (kW)	11.2
PV penetration (%)	47.3
Hours of operation (hours/year)	4,386
Levelized cost of energy ($/kWh)	0.0634

% continued to the next page

Renewable Energy-Integrated Electric Vehicle Charging Infrastructure

Figure 7. Power output characteristics of PV module and wind turbine for EV charging station

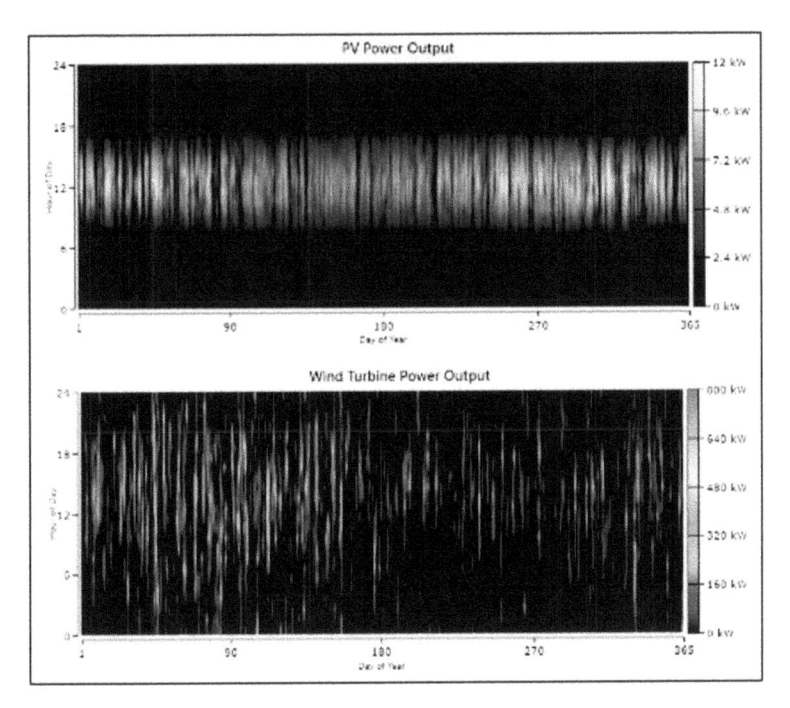

(b). Wind Power Generation: A wind turbine is a renewable energy conversion device that starts by converting wind energy from kinetic to mechanical form. Because wind energy is non-stationary and intermittent, changes in wind speed can quickly lead to system fluctuations, which compromise the wind power systems' ability to reliably produce power. The capital, replacement and O&M cost are $30,000.00, $30,000 and $600, respectively. The life time of the wind turbine is 20 years. Table 6 shows the details about the wind turbine. The total production is 282,738 kWh/year and levelized cost is 0.0115$/kWh. Here, Figures 7-8 shows the wind turbine power output plotted against day of year (x-axis) and corresponding power for Leh, Ladakh site.

Table 6. Wind turbine configuration and power generation results for EV charging station

Operational parameters	Results
Total rated capacity (kW)	660
Mean power output (kW)	32.3
Capacity factor (%)	4.89
Total power production (kWh/year)	282,738
Minimum power output (kW)	0
Maximum power output (kW)	721
Wind penetration (%)	775
Hours of penetration (hours/year)	5,022
Levelized cost of energy ($/kWh)	0.0115

Figure 8. Wind power output daily profile characteristics for EV charging station

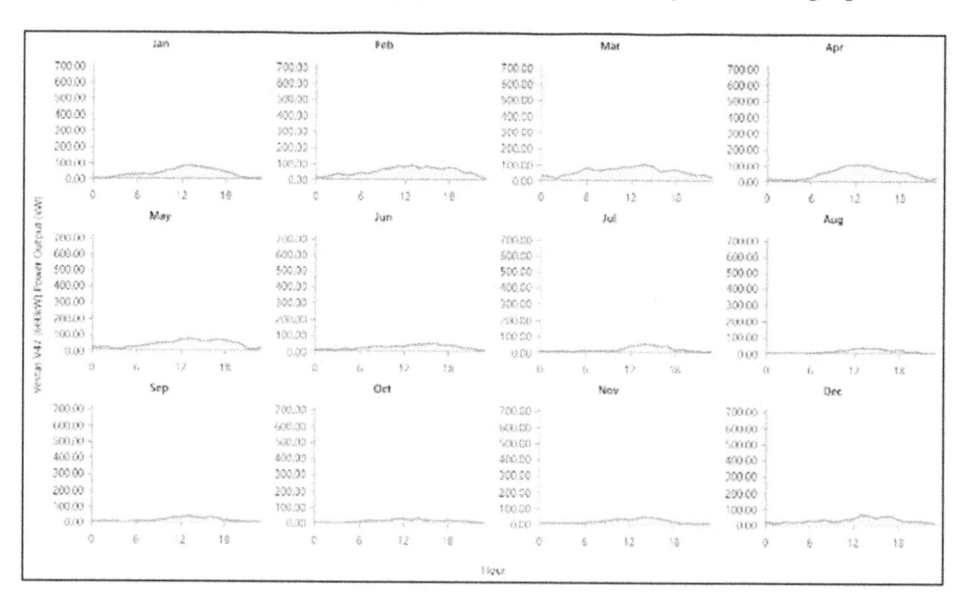

(c). Power Converter Operation: Battery units are used to control the system's energy based on the bidirectional converter. The batteries' maximum storage capacity is not equal to the maximum power at PCC. As a result, a longer battery life at PCC preserves the ideal DC link voltage. Table 7 shows the input and output details for the generic converter. The values from the characteristics curve provide information about the capacity, efficiency, energy production

and consumption of the inverter and rectifier components of the converter system. The inverter converts DC to AC and while rectifier converts AC to DC. In Figure 9, it is shown the details about the generic converter for Leh, Ladakh. The capacity:50kW, capital and replacement: 500.00$, O&M: 1$/year, respectively. The lifetime is 15 years, efficiency is 95% for inverter input. Relative capacity is 100% and efficiency is 95% for rectifier inputs.

Table 7. Power conversion mechanism for EV charging station

Quantity	Inverter	Rectifier
Rated power capacity (kW)	50.0	50.0
Mean power output (kW)	2.33	1.22
Minimum power putput (kW)	0	0
Maximum power output (kW)	5.95	50.0
Capacity factor (%)	4.67	2.45
Hours of operation (hour/year)	5,518	2,089
Energy out (kWh/year)	20,452	10,726
Energy in (kWh/year)	21,528	11,290
Power losses (kWh/year)	1,076	565

% continued to the next page

Figure 9. Operating characteristics of power inverter and rectifier for EV charging station

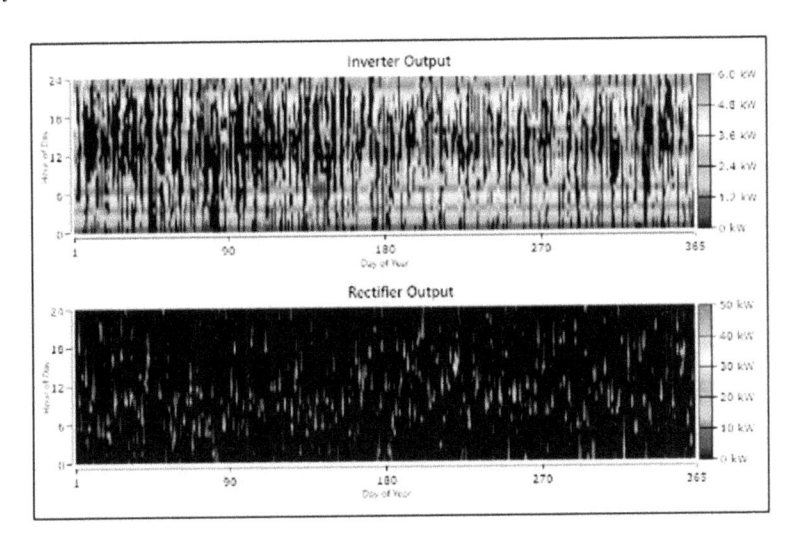

(d). Battery Energy Storage: Renewable energy hybrid system batteries serve as storage devices, storing surplus power generated by renewable sources. In times of power shortages, these batteries release stored energy, ensuring a continuous and reliable power supply. Figure 10 represents the storage specifications for Leh, Ladakh. The figures show the capital 2954$, replacement 2954$, O&M 60$/year cost of the battery. Nominal Voltage V8.4. Nominal capacity is 1.34 kWh andnominal capacity is 160Ah; while, roundtrip efficiency is90%. Maximum charge rate is 0.5 A/Ah. Maximum charge current is 80 A; while, maximum discharge current is 80 A. Table 8 shows the storage graphs details and specification for Leh, Ladakh.

Table 8. Energy storage with battery for EV charging station

Operational Parameters	Results	Operational Parameters	Results
Battery (quantity)	120.00	Lifetime throughput (kWh)	4,320,000
String size	1.00	Expected life (year)	254
Strings in parallel (strings)	120.00	Average energy cost ($/kWh)	0
Bus voltage (V)	8.40	Energy in (kWh/year)	17,912
Autonomy (hour)	38.7	Energy out (kWh/year)	16,139
Storage wear cost ($/kWh)	0.000721	Storage depletion (kWh/year)	19.3
Nominal capacity (kWh)	161	Losses (kWh/year)	1,792
Usable nominal capacity (kWh)	161	Annual throughput (kWh/year)	17,012

Figure 10. Energy storage characteristics of battery for EV charging station

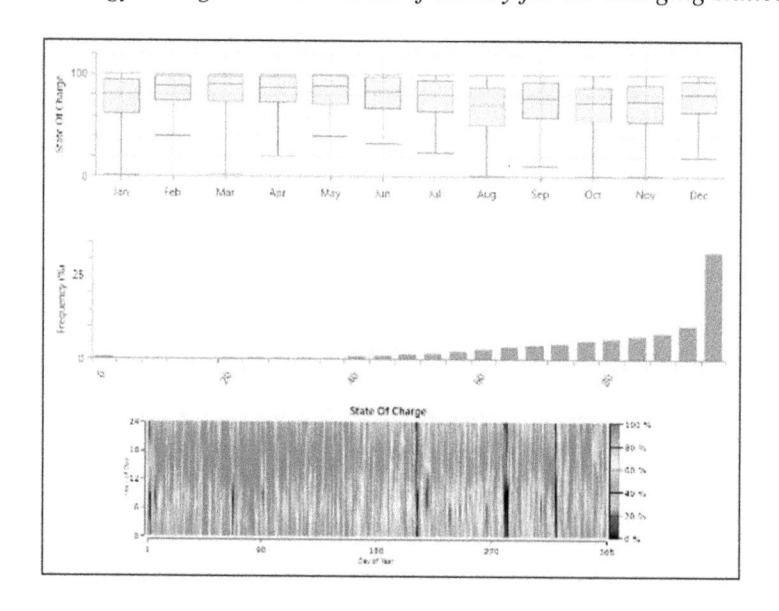

(e). Conventional Energy Production with Genset: A dependable EV charging station is always supported by a fallback power source in case demand exceeds renewable energy production. That backup power source will often be a diesel generator. The benefits of diesel generators include extended operating hours, inexpensive start up costs, convenient fuel access and ease of use. The biggest disadvantages of diesel generators, in spite of all their benefits, are their high operating and maintenance costs as well as their greenhouse gas emissions. The initial capital cost, replacement cost and O&M costare500$, 500$, O&M 0.030 $/op, respectively. The power operating characteristics of diesel generator is shown in Figure 11 and corresponding operational parameters are given in Table 9.

Figure 11. Characteristics of Genset power output for EV charging station at Leh, Ladakh

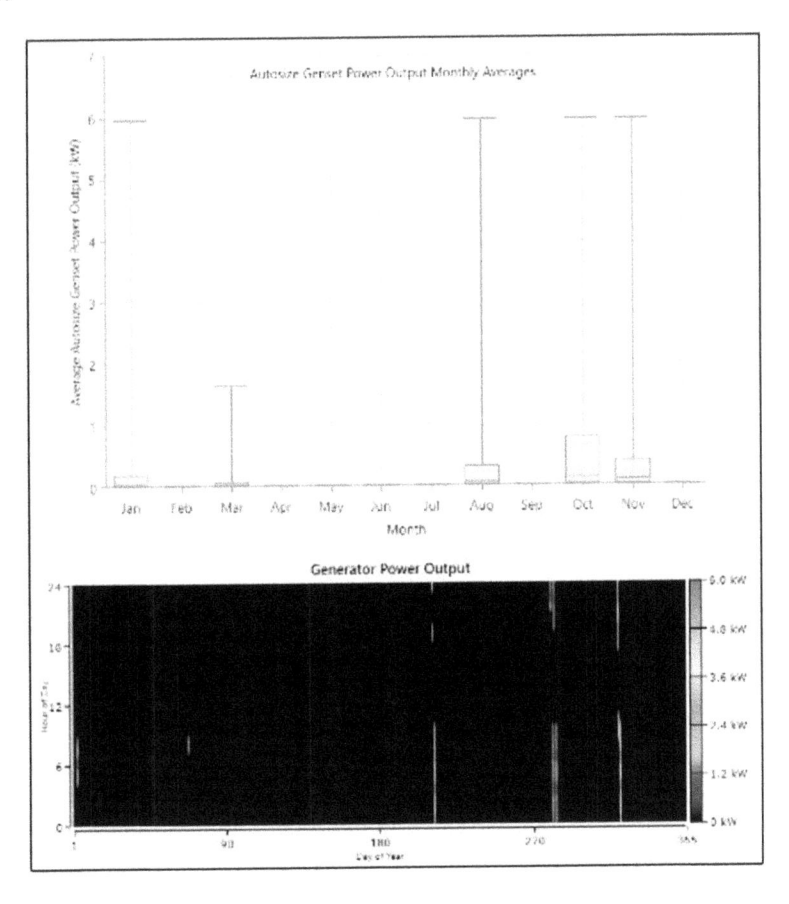

Table 9. Operational parameters of diesel generator for EV charging station regions

Operational Parameters	Results	Operational Parameters	Results
Hours of operation (hour/year)	68.0	Electrical production (kWh/year)	214
Number of starts (starts/hour)	8.00	Mean electrical output (kW)	3.14
Operational life (year)	221	Minimum electrical output (kW)	1.65
Capacity factor (%)	0.370	Maximum electrical output (kW)	5.95
Fixed generation cost ($/hour)	3.37	Fuel consumption (L)	75.5
Marginal generation cost ($/kWh)	1.89	Specific fuel consumption (L/kWh)	0.353
Mean electrical efficiency (%)	28.8	Fuel energy input (kWh/year)	743

(e). DC Operating Capacity Behavior: Here, Figure 12 illustrates the characteristics curve for average DC operating capacity for different months of a year.

Figure 12. DC operating capacity monthly average for the EV charging station

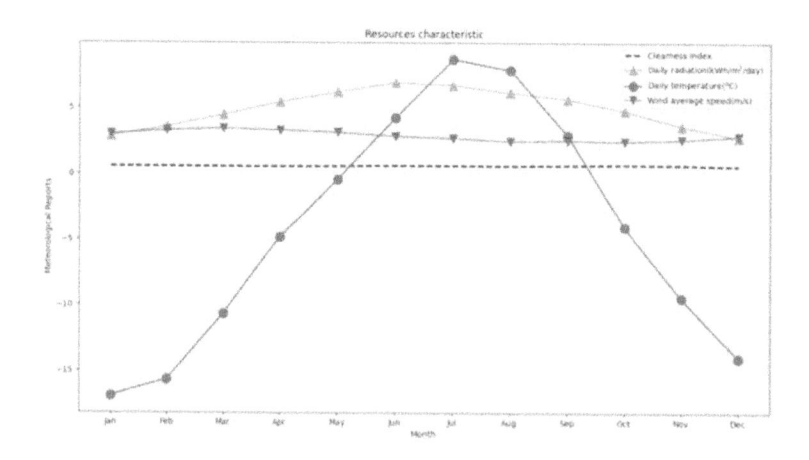

(f). AC Operating Capacity Behavior: In the present study, Figure 13 illustrates the characteristics curve for average AC operating capacity daily profile for a year.

Figure 13. AC operating capacity daily profile for the EV charging station

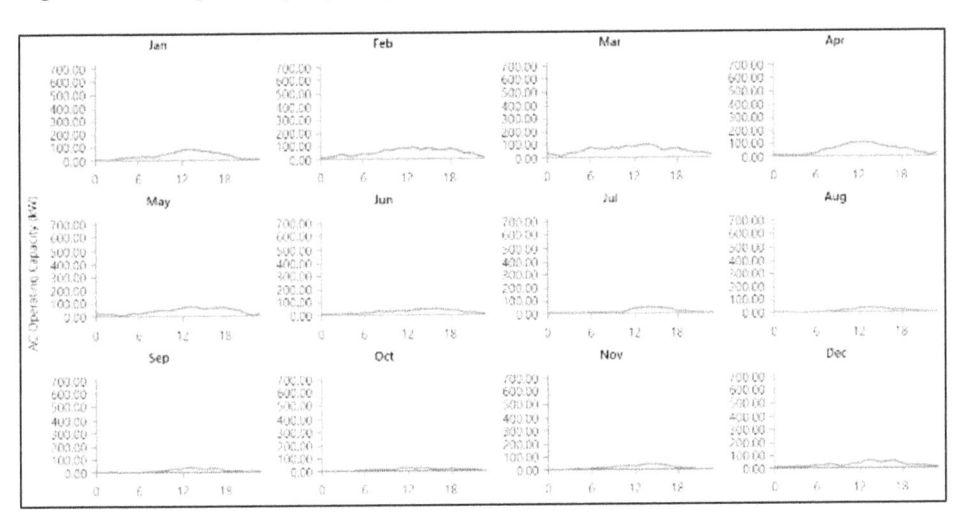

(iii). Economical Aspect of Optimization: Since renewable energy resources can be found in remote areas and because sustainable energy technologies are developing more quickly than fossil fuels, renewable energy is more affordable than fossil fuels. From direct to indirect ways, the green finance options for projects involving the production of renewable energy have evolved. Direct investments were the first to give rise to the taxes channels. Its correspond observation and comparative review have been presented in Table 9-11 and Figure 14-15. Having a twofold influence on the detrimental environmental externalities produced by the energy producing processes may be one of the driving forces for this transition. In reality, this element is receiving more and more attention due to the potential introduction of compensation for the averted CO_2 emissions.

Table 9. Rate chart details of the proposed green energy model for EV charging station

Banking Rate	Value
Discount rate (%)	8.00
Inflation rate (%)	2.00
Annual capacity storage (%)	0.00
Project lifetime (years)	25.00

Figure 14. Economical optimization results for EV charging station

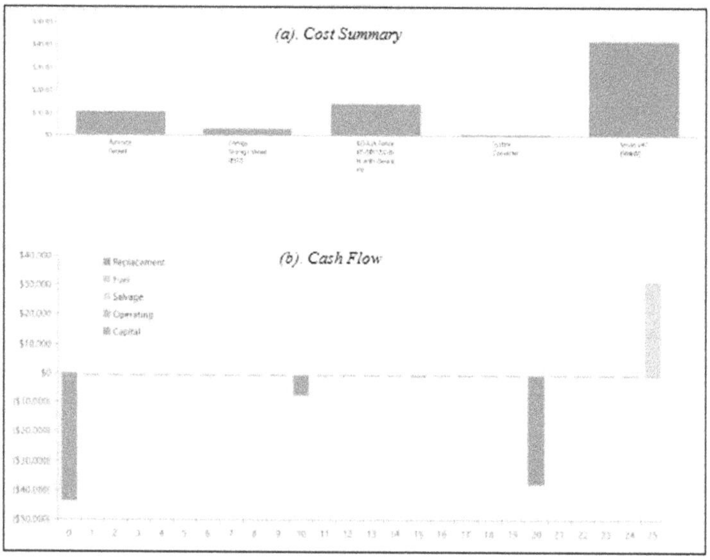

Table 10. Cost summary details of the proposed green energy model for EV charging station

Component	Capital ($)	Replacement ($)	O&M ($)	Salvage ($)	Total ($)
Autosize genset	$3,300	$0.00	$174.06	$700.95	$10,582.15
KF-SPI125K-B-H with Generic PV	$7,000.00	$6,184.00	$1,809.85	$838.45.00	$14,155.46
Energy storage	$2,954.00	$0.00	$775.65	$637.99	$3,091.67
Wind energy	$30,000.00	$9,564.22	$7,756.51	$5,390.05	$41,930.68
System converter	$500.00	$212.14	$12.93	$39.93	$685.14
Overall System	$43,754	$15,960.42	$10,529.00	$7,607.36	$70,445.10

Table 11. Comparative economics analysis results for EV charging station

Metric	Value
Present worth ($)	46,404
Annual worth ($/year)	3,590
Return on investment (%)	55.7
Internal rate of return (%)	63.8
Simple payback (year)	1.56
Discounted payback (year)	1.69s

Figure 15. Comparative optimization results summary for EV charging station at Leh, Ladakh

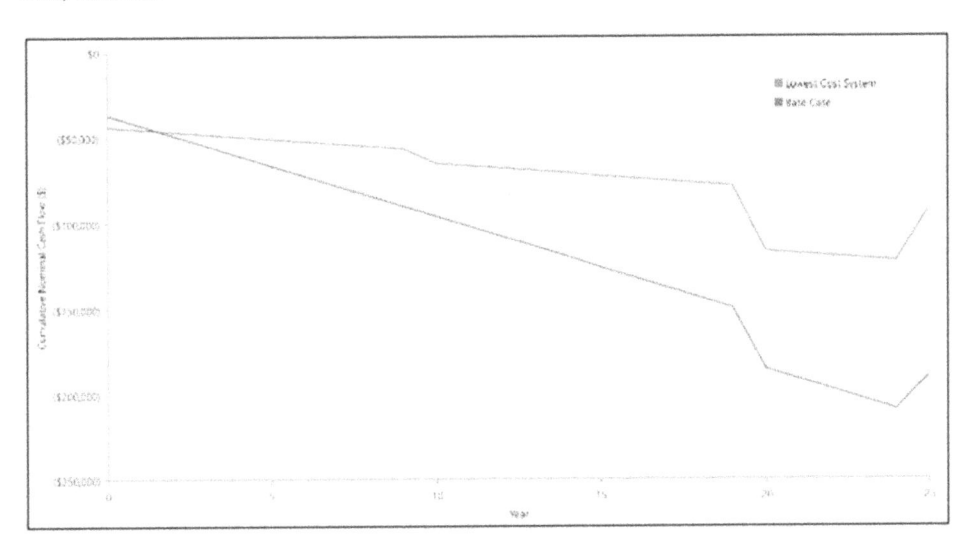

(iv). Environmental Impact Assessment: Apart from the economic factor, which is significant from the standpoint of the community and end user, the environmental aspect is also crucial, particularly in this case when the primary objective of the incentive is to reduce carbon emissions. Numbers of elements influence the pollutant emissions such as innovation, financial development, transportation infrastructure and economic expansion, which in turn impact environmental sustainability. The nation can achieve environmental stability through innovation. Through improved production techniques and low energy-intensive products, it contributes to the reduction of harmful emissions as one of the main ways to cut CO_2, SO_x and NO_x emissions. Since innovation offers new and more effective ways to carry out production processes, it is anticipated that innovation will have an adverse effect on carbon emissions. In this study, Table 12 illustrates the harmful gases emitted to the environment.

Table 12. Emissions analysis results for EV charging station regions

Quantity	Value(kg/year)
Carbon Dioxide	198
Carbon Monoxide	1.25
Unburned Hydrocarbons	0.0544
Particular Matter	0.00755
Sulfur Dioxide	0.484
Nitrogen Oxides	1.17

CONCLUSION

This study presents an innovative strategy for efficient and ecological transportation considering energy harvesting and management technologies into smart roadways designed for electric vehicles. Although smart roads with energy collecting capabilities may have high implementation costs at first, there are considerable long-term economic benefits. In the larger framework of sustainable urban development, these systems are economically feasible due to the decrease in conventional energy usage as well as the possibility of generating income through energy sales to the grid or electric vehicle charging. Throughout this study, it has been looked at a number of renewable energy sources and technologies for maximizing the greener way to power EVs along the road infrastructure. Energy harvesting technologies, including PV panels, wind turbine, etc. have shown promising approach in capturing and converting ambient energy from sunlight, wind and moving vehicles in daily life. This makes it possible for self-sustaining ecosystems to develop on smart highways. In future perspective, the incorporation of energy harvesting and management technologies into electric vehicle-specific smart roadways offers a revolutionary chance to develop intelligent, efficient, and sustainable transportation networks.

REFERENCES

Abid, M. S., Ahshan, R., Al Abri, R., Al-Badi, A., & Albadi, M. (2024). Techno-economic and environmental assessment of renewable energy sources, virtual synchronous generators, and electric vehicle charging stations in microgrids. *Applied Energy, 353*, 122028. doi:10.1016/j.apenergy.2023.122028

AbuElrub, A., Hamed, F., & Saadeh, O. (2020). Microgrid integrated electric vehicle charging algorithm with photovoltaic generation. *Journal of Energy Storage, 32*, 101858. doi:10.1016/j.est.2020.101858

Al Wahedi, A., & Bicer, Y. (2022). Techno-economic optimization of novel stand-alone renewables-based electric vehicle charging stations in Qatar. *Energy, 243*, 123008. doi:10.1016/j.energy.2021.123008

AltEnergyMag. (2018). *Piezoelectric Power Generation Automotive Tires.* https://www.altenergymag.com/article/2017/12/1-article-for-2018-piezoelectric-power-generation-in-automotive-tires/27642/

Bai, S., & Liu, C. (2021). Overview of energy harvesting and emission reduction technologies in hybrid electric vehicles. *Renewable & Sustainable Energy Reviews, 147*, 111188. doi:10.1016/j.rser.2021.111188

Bilal, M., Ahmad, F., & Rizwan, M. (2023). Techno-economic assessment of grid and renewable powered electric vehicle charging stations in India using a modified metaheuristic technique. *Energy Conversion and Management, 284*, 116995. doi:10.1016/j.enconman.2023.116995

Chen, C., Xu, T. B., Yazdani, A., & Sun, J. Q. (2021). A high density piezoelectric energy harvesting device from highway traffic - System design and road test. *Applied Energy, 299*, 117331. doi:10.1016/j.apenergy.2021.117331

Cunha, Á., Brito, F. P., Martins, J., Rodrigues, N., Monteiro, V., Afonso, J. L., & Ferreira, P. (2016). Assessment of the use of vanadium redox flow batteries for energy storage and fast charging of electric vehicles in gas stations. *Energy, 115*, 1478–1494. doi:10.1016/j.energy.2016.02.118

Electrek, (2022). *BMW designs EV suspension system that turns bumpy roads into usable power.* https://electrek.co/2022/12/01/bmw-designs-ev-suspension-system-that-generates-usable-energy/

Engineering New-Record. (2022). *Mixed Results as Smart Road Testing Begins.* https://www.enr.com/articles/54723-q3-tech-focus-mixed-results-as-smart-roading-testing-begins

Ghosh, A., & Aggarwal, V. (2018). Menu-based pricing for charging of electric vehicles with vehicle-to-grid service. *IEEE Transactions on Vehicular Technology, 67*(11), 10268–10280. doi:10.1109/TVT.2018.2865706

Government Technology. (2016). *California to Test Road Vibrations as an Energy Source*. https://www.govtech.com/fs/california-to-test-road-vibrations-as-an-energy-source.html

Hasan, S., Zeyad, M., Ahmed, S. M., Mahmud, D. M., Anubhove, M. S. T., & Hossain, E. (2023). Techno-economic feasibility analysis of an electric vehicle charging station for an International Airport in Chattogram, Bangladesh. *Energy Conversion and Management, 293*.

Jeon, D. H., Cho, J. Y., Jhun, J. P., Ahn, J. H., Jeong, S., Jeong, S. Y., Kumar, A., Ryu, C. H., Hwang, W., Park, H., Chang, C., Lee, H., & Sung, T. H. (2021). A lever-type piezoelectric energy harvester with deformation-guiding mechanism for electric vehicle charging station on smart road. *Energy, 218*, 119540. doi:10.1016/j.energy.2020.119540

Joseph, P. K., & Devaraj, E. (2019). Design of hybrid forward boost converter for renewable energy powered electric vehicle charging applications. *IET Power Electronics, 12*(8), 2015–2021. doi:10.1049/iet-pel.2019.0151

Kang, T., Li, H., Zheng, L., Li, J., Xia, D., Ji, L., Shi, Y., Wang, H., & Chen, M. (2023). Distributed plug-in electric vehicles charging strategy considering driver behaviours and load constraints. *Electric Power Systems Research, 220*, 109367. doi:10.1016/j.epsr.2023.109367

Khan, K. A., Quamar, M. M., Al-Qahtani, F. H., Asif, M., Alqahtani, M., & Khalid, M. (2023). Smart grid infrastructure and renewable energy deployment: A conceptual review of Saudi Arabia. *Energy Strategy Reviews., 50*, 101247. doi:10.1016/j.esr.2023.101247

Khan, S., Sudhakar, K., Yusof, M. H. B., Azmi, W. H., & Ali, H. M. (2023). Roof integrated photovoltaic for electric vehicle charging towards net zero residential buildings in Australia. *Energy for Sustainable Development, 73*, 340–354. doi:10.1016/j.esd.2023.02.005

Kim, H., Tai, W. C., Parker, J., & Zuo, L. (2019). Self-tuning stochastic resonance energy harvesting for rotating systems under modulated noise and its application to smart tires. *Mechanical Systems and Signal Processing, 122*, 769–785. doi:10.1016/j.ymssp.2018.12.040

Li, K., Chen, J., Sun, X., Lei, G., Cai, Y., & Chen, L. (2023). Application of wireless energy transmission technology in electric vehicles. *Renewable & Sustainable Energy Reviews, 184*, 113569. doi:10.1016/j.rser.2023.113569

Li, K., Chen, J., Sun, X., Lei, G., Cai, Y., & Chen, L. (2023). Transportation systems management considering dynamic wireless charging electric vehicles: Review and prospects. *Transportation Research Part E, Logistics and Transportation Review*, *163*, 102761.

Li, Z., Xu, Z., Chen, Z., Xie, C., Chen, G., & Zhong, M. (2023). An empirical analysis of electric vehicles' charging patterns. *Transportation Research Part D, Transport and Environment*, *117*, 103651. doi:10.1016/j.trd.2023.103651

Lin, H., Zhou, Y., Li, Y., & Zheng, H. (2024). Aggregator pricing and electric vehicles charging strategy based on a two-layer deep learning model. *Electric Power Systems Research*, *227*, 109971. doi:10.1016/j.epsr.2023.109971

Lyu, F., Cai, T., & Huang, F. (2023). A universal wireless charging platform with novel bulged-structure transmitter design for multiple heterogeneous autonomous underwater vehicles (AUVs). *IET Power Electronics*, *16*(13), 2162–2177. doi:10.1049/pel2.12536

Mehrjerdi, H. (2020). Dynamic and multi-stage capacity expansion planning in microgrid integrated with electric vehicle charging station. *Journal of Energy Storage*, *29*, 101351. doi:10.1016/j.est.2020.101351

Nasr, T., Torabi, S., Bou-Harb, E., Fachkha, C., & Assi, C. (2022). Power jacking your station: In-depth security analysis of electric vehicle charging station management systems. *Computers & Security*, *112*, 102511. doi:10.1016/j.cose.2021.102511

Next Move Strategy Consulting. (2023). https://www.nextmsc.com/report/india-electric-vehicle-ev-cha rging-market

Ozkan, H. A., & Erol-Kantarci, M. (2022). A novel Electric Vehicle Charging/Discharging Scheme with incentivization and complementary energy sources. *Journal of Energy Storage*, *51*, 104493. doi:10.1016/j.est.2022.104493

Pei, J., Guo, F., Zhang, J., Zhou, B., Bi, Y., & Li, R. (2021). Review and analysis of energy harvesting technologies in roadway transportation. *Journal of Cleaner Production*, *288*, 125338. doi:10.1016/j.jclepro.2020.125338

Qin, Y., Kishk, M. A., & Alouini, M. S. (2022). Performance Analysis of Charging Infrastructure Sharing in UAV and EV-involved Networks. *IEEE Transactions on Vehicular Technology*, *72*(3), 3973–3988. doi:10.1109/TVT.2022.3219764

Sadeghian, O., Oshnoei, A., Mohammadi-Ivatloo, B., Vahidinasab, V., & Anvari-Moghaddam, A. (2022). A comprehensive review on electric vehicles smart charging: Solutions, strategies, technologies, and challenges. *Journal of Energy Storage, 54*, 105241. doi:10.1016/j.est.2022.105241

Sagaria, S., Duarte, G., Neves, D., & Baptista, P. (2022). Photovoltaic integrated electric vehicles: Assessment of synergies between solar energy, vehicle types and usage patterns. *Journal of Cleaner Production, 348*, 131402. doi:10.1016/j.jclepro.2022.131402

Solar Power World. (2019). *4 ways solar is contributing to smart road technologies.* https://www.solarpowerworldonline.com/2019/03/4-ways-solar-is-contributing-to-smart-road-technologies/

Sun, M., Wang, W., Zheng, P., Luo, D., & Zhang, Z. (2021). A novel road energy harvesting system based on a spatial double V-shaped mechanism for near-zero-energy toll stations on expressways. *Sensors and Actuators. A, Physical, 323*, 112648. doi:10.1016/j.sna.2021.112648

The Engineer. (2019). *Smart infrastructure to harvest energy from roads.* https://www.theengineer.co.uk/content/news/smart-infrastructure-to-harvest-energy-from-roads/

The Times of India. (2023). *Why India-specific EV charging and integrated solutions are the key to success for green mobility in India.* https://timesofindia.indiatimes.com/blogs/voices/why-india-specific-ev-charging-and-integrated-solutions-are-the-key-to-success-for-green-mobility-in-india/

TheNextWeb. (2022). *Can vehicle charging-roads power our electric future?* https://thenextweb.com/news/energy-generating-roads-can-charge-electric-vehicles

Venugopal, P., Shekhar, A., Visser, E., Scheele, N., Mouli, G. R. C., Bauer, P., & Silvester, S. (2018). Roadway to self-healing highways with integrated wireless electric vehicle charging and sustainable energy harvesting technologies. *Applied Energy, 212*, 1226–1239. doi:10.1016/j.apenergy.2017.12.108

Wang, C., Li, J., Yang, Y., & Ye, F. (2017). Combining solar energy harvesting with wireless charging for hybrid wireless sensor networks. *IEEE Transactions on Mobile Computing, 17*(3), 560–576. doi:10.1109/TMC.2017.2732979

Wu, Y., Zhang, J., Ravey, A., Chrenko, D., & Miraoui, A. (2020). Real-time energy management of photovoltaic-assisted electric vehicle charging station by markov decision process. *Journal of Power Sources*, *476*, 228504. doi:10.1016/j.jpowsour.2020.228504

Yang, C. H., Song, Y., Woo, M. S., Eom, J. H., Song, G. J., Kim, J. H., Kim, J., Lee, T. H., Choi, J. Y., & Sung, T. H. (2017). Feasibility study of impact-based piezoelectric road energy harvester for wireless sensor networks in smart highways. *Sensors and Actuators. A, Physical*, *261*, 317–324. doi:10.1016/j.sna.2017.04.025

Yap, K. Y., Chin, H. H., & Klemeš, J. J. (2022). Solar Energy-Powered Battery Electric Vehicle charging stations: Current development and future prospect review. *Renewable & Sustainable Energy Reviews*, *169*, 112862. doi:10.1016/j.rser.2022.112862

Zhang, Y., & Cai, L. (2018). Dynamic charging scheduling for EV parking lots with photovoltaic power system. *IEEE Access : Practical Innovations, Open Solutions*, *6*, 56995–57005. doi:10.1109/ACCESS.2018.2873286

Zou, H. X., Zhao, L. C., Gao, Q. H., Zuo, L., Liu, F. R., Tan, T., Wei, K. X., & Zhang, W. M. (2019). Mechanical modulations for enhancing energy harvesting: Principles, methods and applications. *Applied Energy*, *255*, 113871. doi:10.1016/j.apenergy.2019.113871

Zou, H. X., Zhu, Q. W., He, J. Y., Zhao, L. C., Wei, K. X., Zhang, W. M., Du, R. H., & Liu, S. (2024). Energy harvesting floor using sustained-release regulation mechanism for self-powered traffic management. *Applied Energy*, *353*, 122082. doi:10.1016/j.apenergy.2023.122082

Chapter 3

Intelligent Gas Detection Systems:
Leveraging IoT and Machine Learning for Early Warning of Hazardous Gas and LPG Leakage

Kondireddy Muni Sankar
Department of Information Technology, Vel's University (VISTAS), Chennai, India

B. Booba
Department of Information Technology, Vel's University (VISTAS), Chennai, India

ABSTRACT

The chapter explores the integration of internet of things (IoT) and machine learning (ML) algorithms in hazardous gas detection systems, specifically LPG leakage. The gas detection systems industry is undergoing a significant transformation due to technological advancements, innovative sensor technologies, and intelligent solutions. This shift is crucial for safety in residential and industrial environments. The integration of gas sensing technologies with IoT devices has led to real-time monitoring, data analytics, and smart automation. Smart homes and industrial facilities benefit from interconnected gas detection systems that detect anomalies and trigger automated responses. Regulatory bodies are implementing standards to integrate advanced gas detection technologies into the automobile sector, focusing on sensor capabilities, data analytics, and automation integration. This integration will create a safer environment, prioritizing safety in connected living and working spaces. The chapter also discusses real-world case studies using automobiles.

DOI: 10.4018/978-1-6684-9214-7.ch003

INTRODUCTION

The ever-increasing demand for energy resources, coupled with the widespread use of liquefied petroleum gas (LPG), has necessitated the development of robust and intelligent gas detection systems. This chapter introduces the concept of Intelligent Gas Detection Systems, emphasizing their pivotal role in enhancing safety and preventing potential hazards associated with the leakage of hazardous gases, particularly LPG. As the consequences of gas leaks can be severe, ranging from environmental pollution to life-threatening situations, the need for advanced technologies to provide early warnings becomes paramount (Baballe & Bello, 2022).

LPG, widely used in residential, commercial, and industrial sectors, poses inherent risks due to its flammable nature. Intelligent gas detection systems can mitigate these risks and safeguard lives and property. Traditional systems often lack timely alerts, necessitating a paradigm shift towards intelligent systems for better response times and incident severity. The paragraph highlights the integration of IoT and Machine Learning technologies in intelligent gas detection systems, enabling real-time monitoring and data collection, and improving detection accuracy, offering a dynamic approach to gas leak prevention (Tang et al., 2020).

It delves into the fundamentals of gas sensing, discussing various types of sensors, their applications, and their working principles, providing insights into gas detection mechanisms. The introduction introduces the chapter, detailing IoT integration in gas detection systems, ML algorithms for gas leak prediction, practical implementation considerations, and real-world case studies. It provides a comprehensive understanding of intelligent gas detection systems. Hazardous gas detection is crucial for preventing potential disasters and promoting safety in various settings, including industrial facilities and residential spaces. It is essential to detect hazardous gases to prevent fires, explosions, and health hazards, ensuring a secure living and working environment (Ha et al., 2020).

The paragraph discusses hazardous gas detection's objectives, including preventing accidents and providing early warnings for evacuation. It emphasizes the need for sensitivity and specificity to minimize false alarms. The evolution of gas detection technologies, including advanced sensors, communication technologies, and data analytics, has transformed it into a proactive field (Khan, 2020).

This summary delves into hazardous gas detection in various industries like manufacturing, petrochemicals, and utilities, emphasizing the significance of regulatory standards in shaping the industry landscape, mitigating risks, and fostering a safety culture, while providing a foundation for understanding hazardous gas detection technologies.

Gap Identification

- The introduction discusses the integration of IoT and Machine Learning technologies in intelligent gas detection systems, but it lacks a comprehensive analysis of potential challenges and limitations, suggesting further research is needed.
- The paragraph discusses practical implementation challenges of intelligent gas detection systems, such as real-time monitoring and data collection, but lacks a comprehensive exploration of the barriers faced in real-world scenarios, suggesting that identifying and addressing these issues could bridge the gap between theory and practice.
- The introduction emphasizes the importance of sensitivity and specificity in gas detection systems to reduce false alarms, but lacks performance metrics or evaluation methods, suggesting the need for clear performance criteria.
- The introduction of hazardous gas detection lacks a comprehensive understanding of industry-specific challenges, indicating the need for a broader exploration of gas detection needs across industries, and lacks specific details on compliance issues.

Importance of Early Detection in LPG Leakage

Liquefied Petroleum Gas (LPG) is a versatile energy source, but its flammable nature poses risks in leaks. Early detection is crucial to prevent minor incidents and catastrophic consequences like explosions or fires. Delayed detection can have cascading effects on safety, property, and the environment. Moreover, this part explores the economic impact of early detection, emphasizing that swift responses can minimize damage, reduce downtime, and lower recovery costs. By preventing large-scale disasters, early detection contributes to preserving assets, protecting investments, and ensuring the continuity of operations in both residential and industrial settings. This section discusses the environmental impact of LPG leakage, emphasizing the need for early detection systems to prevent environmental damage and promote sustainability. It also explores the potential of advanced technologies like IoT and Machine Learning to enhance gas detection systems (Rathor et al., 2023).

Role of IoT and Machine Learning in Enhancing Gas Detection Systems

The integration of IoT and Machine Learning has significantly improved gas detection systems, enhancing accuracy, efficiency, and adaptability. IoT connects gas sensors via wireless networks, enabling real-time monitoring and data collection.

Machine learning (ML) algorithms in gas detection allow systems to learn and adapt based on data patterns. Supervised learning algorithms can identify gas signatures accurately, while unsupervised learning methods excel in anomaly detection. The synergy between IoT and ML results in predictive analysis, allowing proactive measures to be taken before critical situations arise. ML's adaptive nature ensures that detection algorithms evolve over time, optimizing performance in response to changing conditions (Mohammed et al., 2021; Roshani et al., 2021).

It emphasizes the integration of IoT and ML in gas detection systems, enhancing their responsiveness, accuracy, and adaptability. It will delve into the technical aspects of implementing these technologies, providing practical applications and benefits (Miller et al., 2023).

TYPES OF GAS SENSORS

Gas sensors play a pivotal role in detecting and identifying various gases in diverse environments. Understanding the working principles and factors influencing gas detection is essential for designing effective and reliable gas sensing systems. This section delves into the working principles of seven distinct types of gas sensors, highlighting their distinctive features and their various applications (Song et al., 2021). The figure 1 depicts various types of gas detection sensors.

Catalytic Bead Sensors

Working Principle: These sensors operate on the principle of catalysis. When a combustible gas interacts with a catalyst bead, it causes a chemical reaction that increases the temperature of the bead. The change in temperature is measured and correlated to the gas concentration (Ha et al., 2020).

Applications: Catalytic bead sensors are commonly used for detecting flammable gases such as methane and propane in industrial settings.

Electrochemical Sensors

Working Principle: Electrochemical sensors utilize the electrochemical reaction between a gas and an electrode, producing an electric current that is directly proportional to the gas concentration.

Applications: Commonly used for detecting toxic gases like carbon monoxide and hydrogen sulfide, electrochemical sensors find applications in industrial safety and indoor air quality monitoring.

Infrared (IR) Gas Sensors

Working Principle: IR sensors detect gases based on their absorption of infrared radiation. Each gas absorbs specific wavelengths of infrared light, and by measuring the attenuation of the light passing through a gas sample, the sensor can identify and quantify the gas.

Applications: Infrared sensors are versatile and used for detecting a wide range of gases, including methane, carbon dioxide, and refrigerants.

Photoionization Detectors (PID)

Working Principle: PID sensors operate by ionizing gas molecules when exposed to ultraviolet light. The generated ions produce a measurable current that is proportional to the gas concentration.

Applications: PID sensors are effective in detecting volatile organic compounds (VOCs) and are widely used in environmental monitoring and industrial safety.

Metal Oxide Semiconductor (MOS) Sensors

Working Principle: MOS sensors detect gases based on the changes in conductivity of a metal oxide semiconductor when it comes into contact with a target gas. Gas adsorption alters the conductivity, and this change is measured.

Applications: MOS sensors are used for detecting a variety of gases, including methane, ammonia, and volatile organic compounds, making them suitable for applications in residential and industrial environments.

Semiconductor Gas Sensors

Working Principle: Semiconductor sensors operate on the principle of the gas-induced change in conductivity of a semiconductor material. The presence of a specific gas alters the electrical properties of the semiconductor.

Applications: Widely used for detecting gases like carbon monoxide, hydrogen, and methane, semiconductor sensors are employed in industrial safety and automotive applications.

Acoustic Wave Sensors

Working Principle: These sensors rely on the propagation of acoustic waves through a material. The presence of a gas alters the acoustic properties, and the changes are measured to determine the gas concentration.

Intelligent Gas Detection Systems

Applications: Acoustic wave sensors are employed for the detection of various gases, including volatile organic compounds, and find applications in environmental monitoring and industrial safety.

Factors influencing gas detection across these sensor types include environmental conditions (temperature, humidity), interference from other gases, sensor sensitivity, response time, and maintenance requirements. Selecting the appropriate sensor type depends on the specific gas to be detected, the application requirements, and the environmental conditions in which the sensor will operate.

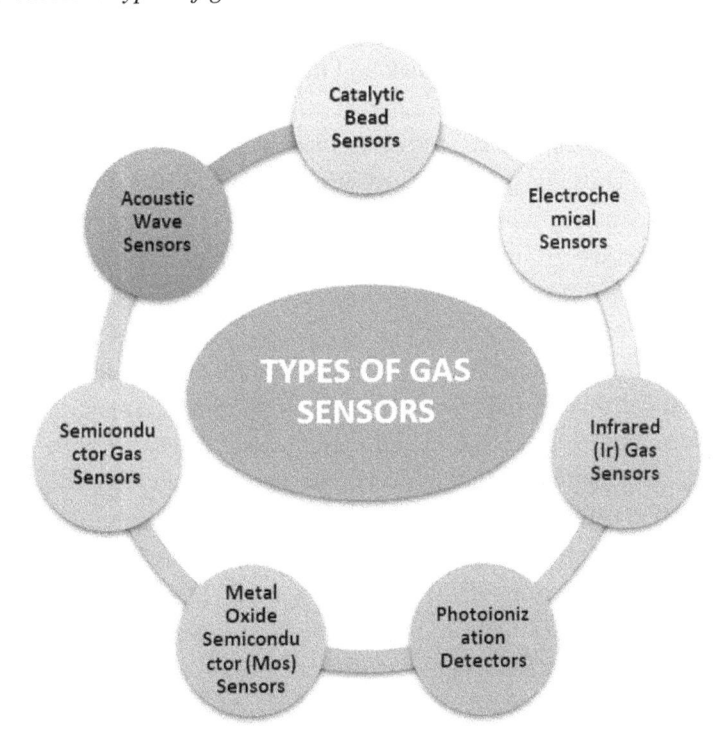

Figure 1. Variable types of gas detection Sensors

IOT IN GAS DETECTION SYSTEMS

Integration of IoT Devices for Real-Time Monitoring

The integration of Internet of Things (IoT) devices has revolutionized gas detection systems, propelling them into an era of real-time monitoring and dynamic responsiveness. This section explores how the synergy between gas detection systems and IoT contributes to enhanced safety measures and improved operational efficiency (Agrawal et al., 2023; Maguluri, Ananth, et al., 2023).

IoT devices, including wireless sensors and communication modules, enable the seamless integration of gas detection systems into interconnected networks. These networks facilitate real-time data transmission and monitoring, allowing for instantaneous detection of gas leaks or anomalies. The real-time aspect is particularly crucial in scenarios where rapid responses are paramount to prevent potential hazards (Kavitha et al., 2023; Venkateswaran, Kumar, et al., 2023). One key advantage of IoT integration is the ability to remotely monitor gas levels across diverse environments. This remote accessibility provides flexibility in monitoring large and geographically dispersed areas, making it well-suited for applications in industrial complexes, smart cities, and critical infrastructure. Operators can access real-time data through centralized dashboards, ensuring quick and informed decision-making (Hussain et al., 2023; Pachiappan et al., 2023; Rahamathunnisa, Sudhakar, Padhi, et al., 2023).

Furthermore, IoT-enabled gas detection systems offer continuous and automated data collection. This constant flow of information allows for the creation of comprehensive datasets, which, when coupled with advanced analytics, empower Machine Learning algorithms to discern patterns and predict potential gas leakages. This predictive capability enhances the overall effectiveness of the gas detection system, enabling proactive measures to be taken before critical situations arise. In addition to real-time monitoring, IoT integration supports predictive maintenance strategies (Karthik et al., 2023; Kumar B et al., 2024; Satav, Hasan, et al., 2024). By continuously monitoring the health and performance of gas sensors, maintenance needs can be anticipated, reducing downtime and ensuring the reliability of the entire system. This predictive maintenance approach contributes to cost savings and enhances the longevity of the gas detection infrastructure (Boopathi, 2023; Hema et al., 2023; Koshariya et al., 2023; Syamala et al., 2023).

The integration of IoT devices in gas detection systems revolutionizes safety and monitoring by providing real-time data, remote accessibility, and predictive capabilities. These systems not only detect gas leaks promptly but also optimize performance and reduce operational risks. As industries and urban environments adopt IoT technologies, intelligent gas detection systems are making significant strides towards safer and more resilient spaces.

Wireless Sensor Networks for Scalability

The implementation of Wireless Sensor Networks (WSNs) plays a pivotal role in enhancing the scalability of gas detection systems. This section explores how WSNs contribute to scalability by providing a flexible and interconnected framework for gas sensors, enabling seamless expansion and comprehensive coverage (Agrawal et al., 2023; Koshariya et al., 2023).

Scalability in gas detection systems refers to their ability to accommodate an increasing number of sensors and effectively cover larger geographical areas. WSNs offer a wireless communication infrastructure that allows sensors to communicate with each other and with a central monitoring station without the need for extensive cabling or fixed connections. This wireless communication paradigm significantly simplifies the deployment and expansion of gas detection networks. WSNs are highly adaptable to diverse environments, allowing them to cover additional zones without significant infrastructure changes. This scalability is particularly beneficial in dynamic settings like industrial facilities or smart cities, where gas detection requirements may evolve over time, making them a valuable tool.

Moreover, WSNs facilitate cost-effective scalability. Traditional wired systems can be expensive and cumbersome to expand, involving significant labor and material costs. In contrast, WSNs eliminate the need for extensive wiring, reducing installation and maintenance expenses. The wireless nature of the network also allows for quick and efficient reconfiguration as operational needs change. The distributed nature of WSNs enhances reliability and fault tolerance. In the event of a sensor failure or network disruption, the remaining sensors can continue to operate, ensuring that gas detection capabilities are not compromised. This redundancy contributes to the robustness of the overall system, a crucial factor in ensuring continuous and reliable gas monitoring (Vennila et al., 2023).

Additionally, the scalability of WSNs aligns with the requirements of complex and large-scale industrial processes. In scenarios where multiple sensors are required to cover extensive facilities, such as refineries or manufacturing plants, WSNs provide a scalable solution that can be easily expanded to match the evolving needs of the operational environment. In conclusion, Wireless Sensor Networks offer a scalable foundation for gas detection systems, enabling them to evolve alongside changing requirements and expanding spatial considerations. The wireless communication, adaptability to diverse environments, cost-effectiveness, and fault tolerance make WSNs a fundamental component in achieving scalable and efficient gas monitoring solutions. As industries continue to embrace the benefits of scalable WSNs, the evolution of intelligent and expansive gas detection networks advances towards safer and more resilient applications (Vennila et al., 2023).

Cloud Connectivity for Data Storage and Analysis

Cloud connectivity plays a pivotal role in advancing the capabilities of gas detection systems by providing a robust platform for data storage, analysis, and remote accessibility. This section explores how cloud integration enhances the efficiency, scalability, and intelligence of gas detection systems (Satav, Lamani, et al., 2024; M. Sharma et al., 2023; Srinivas et al., 2023).

Efficient Data Storage: Cloud platforms offer virtually unlimited storage capacity, eliminating the constraints imposed by local storage solutions. Gas detection systems generate large volumes of data from multiple sensors in real-time. Cloud storage allows this data to be securely stored, ensuring that historical information is readily available for analysis, compliance reporting, and auditing purposes.

Real-Time Data Analysis: Cloud-based solutions enable real-time analysis of gas concentration data. By leveraging cloud computing resources, gas detection systems can employ advanced analytics and machine learning algorithms to process incoming data streams. This analysis can identify patterns, anomalies, and trends, providing valuable insights into potential gas leakages or environmental changes (Agrawal et al., 2024; Rahamathunnisa, Sudhakar, Murugan, et al., 2023; Venkateswaran, Vidhya, et al., 2023).

Remote Accessibility: Cloud connectivity allows authorized users to access gas detection system data remotely from any location with internet access. This remote accessibility is particularly valuable for monitoring distributed facilities or conducting analysis without the need for physical presence on-site. Cloud-based dashboards provide real-time information, facilitating prompt decision-making.

Scalability And Flexibility: Cloud platforms inherently offer scalability, allowing gas detection systems to expand seamlessly as the number of sensors or the scope of monitored areas increases. The on-demand nature of cloud resources ensures that the system can adapt to changing requirements, providing flexibility in deployment and resource utilization.

Integration With IOT Devices: Cloud connectivity enables seamless integration with Internet of Things (IoT) devices, creating a holistic ecosystem for gas detection. The data collected by sensors can be transmitted to the cloud in real-time, fostering a connected environment where gas detection systems, analytics engines, and other IoT devices collaborate to enhance overall operational efficiency and safety (Reddy et al., 2023; Samikannu et al., 2023; Satav, Hasan, et al., 2024).

Data Security And Redundancy: Leading cloud service providers prioritize data security and offer advanced encryption measures to protect sensitive information. Additionally, cloud platforms often implement redundancy measures to ensure data integrity and availability. These security features contribute to the reliability and resilience of gas detection systems.

Cost-Effective Solutions: Cloud-based solutions eliminate the need for extensive on-site infrastructure and maintenance. Gas detection systems can leverage the pay-as-you-go model, allowing organizations to pay only for the resources they consume. This cost-effective approach is particularly advantageous for small to medium-sized enterprises seeking advanced gas detection capabilities without significant upfront investments (Dhanya et al., 2023; Mohanty et al., 2023).

Cloud connectivity is revolutionizing modern gas detection systems by providing efficient data storage, real-time analysis, remote accessibility, scalability, and enhanced integration capabilities. As industries adopt cloud solutions, the synergy between gas detection systems and cloud platforms drives innovation, making them intelligent, adaptable, and accessible for a safer future.

MACHINE LEARNING FOR GAS LEAKAGE PREDICTION

This section delves into the application of Machine Learning (ML) in gas leakage prediction, focusing on model training, anomaly detection, and methods for enhancing accuracy in intelligent gas detection systems (Boopathi & Kanike, 2023; Maheswari et al., 2023; Syamala et al., 2023; Zekrifa et al., 2023).

Training Machine Learning Models: Training ML models for gas leakage prediction involves utilizing historical data to enable the algorithm to learn patterns and correlations between various parameters. Supervised learning is commonly employed in this context. In the training phase, the model is fed with labeled datasets, where the features represent different environmental variables, and the labels indicate whether a gas leakage occurred. During training, the ML model learns to recognize distinctive patterns associated with normal conditions and those indicative of gas leakage. The choice of features is crucial, considering factors such as gas concentration levels, temperature, humidity, and sensor responses. The model is fine-tuned iteratively to optimize its performance in distinguishing between normal and anomalous situations (Ingle et al., 2023; Maheswari et al., 2023; Ramudu et al., 2023).

Anomaly Detection: One of the primary strengths of ML in gas leakage prediction lies in its ability to detect anomalies. Unsupervised learning methods, such as clustering or autoencoders, are often employed for anomaly detection. These algorithms learn the normal behavior of the system during training and can subsequently identify deviations from this learned pattern as anomalies. For example, clustering algorithms group data points based on similarity, and any data point that falls outside the established clusters may be flagged as an anomaly. Autoencoders, a type of neural network, learn to reconstruct input data, and instances where the reconstruction deviates significantly from the original input indicate anomalies (Kumar B et al., 2024).

Methods for Improved Accuracy:

Various techniques are utilized to enhance the precision of machine learning models in predicting gas leakage (Ramudu et al., 2023).

- **Feature Engineering:** Careful selection and engineering of features can significantly impact model performance. Identifying the most relevant variables and transforming or combining them intelligently can improve the model's ability to discern subtle patterns associated with gas leakage.
- **Hyperparameter Tuning:** Fine-tuning the hyperparameters of ML algorithms is essential for optimizing their performance. Adjusting parameters such as learning rates, regularization terms, and model architectures can significantly improve accuracy.
- **Ensemble Learning:** Ensemble methods, such as Random Forests or Gradient Boosting, combine predictions from multiple models to create a more robust and accurate final prediction. This approach is particularly effective in capturing diverse patterns present in complex datasets.
- **Cross-Validation Techniques:** To ensure the generalizability of the model, cross-validation is crucial. Techniques like k-fold cross-validation help assess the model's performance on various subsets of the dataset, reducing the risk of overfitting or underfitting.
- **Imbalanced Data Handling:** In gas detection scenarios, imbalanced datasets where instances of gas leakage are relatively rare can pose challenges. Techniques like oversampling, undersampling, or using specific algorithms designed for imbalanced data help address this issue.

Machine Learning is utilized in gas leakage prediction by training models with historical data, detecting anomalies, and enhancing accuracy through strategies (Boopathi & Kanike, 2023; Ingle et al., 2023; Syamala et al., 2023; Zekrifa et al., 2023). The continuous evolution of ML algorithms, coupled with advancements in feature engineering and model optimization, contributes to the development of highly effective and reliable gas detection systems (Kumar B et al., 2024). As the field progresses, the integration of these methods further enhances the proactive capabilities of intelligent gas detection systems, making them integral components of safety and environmental monitoring.

DATA PREPROCESSING FOR GAS LEAKAGE PREDICTION

This section discusses the process of efficient data preprocessing and feature engineering in creating accurate Machine Learning models for gas leakage prediction, including cleaning sensor data, extracting relevant features, and handling imbalanced datasets (Boopathi, Sureshkumar, et al., 2023a). The data preprocessing for gas leakage prediction is depicted in Figure 2.

Figure 2. Data preprocessing for gas leakage prediction

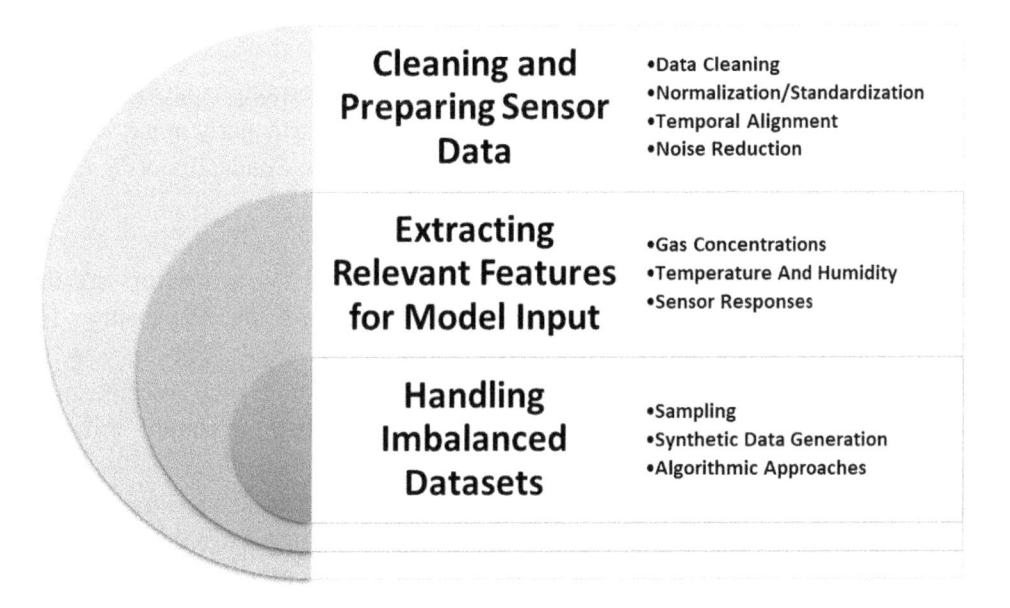

Cleaning and Preparing Sensor Data
- Data Cleaning
- Normalization/Standardization
- Temporal Alignment
- Noise Reduction

Extracting Relevant Features for Model Input
- Gas Concentrations
- Temperature And Humidity
- Sensor Responses

Handling Imbalanced Datasets
- Sampling
- Synthetic Data Generation
- Algorithmic Approaches

Cleaning and Preparing Sensor Data

The performance of Machine Learning models is significantly influenced by the quality of the data they train and test (Vennila et al., 2023).

- **Data Cleaning:** Identifying and handling missing or erroneous data is essential. This may involve imputing missing values, removing outliers, and ensuring data consistency.
- **Normalization/Standardization:** Bringing sensor data to a common scale is crucial for the proper functioning of ML algorithms. Normalization (scaling features between 0 and 1) or standardization (transforming features to have a mean of 0 and a standard deviation of 1) ensures that features contribute

uniformly to model training (Boopathi et al., 2022; Haribalaji et al., 2021; Saravanan et al., 2022).
- **Temporal Alignment:** In gas detection systems, sensor readings may be timestamped. Temporal alignment ensures that readings from different sensors are synchronized, facilitating the creation of coherent datasets for training and testing.
- **Noise Reduction:** Sensor data may contain noise due to environmental factors. Applying filtering techniques or signal processing methods can help reduce noise and enhance the accuracy of gas leakage prediction.

Extracting Relevant Features for Model Input

Feature engineering is the process of selecting and transforming input variables to improve the performance of machine learning models, particularly in gas leakage prediction (Palaniappan et al., 2023; Senthil et al., 2023; Sundaramoorthy et al., 2023; Zekrifa et al., 2023).

- **Gas Concentrations:** The primary data collected by gas sensors are the concentrations of different gases. These values form the core features for predicting gas leakage.
- **Temperature And Humidity:** Environmental conditions such as temperature and humidity can influence gas dispersion. Including these features provides contextual information for the model.
- **Sensor Responses:** The responses of sensors to gas concentrations are valuable features. The unique signatures of different gases influence sensor responses, aiding in the identification of specific gas types.
- **Time-Based Features:** Time-related features, such as time of day, day of the week, or trends over time, can capture temporal patterns in gas leakage occurrences.
- **Spatial Features:** If gas sensors are distributed across different locations, spatial features, such as proximity to potential sources of leakage, can be informative.

Handling Imbalanced Datasets

Imbalanced datasets, where instances of gas leakage are relatively rare compared to normal instances, are common in gas detection scenarios. Several techniques address this imbalance (Dhanya et al., 2023; Rebecca et al., 2023):

- **Oversampling:** Duplicating instances of the minority class (gas leakage) to balance the dataset.
- **Under-Sampling:** Randomly removing instances of the majority class to balance the dataset.
- **Synthetic Data Generation:** Creating synthetic instances of the minority class using techniques like SMOTE (Synthetic Minority Over-sampling Technique).
- **Algorithmic Approaches:** Certain machine learning algorithms have built-in mechanisms to handle imbalanced datasets, such as assigning different weights to classes.

Techniques address imbalanced dataset challenges, enabling machine learning models to accurately predict gas leakages, even in infrequent scenarios. Data preprocessing and feature engineering are crucial in creating accurate gas leakage prediction models (Dhanya et al., 2023; Rebecca et al., 2023). These steps involve cleaning sensor data, extracting relevant features, and handling imbalanced datasets. These steps, combined with advanced Machine Learning algorithms, form the foundation of intelligent gas detection systems with enhanced predictive capabilities.

MODEL DEVELOPMENT AND TRAINING FOR GAS LEAKAGE PREDICTION

This section outlines the process of creating robust and effective machine learning models for gas leakage prediction, which includes selecting suitable models, hyperparameter tuning, and cross-validation techniques, to ensure their effectiveness (Agrawal et al., 2024; Boopathi, Sureshkumar, et al., 2023b; Dhanya et al., 2023; Kumar Reddy R. et al., 2023).

Selection of Machine Learning Models

The selection of the appropriate machine learning model is crucial for the performance of a gas leakage prediction system, requiring consideration of the dataset's characteristics, problem nature, and desired outcomes (Kumar B et al., 2024; Satav, Hasan, et al., 2024; Satav, Lamani, et al., 2024).

- **Logistic Regression:** Suitable for binary classification tasks, logistic regression models the probability of an event occurring.

- **Random Forest:** An ensemble learning method that constructs a multitude of decision trees and combines their predictions, providing robustness and accuracy.
- **Support Vector Machines (Svm):** Effective for both binary and multiclass classification, SVM aims to find a hyperplane that best separates classes in feature space.
- **Gradient Boosting Algorithms (E.G., Xgboost, Lightgbm):** These algorithms build multiple weak models sequentially, with each model correcting the errors of the previous one.
- **Neural Networks:** Deep learning models, particularly recurrent neural networks (RNNs) or long short-term memory networks (LSTMs), can capture complex patterns in sequential data, making them suitable for time-series prediction.

The model selection is influenced by the size, dimensionality, and complexity of the underlying relationships in the dataset.

Hyperparameter Tuning

Hyperparameters are external configuration settings for machine learning models, set before training, crucial for optimizing model performance through a procedure that involves (Boopathi, Sureshkumar, et al., 2023a):

- **Define Hyperparameter Search Space:** Identify the hyperparameters that significantly impact model performance and define a range of values for each.
- **Select Optimization Technique:** Choose an optimization technique, such as grid search, random search, or Bayesian optimization, to explore the hyperparameter space efficiently.
- **Evaluate Performance:** For each set of hyperparameters, train the model on a subset of the data and evaluate its performance using metrics like accuracy, precision, recall, or area under the ROC curve.
- **Iterate And Refine:** Based on performance results, refine the search space and iterate the process until an optimal set of hyperparameters is found.

Hyperparameter tuning is a technique used to refine a model's predictive capabilities by extracting the most valuable information from the data.

Cross-Validation Techniques

Cross-validation is a crucial method for evaluating a model's generalization performance and reducing the risk of overfitting (Rebecca et al., 2023).

- **Data Splitting:** Divide the dataset into training and testing sets. In k-fold cross-validation, the dataset is divided into k subsets.
- **Training And Validation:** Train the model on k-1 subsets and validate it on the remaining subset. Repeat this process k times, using a different subset as the validation set each time.
- **Performance Metrics Aggregation:** Aggregate the performance metrics (e.g., accuracy, precision, recall) obtained from each iteration to assess the model's overall performance.
- **Model Selection:** Choose the model with the best average performance across all folds for deployment.

Cross-validation enhances the model's performance estimation reliability by minimizing the likelihood of overfitting to a specific data subset.

The development and training of gas leakage prediction models involve careful selection of machine learning models, fine-tuning of hyperparameters, and cross-validation techniques. These steps create accurate, robust models that can generalize well to new data. The iterative nature of these processes ensures continuous refinement and optimization, improving gas detection systems' predictive capabilities.

INTEGRATION OF GAS SENSORS WITH IOT DEVICES

The integration of gas sensors with IoT devices enhances gas detection systems by enabling real-time monitoring, data collection, and connectivity. This involves hardware requirements, sensor networks, and power management, and can be achieved through a procedural idea (Agrawal et al., 2023; Boopathi, 2023; Hussain et al., 2023; Satav, Hasan, et al., 2024). The integration of gas sensors with IoT devices is depicted in Figure3.

Hardware Requirements

- **Gas Sensors:** Choose appropriate gas sensors based on the types of gases to be detected. Common sensors include catalytic bead sensors for combustible gases, electrochemical sensors for toxic gases, and infrared sensors for specific gas types. Ensure compatibility with IoT communication protocols for seamless integration.

- **Microcontrollers/Microprocessors:** Select microcontrollers or microprocessors (e.g., Arduino, Raspberry Pi) to interface with the gas sensors. These devices will process sensor data, manage communication, and control sensor operations.
- **Communication Modules:** Integrate communication modules such as Wi-Fi, Bluetooth, or Zigbee to enable data transmission between gas sensors and IoT devices. Consider the range and data rate requirements when selecting communication protocols.
- **Power Supply:** Choose a reliable power supply source for both gas sensors and IoT devices. Battery-powered solutions may be suitable for mobile or remote deployments, while wired solutions may be used for continuous power.
- **Enclosures:** Use protective enclosures to shield the gas sensors and IoT devices from environmental factors such as dust, moisture, and temperature extremes. Ensure the enclosures do not interfere with sensor readings.

Figure 3. Integration of gas sensors with IoT devices

Sensor Networks

- **Wireless Sensor Networks (WSNs):** Deploy gas sensors in a networked configuration to cover larger areas. WSNs facilitate communication between sensors and the central IoT device. Ensure the wireless network topology is suitable for the specific deployment environment, considering factors like signal range and interference (Agrawal et al., 2023; Koshariya et al., 2023).
- **Data Aggregation:** Implement a data aggregation strategy where data from multiple gas sensors is collected and transmitted to the IoT device. This reduces the amount of data transmitted and conserves power.
- **Mesh Networking:** Implement mesh networking for increased reliability and redundancy. Mesh networks allow sensors to relay data to neighboring sensors, ensuring robust communication even if some sensors experience connectivity issues.

Power Management

- **Sleep Modes:** Implement sleep modes for gas sensors and IoT devices during periods of inactivity. This conserves power and extends the lifespan of batteries in battery-powered systems (Chandrika et al., 2023; Domakonda et al., 2023; Gnanaprakasam et al., 2023; Sundaramoorthy et al., 2023).
- **Dynamic Power Allocation:** Employ dynamic power allocation based on the urgency of data transmission. For instance, during normal operating conditions, sensors and devices can operate in low-power modes, with increased power allocated when a gas leakage event is detected.
- **Energy Harvesting:** Explore energy harvesting solutions, such as solar panels or vibration-based harvesters, to supplement or replace traditional power sources. This is particularly useful for remote or inaccessible locations.
- **Battery Monitoring:** Implement battery monitoring systems to track the health of the power source. This enables proactive replacement or recharging to avoid unexpected downtime.

The integration of gas sensors with IoT devices can create a robust, efficient, and scalable gas detection system, enabling real-time monitoring and timely responses to leakage events, thereby enhancing safety and environmental protection.

SCALABILITY AND DEPLOYMENT CHALLENGES IN GAS DETECTION SYSTEMS

Gas detection systems' scalability and deployment present both opportunities and challenges, necessitating adaptation to diverse environments, understanding their strengths and weaknesses, and ensuring robust deploymen t(Ramudu et al., 2023; D. M. Sharma et al., 2024).

Adapting Systems for Scalability

Sensor Network Design

- **Challenge:** Designing sensor networks that can scale with the size and complexity of the monitored area.
- **Solution:** Implement wireless sensor networks (WSNs) for flexible deployment. Use mesh networking to extend coverage and adapt to changing spatial requirements.

Data Handling

- **Challenge:** Managing and processing large volumes of data generated by an increasing number of sensors.
- **Solution:** Leverage cloud computing for scalable and on-demand data storage and processing. Implement edge computing to process data closer to the source, reducing latency.

Communication Protocols

- **Challenge:** Ensuring that communication protocols can handle the increasing data traffic in a scalable manner.
- **Solution:** Choose efficient and scalable communication protocols such as MQTT or CoAP. Optimize data transmission by aggregating and compressing sensor data.

Positives and Negatives

Positives

- **Scalability:** Gas detection systems can scale to cover larger areas and accommodate more sensors, providing comprehensive coverage.

- **Cost-Effectiveness:** Scalable solutions often lead to cost-effective deployments, as the infrastructure can adapt to changing requirements without significant overhauls.
- **Improved Accuracy:** More sensors enable finer granularity in monitoring, leading to improved accuracy in detecting gas leaks.

Negatives

- **Complexity:** As the system scales, the complexity of managing numerous sensors, data streams, and communication paths increases.
- **Data Overload:** A higher number of sensors may generate vast amounts of data, leading to potential challenges in storage, processing, and analysis.
- **Interference:** Increased density of sensors may result in interference, affecting the accuracy of readings.

Ensuring Robustness in Deployment

Redundancy

- **Strategy:** Implement redundancy in critical components such as communication pathways and power sources.
- **Benefits:** Redundancy ensures system robustness by minimizing the impact of failures on overall operations.

Regular Maintenance

- **Strategy:** Establish a routine maintenance schedule to ensure sensors are calibrated, communication devices are functioning, and power sources are operational.
- **Benefits:** Regular maintenance enhances the longevity and reliability of the gas detection system.

Adaptive Algorithms

- **Strategy:** Implement adaptive algorithms that can self-adjust based on changing environmental conditions and sensor characteristics.
- **Benefits:** Adaptive algorithms ensure that the system remains effective in different scenarios without constant manual adjustments.

Security Measures

- **Strategy:** Implement robust security measures to protect the system from unauthorized access or tampering.
- **Benefits:** Ensuring the security of the gas detection system is crucial for maintaining the integrity of collected data and preventing malicious interference.

User Training

- **Strategy:** Train end-users and maintenance personnel on system operations, troubleshooting, and emergency response procedures.
- **Benefits:** Well-trained users contribute to the efficient operation of the system and facilitate quick responses to gas leakage incidents.

To optimize gas detection systems for scalability, a strategic approach is needed to address complexities, effectively manage data, and address potential negatives. A well-designed, scalable system enhances safety, environmental protection, and operational efficiency. Regular monitoring, maintenance, and adaptation to evolving technologies are crucial for system success.

Figure 4. Integrating an intelligent gas detection system into automobiles

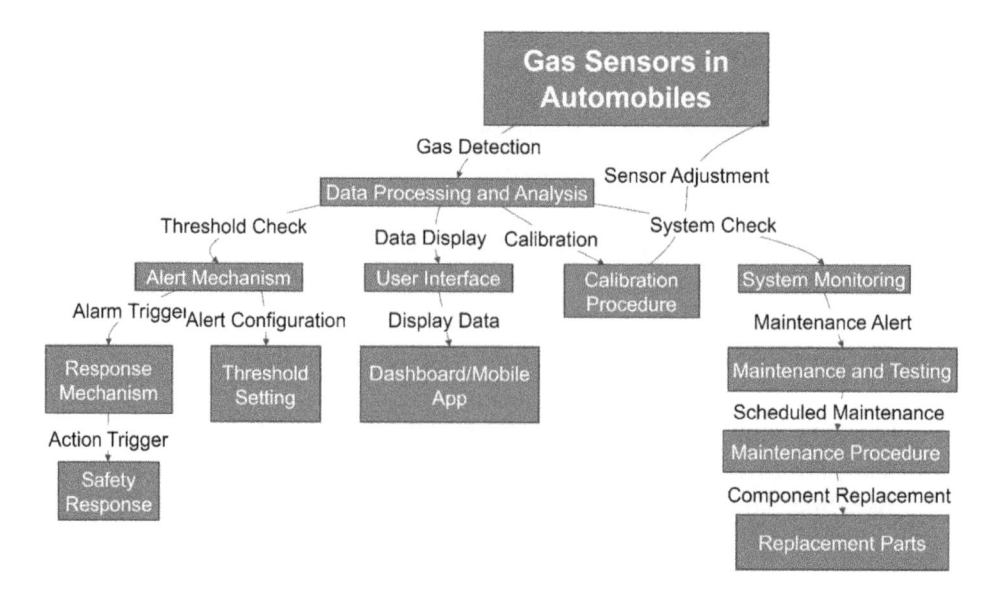

INTEGRATING AN INTELLIGENT GAS DETECTION SYSTEM INTO AUTOMOBILES

Integrating an intelligent gas detection system into automobiles involves several procedures and methods to ensure its effectiveness and reliability (Figure 4). The following procedures outline the implementation of intelligent gas detection systems in automobiles, which are crucial for detecting and alerting occupants to harmful gases like CO and CH4, and can be effectively utilized.

System Components: These systems typically consist of gas sensors, a processing unit (microcontroller or microprocessor), an alarm mechanism (audible and/or visual), and possibly a communication interface.

Gas Sensors Installation: Gas sensors are strategically placed within the vehicle cabin to ensure efficient detection. Common locations include near the exhaust system, within the engine compartment, and inside the cabin itself.

Calibration: Before deploying the system, gas sensors should be calibrated to ensure accurate detection. This involves exposing the sensors to known concentrations of gases and adjusting them accordingly.

Threshold Setting: Threshold levels for each gas being monitored should be set based on safety standards and regulations. Once these levels are exceeded, the system will trigger an alert.

Data Processing and Analysis: The processing unit continuously monitors the sensor readings. It compares the readings to the preset thresholds and analyzes the data to determine if there's a potential gas leak or accumulation.

Alert Mechanism: When gas levels exceed the preset thresholds, the system activates the alarm mechanism. This could include sounding an audible alarm, flashing lights, or even sending alerts to the vehicle's dashboard or mobile app.

Response Mechanism: In addition to alerting occupants, the system may also trigger specific response actions such as activating ventilation systems to clear the air, shutting down the engine, or opening windows.

User Interface: The system may include a user interface for configuration and monitoring purposes. This could be a dashboard display within the vehicle or a smartphone app that allows users to view real-time gas levels and system status.

Maintenance and Testing: Regular maintenance and testing are essential to ensure the system remains functional. This includes checking sensor calibration, testing the alarm mechanism, and replacing sensors or components as needed.

Integration with Other Safety Systems: Intelligent gas detection systems can be integrated with other safety systems within the vehicle, such as airbag deployment or automatic emergency braking, to provide a comprehensive safety network.

Compliance: Ensure that the system complies with relevant safety standards and regulations, such as those set by automotive industry organizations or government agencies.

Intelligent gas detection systems can enhance safety in automobiles by detecting and alerting occupants to harmful gases. By following specific procedures and methods, these systems can be effectively integrated into vehicles, ensuring the protection of occupants.

FUTURE TRENDS AND INNOVATIONS IN GAS DETECTION SYSTEMS

Advancements in Gas Sensing Technologies

As technology continues to evolve, gas sensing technologies are undergoing significant advancements, paving the way for more accurate, sensitive, and versatile gas detection systems. Emerging trends include the development of miniaturized sensors with improved selectivity, allowing for the detection of specific gases with higher precision. Nanotechnology is playing a key role in enhancing sensor performance, enabling the creation of sensors that are not only smaller but also more sensitive to low concentrations of gases (Boopathi, Khare, et al., 2023; Das et al., 2024; Pramila et al., 2023).

The integration of novel materials, such as graphene and metal-organic frameworks, into gas sensors is opening new possibilities. These materials offer unique properties that enhance the sensitivity and responsiveness of sensors, making them capable of detecting even trace amounts of gases. Additionally, advancements in sensor manufacturing techniques, such as 3D printing and microfabrication, contribute to the production of cost-effective and highly customizable sensors (Palaniappan et al., 2023; Senthil et al., 2023). These innovations are reshaping the landscape of gas detection, providing solutions that are not only more efficient but also more adaptable to diverse applications, ranging from industrial settings to smart homes.

Integration With Smart Home and Industrial Automation

The integration of gas detection systems with smart home and industrial automation platforms is a transformative trend that enhances the overall safety and efficiency of spaces. In smart homes, gas sensors are becoming integral components of connected ecosystems, capable of communicating with other devices. For instance, in the event of a gas leak, smart gas detectors can trigger automatic responses such as shutting off gas supplies, activating ventilation systems, and sending real-time alerts to homeowners' smartphones (Chandrika et al., 2023; Nishanth et al., 2023; Vijayakumar et al., 2023). This integration ensures swift responses to potential hazards, preventing accidents and minimizing the impact of gas-related incidents.

In industrial settings, the convergence of gas detection with automation technologies is optimizing operational workflows. Smart sensors, when integrated into industrial automation systems, enable real-time monitoring and control. These systems can autonomously adjust ventilation, activate safety protocols, and provide data for predictive maintenance. The seamless integration of gas detection into automation frameworks not only enhances safety but also contributes to operational efficiency by reducing downtime, improving resource utilization, and enabling data-driven decision-making (Gnanaprakasam et al., 2023).

Potential Impacts on Safety Regulations

The evolution of gas detection technologies and their integration into smart systems is expected to influence safety regulations across various sectors. Regulatory bodies are likely to adapt to the capabilities of advanced gas detection systems by incorporating new standards that reflect the advancements in sensor technologies and the integration with automation. Stricter regulations and guidelines may be established to ensure the deployment of state-of-the-art gas detection solutions in critical environments such as industrial plants, laboratories, and residential spaces (Boopathi, 2023; Maguluri, Arularasan, et al., 2023).

Furthermore, the increasing reliance on data analytics and machine learning in gas detection systems may prompt regulatory bodies to set guidelines for the responsible use of these technologies. Standards for data privacy, security, and algorithmic transparency may become integral aspects of safety regulations, ensuring that the adoption of advanced technologies aligns with ethical and regulatory considerations. Overall, the growing sophistication of gas detection systems is likely to contribute to a paradigm shift in safety regulations, emphasizing the need for intelligent, connected, and data-driven approaches to ensure the well-being of individuals and the environment (Karthik et al., 2023; Srinivas et al., 2023).

The future of gas detection systems is characterized by advancements in sensing technologies, integration with smart systems, and potential safety regulations impacts. These trends lead to more responsive, adaptable, and effective gas detection solutions, promoting safety and technological enrichment in residential, industrial, and regulatory sectors.

CONCLUSION

The gas detection systems industry is undergoing a transformation due to technological advancements, innovative sensor technologies, and intelligent solutions. This is crucial for ensuring safety in residential and industrial environments. The transition from traditional methods to modern, intelligent systems involves sensor advancements, IoT integration, machine learning applications, and scalability and robust deployment.

Advancements in gas sensing technologies, marked by the use of nanomaterials, miniaturization, and novel fabrication techniques, have elevated the precision and sensitivity of gas sensors. The integration of these sensors with Internet of Things (IoT) devices has ushered in an era of real-time monitoring, data analytics, and smart automation. Smart homes and industrial facilities are benefiting from interconnected gas detection systems that not only detect anomalies promptly but also trigger automated responses to mitigate potential hazards.

The future trajectory of gas detection systems also intersects with potential impacts on safety regulations. Regulatory bodies are likely to adapt to the capabilities of advanced gas detection technologies by establishing standards that reflect the evolving landscape of sensor capabilities, data analytics, and automation integration. Stricter guidelines may be implemented to ensure the responsible use of machine learning algorithms, data privacy, and overall system transparency, aligning safety regulations with the ethical and regulatory considerations of evolving technologies.

The integration of gas detection systems with automobile sectors is expected to create a safer and more responsive environment. The ability to detect gas leakages promptly, combined with intelligent automation and data-driven decision-making, represents a paradigm shift in safety practices. This evolution contributes to a future where safety is a priority and integrated into connected living and working spaces.

REFERENCES

Agrawal, A. V., Magulur, L. P., Priya, S. G., Kaur, A., Singh, G., & Boopathi, S. (2023). Smart Precision Agriculture Using IoT and WSN. In Advances in Information Security, Privacy, and Ethics (pp. 524–541). IGI Global. doi:10.4018/978-1-6684-8145-5.ch026

Agrawal, A. V., Shashibhushan, G., Pradeep, S., Padhi, S. N., Sugumar, D., & Boopathi, S. (2024). Synergizing Artificial Intelligence, 5G, and Cloud Computing for Efficient Energy Conversion Using Agricultural Waste. In B. K. Mishra (Ed.), Practice, Progress, and Proficiency in Sustainability. IGI Global. doi:10.4018/979-8-3693-1186-8.ch026

Baballe, M. A., & Bello, M. I. (2022). Gas leakage detection system with alarming system. *Review of Computer Engineering Research*, 9(1), 30–43. doi:10.18488/76.v9i1.2984

Boopathi, S. (2023). Securing Healthcare Systems Integrated With IoT: Fundamentals, Applications, and Future Trends. In A. Suresh Kumar, U. Kose, S. Sharma, & S. Jerald Nirmal Kumar (Eds.), Advances in Healthcare Information Systems and Administration. IGI Global. doi:10.4018/978-1-6684-6894-4.ch010

Boopathi, S., & Kanike, U. K. (2023). Applications of Artificial Intelligent and Machine Learning Techniques in Image Processing. In B. K. Pandey, D. Pandey, R. Anand, D. S. Mane, & V. K. Nassa (Eds.), Advances in Computational Intelligence and Robotics. IGI Global. doi:10.4018/978-1-6684-8618-4.ch010

Boopathi, S., Khare, R., Jaya Christiyan, K. G., Muni, T. V., & Khare, S. (2023). Additive Manufacturing Developments in the Medical Engineering Field. In R. Keshavamurthy, V. Tambrallimath, & J. P. Davim (Eds.), Advances in Chemical and Materials Engineering. IGI Global. doi:10.4018/978-1-6684-6009-2.ch006

Boopathi, S., Sureshkumar, M., & Sathiskumar, S. (2022). Parametric Optimization of LPG Refrigeration System Using Artificial Bee Colony Algorithm. *International Conference on Recent Advances in Mechanical Engineering Research and Development*, 97–105.

Boopathi, S., Sureshkumar, M., & Sathiskumar, S. (2023). Parametric Optimization of LPG Refrigeration System Using Artificial Bee Colony Algorithm. In S. Tripathy, S. Samantaray, J. Ramkumar, & S. S. Mahapatra (Eds.), *Recent Advances in Mechanical Engineering* (pp. 97–105). Springer Nature Singapore. doi:10.1007/978-981-19-9493-7_10

Chandrika, V. S., Sivakumar, A., Krishnan, T. S., Pradeep, J., Manikandan, S., & Boopathi, S. (2023). Theoretical Study on Power Distribution Systems for Electric Vehicles. In B. K. Mishra (Ed.), Advances in Civil and Industrial Engineering. IGI Global. doi:10.4018/979-8-3693-0044-2.ch001

Das, S., Lekhya, G., Shreya, K., Lydia Shekinah, K., Babu, K. K., & Boopathi, S. (2024). Fostering Sustainability Education Through Cross-Disciplinary Collaborations and Research Partnerships: Interdisciplinary Synergy. In P. Yu, J. Mulli, Z. A. S. Syed, & L. Umme (Eds.), Advances in Higher Education and Professional Development. IGI Global. doi:10.4018/979-8-3693-0487-7.ch003

Dhanya, D., Kumar, S. S., Thilagavathy, A., Prasad, D. V. S. S. S. V., & Boopathi, S. (2023). Data Analytics and Artificial Intelligence in the Circular Economy: Case Studies. In B. K. Mishra (Ed.), Advances in Civil and Industrial Engineering. IGI Global. doi:10.4018/979-8-3693-0044-2.ch003

Domakonda, V. K., Farooq, S., Chinthamreddy, S., Puviarasi, R., Sudhakar, M., & Boopathi, S. (2023). Sustainable Developments of Hybrid Floating Solar Power Plants: Photovoltaic System. In P. Vasant, R. Rodríguez-Aguilar, I. Litvinchev, & J. A. Marmolejo-Saucedo (Eds.), Advances in Environmental Engineering and Green Technologies. IGI Global. doi:10.4018/978-1-6684-4118-3.ch008

Gnanaprakasam, C., Vankara, J., Sastry, A. S., Prajval, V., Gireesh, N., & Boopathi, S. (2023). Long-Range and Low-Power Automated Soil Irrigation System Using Internet of Things: An Experimental Study. In G. S. Karthick (Ed.), Advances in Environmental Engineering and Green Technologies. IGI Global. doi:10.4018/978-1-6684-7879-0.ch005

Ha, N., Xu, K., Ren, G., Mitchell, A., & Ou, J. Z. (2020). Machine learning-enabled smart sensor systems. *Advanced Intelligent Systems*, *2*(9), 2000063. doi:10.1002/aisy.202000063

Haribalaji, V., Boopathi, S., & Asif, M. M. (2021). Optimization of friction stir welding process to join dissimilar AA2014 and AA7075 aluminum alloys. *Materials Today: Proceedings*, *50*, 2227–2234. doi:10.1016/j.matpr.2021.09.499

Hema, N., Krishnamoorthy, N., Chavan, S. M., Kumar, N. M. G., Sabarimuthu, M., & Boopathi, S. (2023). A Study on an Internet of Things (IoT)-Enabled Smart Solar Grid System. In P. Swarnalatha & S. Prabu (Eds.), Advances in Computational Intelligence and Robotics. IGI Global. doi:10.4018/978-1-6684-8098-4.ch017

Hussain, Z., Babe, M., Saravanan, S., Srimathy, G., Roopa, H., & Boopathi, S. (2023). Optimizing Biomass-to-Biofuel Conversion: IoT and AI Integration for Enhanced Efficiency and Sustainability. In N. Cobîrzan, R. Muntean, & R.-A. Felseghi (Eds.), Advances in Finance, Accounting, and Economics. IGI Global. doi:10.4018/978-1-6684-8238-4.ch009

Ingle, R. B., Swathi, S., Mahendran, G., Senthil, T. S., Muralidharan, N., & Boopathi, S. (2023). Sustainability and Optimization of Green and Lean Manufacturing Processes Using Machine Learning Techniques. In N. Cobîrzan, R. Muntean, & R.-A. Felseghi (Eds.), Advances in Finance, Accounting, and Economics. IGI Global. doi:10.4018/978-1-6684-8238-4.ch012

Karthik, S. A., Hemalatha, R., Aruna, R., Deivakani, M., Reddy, R. V. K., & Boopathi, S. (2023). Study on Healthcare Security System-Integrated Internet of Things (IoT). In M. K. Habib (Ed.), Advances in Systems Analysis, Software Engineering, and High Performance Computing. IGI Global. doi:10.4018/978-1-6684-7684-0.ch013

Kavitha, C. R., Varalatchoumy, M., Mithuna, H. R., Bharathi, K., Geethalakshmi, N. M., & Boopathi, S. (2023). Energy Monitoring and Control in the Smart Grid: Integrated Intelligent IoT and ANFIS. In M. Arshad (Ed.), Advances in Bioinformatics and Biomedical Engineering. IGI Global. doi:10.4018/978-1-6684-6577-6.ch014

Khan, M. M. (2020). Sensor-based gas leakage detector system. *Engineering Proceedings*, *2*(1), 28.

Koshariya, A. K., Kalaiyarasi, D., Jovith, A. A., Sivakami, T., Hasan, D. S., & Boopathi, S. (2023). AI-Enabled IoT and WSN-Integrated Smart Agriculture System. In R. K. Gupta, A. Jain, J. Wang, S. K. Bharti, & S. Patel (Eds.), Practice, Progress, and Proficiency in Sustainability. IGI Global. doi:10.4018/978-1-6684-8516-3.ch011

Kumar, B. M., Kumar, K. K., Sasikala, P., Sampath, B., Gopi, B., & Sundaram, S. (2024). Sustainable Green Energy Generation From Waste Water: IoT and ML Integration. In B. K. Mishra (Ed.), Practice, Progress, and Proficiency in Sustainability. IGI Global. doi:10.4018/979-8-3693-1186-8.ch024

Kumar Reddy, R. V., Rahamathunnisa, U., Subhashini, P., Aancy, H. M., Meenakshi, S., & Boopathi, S. (2023). Solutions for Software Requirement Risks Using Artificial Intelligence Techniques. In Advances in Information Security, Privacy, and Ethics (pp. 45–64). IGI Global. doi:10.4018/978-1-6684-8145-5.ch003

Maguluri, L. P., Ananth, J., Hariram, S., Geetha, C., Bhaskar, A., & Boopathi, S. (2023). Smart Vehicle-Emissions Monitoring System Using Internet of Things (IoT). In P. Srivastava, D. Ramteke, A. K. Bedyal, M. Gupta, & J. K. Sandhu (Eds.), Practice, Progress, and Proficiency in Sustainability. IGI Global. doi:10.4018/978-1-6684-8117-2.ch014

Maguluri, L. P., Arularasan, A. N., & Boopathi, S. (2023). Assessing Security Concerns for AI-Based Drones in Smart Cities. In R. Kumar, A. B. Abdul Hamid, & N. I. Binti Ya'akub (Eds.), Advances in Computational Intelligence and Robotics. IGI Global. doi:10.4018/978-1-6684-9151-5.ch002

Maheswari, B. U., Imambi, S. S., Hasan, D., Meenakshi, S., Pratheep, V. G., & Boopathi, S. (2023). Internet of Things and Machine Learning-Integrated Smart Robotics. In M. K. Habib (Ed.), Advances in Computational Intelligence and Robotics. IGI Global. doi:10.4018/978-1-6684-7791-5.ch010

Miller, K., Reichert, C. L., & Schmid, M. (2023). Biogenic amine detection systems for intelligent packaging concepts: Meat and Meat Products. *Food Reviews International*, *39*(5), 2543–2567. doi:10.1080/87559129.2021.1961270

Mohammed, B. K., Mortatha, M. B., Abdalrada, A. S., & ALRikabi, H. T. H. S. (2021). A comprehensive system for detection of flammable and toxic gases using IoT. *Periodicals of Engineering and Natural Sciences*, *9*(2), 702–711. doi:10.21533/pen.v9i2.1894

Mohanty, A., Venkateswaran, N., Ranjit, P. S., Tripathi, M. A., & Boopathi, S. (2023). Innovative Strategy for Profitable Automobile Industries: Working Capital Management. In Y. Ramakrishna & S. N. Wahab (Eds.), Advances in Finance, Accounting, and Economics. IGI Global. doi:10.4018/978-1-6684-7664-2.ch020

Nishanth, J. R., Deshmukh, M. A., Kushwah, R., Kushwaha, K. K., Balaji, S., & Sampath, B. (2023). Particle Swarm Optimization of Hybrid Renewable Energy Systems. In B. K. Mishra (Ed.), Advances in Civil and Industrial Engineering. IGI Global. doi:10.4018/979-8-3693-0044-2.ch016

Pachiappan, K., Anitha, K., Pitchai, R., Sangeetha, S., Satyanarayana, T. V. V., & Boopathi, S. (2023). Intelligent Machines, IoT, and AI in Revolutionizing Agriculture for Water Processing. In B. B. Gupta & F. Colace (Eds.), Advances in Computational Intelligence and Robotics. IGI Global. doi:10.4018/978-1-6684-9999-3.ch015

Palaniappan, M., Tirlangi, S., Mohamed, M. J. S., Moorthy, R. M. S., Valeti, S. V., & Boopathi, S. (2023). Fused Deposition Modelling of Polylactic Acid (PLA)-Based Polymer Composites: A Case Study. In R. Keshavamurthy, V. Tambrallimath, & J. P. Davim (Eds.), Advances in Chemical and Materials Engineering. IGI Global. doi:10.4018/978-1-6684-6009-2.ch005

Pramila, P. V., Amudha, S., Saravanan, T. R., Sankar, S. R., Poongothai, E., & Boopathi, S. (2023). Design and Development of Robots for Medical Assistance: An Architectural Approach. In G. S. Karthick & S. Karupusamy (Eds.), Advances in Healthcare Information Systems and Administration. IGI Global. doi:10.4018/978-1-6684-8913-0.ch011

Rahamathunnisa, U., Sudhakar, K., Murugan, T. K., Thivaharan, S., Rajkumar, M., & Boopathi, S. (2023). Cloud Computing Principles for Optimizing Robot Task Offloading Processes. In S. Kautish, N. K. Chaubey, S. B. Goyal, & P. Whig (Eds.), Advances in Computational Intelligence and Robotics. IGI Global. doi:10.4018/978-1-6684-8171-4.ch007

Rahamathunnisa, U., Sudhakar, K., Padhi, S. N., Bhattacharya, S., Shashibhushan, G., & Boopathi, S. (2023). Sustainable Energy Generation From Waste Water: IoT Integrated Technologies. In A. S. Etim (Ed.), Advances in Human and Social Aspects of Technology. IGI Global. doi:10.4018/978-1-6684-5347-6.ch010

Ramudu, K., Mohan, V. M., Jyothirmai, D., Prasad, D. V. S. S. S. V., Agrawal, R., & Boopathi, S. (2023). Machine Learning and Artificial Intelligence in Disease Prediction: Applications, Challenges, Limitations, Case Studies, and Future Directions. In G. S. Karthick & S. Karupusamy (Eds.), Advances in Healthcare Information Systems and Administration. IGI Global. doi:10.4018/978-1-6684-8913-0.ch013

Rathor, K., Vidya, S., Jeeva, M., Karthivel, M., Ghate, S. N., & Malathy, V. (2023). Intelligent System for ATM Fraud Detection System using C-LSTM Approach. *2023 4th International Conference on Electronics and Sustainable Communication Systems (ICESC)*, 1439–1444.

Rebecca, B., Kumar, K. P. M., Padmini, S., Srivastava, B. K., Halder, S., & Boopathi, S. (2023). Convergence of Data Science-AI-Green Chemistry-Affordable Medicine: Transforming Drug Discovery. In B. B. Gupta & F. Colace (Eds.), Advances in Computational Intelligence and Robotics. IGI Global. doi:10.4018/978-1-6684-9999-3.ch014

Reddy, M. A., Reddy, B. M., Mukund, C. S., Venneti, K., Preethi, D. M. D., & Boopathi, S. (2023). Social Health Protection During the COVID-Pandemic Using IoT. In F. P. C. Endong (Ed.), Advances in Electronic Government, Digital Divide, and Regional Development. IGI Global. doi:10.4018/978-1-7998-8394-4.ch009

Roshani, M., Phan, G., Faraj, R. H., Phan, N.-H., Roshani, G. H., Nazemi, B., Corniani, E., & Nazemi, E. (2021). Proposing a gamma radiation based intelligent system for simultaneous analyzing and detecting type and amount of petroleum by-products. *Nuclear Engineering and Technology*, 53(4), 1277–1283. doi:10.1016/j.net.2020.09.015

Samikannu, R., Koshariya, A. K., Poornima, E., Ramesh, S., Kumar, A., & Boopathi, S. (2023). Sustainable Development in Modern Aquaponics Cultivation Systems Using IoT Technologies. In P. Vasant, R. Rodríguez-Aguilar, I. Litvinchev, & J. A. Marmolejo-Saucedo (Eds.), Advances in Environmental Engineering and Green Technologies. IGI Global. doi:10.4018/978-1-6684-4118-3.ch006

Saravanan, M., Vasanth, M., Boopathi, S., Sureshkumar, M., & Haribalaji, V. (2022). Optimization of Quench Polish Quench (QPQ) Coating Process Using Taguchi Method. *Key Engineering Materials*, 935, 83–91. doi:10.4028/p-z569vy

Satav, S. D., Hasan, D. S., Pitchai, R., Mohanaprakash, T. A., Sultanuddin, S. J., & Boopathi, S. (2024). Next Generation of Internet of Things (NGIoT) in Healthcare Systems. In B. K. Mishra (Ed.), Practice, Progress, and Proficiency in Sustainability. IGI Global. doi:10.4018/979-8-3693-1186-8.ch017

Satav, S. D., Lamani, D. K. G., H., Kumar, N. M. G., Manikandan, S., & Sampath, B. (2024). Energy and Battery Management in the Era of Cloud Computing: Sustainable Wireless Systems and Networks. In B. K. Mishra (Ed.), Practice, Progress, and Proficiency in Sustainability (pp. 141–166). IGI Global. doi:10.4018/979-8-3693-1186-8.ch009

Senthil, T. S., Ohmsakthi Vel, R., Puviyarasan, M., Babu, S. R., Surakasi, R., & Sampath, B. (2023). Industrial Robot-Integrated Fused Deposition Modelling for the 3D Printing Process. In R. Keshavamurthy, V. Tambrallimath, & J. P. Davim (Eds.), Advances in Chemical and Materials Engineering. IGI Global. doi:10.4018/978-1-6684-6009-2.ch011

Sharma, D. M., Venkata Ramana, K., Jothilakshmi, R., Verma, R., Uma Maheswari, B., & Boopathi, S. (2024). Integrating Generative AI Into K-12 Curriculums and Pedagogies in India: Opportunities and Challenges. In P. Yu, J. Mulli, Z. A. S. Syed, & L. Umme (Eds.), Advances in Higher Education and Professional Development. IGI Global. doi:10.4018/979-8-3693-0487-7.ch006

Sharma, M., Sharma, M., Sharma, N., & Boopathi, S. (2023). Building Sustainable Smart Cities Through Cloud and Intelligent Parking System. In B. B. Gupta & F. Colace (Eds.), Advances in Computational Intelligence and Robotics. IGI Global. doi:10.4018/978-1-6684-9999-3.ch009

Song, Z., Ye, W., Chen, Z., Chen, Z., Li, M., Tang, W., Wang, C., Wan, Z., Poddar, S., Wen, X., Pan, X., Lin, Y., Zhou, Q., & Fan, Z. (2021). Wireless self-powered high-performance integrated nanostructured-gas-sensor network for future smart homes. *ACS Nano*, *15*(4), 7659–7667. doi:10.1021/acsnano.1c01256 PMID:33871965

Srinivas, B., Maguluri, L. P., Naidu, K. V., Reddy, L. C. S., Deivakani, M., & Boopathi, S. (2023). Architecture and Framework for Interfacing Cloud-Enabled Robots: In T. Murugan & N. E. (Eds.), Advances in Information Security, Privacy, and Ethics (pp. 542–560). IGI Global. doi:10.4018/978-1-6684-8145-5.ch027

Sundaramoorthy, K., Singh, A., Sumathy, G., Maheshwari, A., Arunarani, A. R., & Boopathi, S. (2023). A Study on AI and Blockchain-Powered Smart Parking Models for Urban Mobility. In B. B. Gupta & F. Colace (Eds.), Advances in Computational Intelligence and Robotics. IGI Global. doi:10.4018/978-1-6684-9999-3.ch010

Syamala, M. C. R., K., Pramila, P. V., Dash, S., Meenakshi, S., & Boopathi, S. (2023). Machine Learning-Integrated IoT-Based Smart Home Energy Management System. In P. Swarnalatha & S. Prabu (Eds.), Advances in Computational Intelligence and Robotics (pp. 219–235). IGI Global. doi:10.4018/978-1-6684-8098-4.ch013

Tang, S., Chen, W., Jin, L., Zhang, H., Li, Y., Zhou, Q., & Zen, W. (2020). SWCNTs-based MEMS gas sensor array and its pattern recognition based on deep belief networks of gases detection in oil-immersed transformers. *Sensors and Actuators. B, Chemical*, *312*, 127998. doi:10.1016/j.snb.2020.127998

Venkateswaran, N., Kumar, S. S., Diwakar, G., Gnanasangeetha, D., & Boopathi, S. (2023). Synthetic Biology for Waste Water to Energy Conversion: IoT and AI Approaches. In M. Arshad (Ed.), Advances in Bioinformatics and Biomedical Engineering. IGI Global. doi:10.4018/978-1-6684-6577-6.ch017

Venkateswaran, N., Vidhya, K., Ayyannan, M., Chavan, S. M., Sekar, K., & Boopathi, S. (2023). A Study on Smart Energy Management Framework Using Cloud Computing. In P. Ordóñez De Pablos & X. Zhang (Eds.), Practice, Progress, and Proficiency in Sustainability. IGI Global. doi:10.4018/978-1-6684-8634-4.ch009

Vennila, T., Karuna, M. S., Srivastava, B. K., Venugopal, J., Surakasi, R., & B., S. (2023). New Strategies in Treatment and Enzymatic Processes: Ethanol Production From Sugarcane Bagasse. In P. Vasant, R. Rodríguez-Aguilar, I. Litvinchev, & J. A. Marmolejo-Saucedo (Eds.), *Advances in Environmental Engineering and Green Technologies* (pp. 219–240). IGI Global. doi:10.4018/978-1-6684-4118-3.ch011

Vijayakumar, G. N. S., Domakonda, V. K., Farooq, S., Kumar, B. S., Pradeep, N., & Boopathi, S. (2023). Sustainable Developments in Nano-Fluid Synthesis for Various Industrial Applications. In A. S. Etim (Ed.), Advances in Human and Social Aspects of Technology. IGI Global. doi:10.4018/978-1-6684-5347-6.ch003

Zekrifa, D. M. S., Kulkarni, M., Bhagyalakshmi, A., Devireddy, N., Gupta, S., & Boopathi, S. (2023). Integrating Machine Learning and AI for Improved Hydrological Modeling and Water Resource Management. In V. Shikuku (Ed.), Advances in Environmental Engineering and Green Technologies. IGI Global. doi:10.4018/978-1-6684-6791-6.ch003

Chapter 4

The SWIPT-Enabled Cooperative Full-Duplex Relaying Communication for 6G Radio Networks

Rajeev Kumar
https://orcid.org/0000-0001-6465-4247
Indian Institute of Information Technology, Sri City, India

ABSTRACT

This chapter presents the simultaneously wireless information and power transfer (SWIPT)-enabled cooperative communication for 6th generation (6G) radio networks. These networks integrate many applications and use tiny internet-of-thing (IoT) devices that need large amounts of power to maintain energy threshold level of the networks for a long time. Therefore, the energy harvesting (EH) is a promising technique to provide sufficient power to small IoT devices and energy constrained user terminals in the 6G radio networks. For this purpose, the authors consider two-way full-duplex (FD) relaying network to derive closed-form expressions of end-to-end ergodic capacity by exploiting power splitting (PS) and time switching (TS) protocols for Rayleigh fading channels. From the numerical results, they observe that the proposed FD-based EH policies perform better as compared to half-duplex EH policies. Finally, this chapter provides future research directions.

DOI: 10.4018/978-1-6684-9214-7.ch004

The SWIPT-Enabled Cooperative Full-Duplex Relaying Communication

INTRODUCTION

Cooperative communication has been involved more in integration of energy harvesting (EH) and information processing (IP) with assistance of the relay in recent years. The key ideas of enabling advanced technologies such as millimeter wave, ultra MIMO, full-duplex, Terahertz wave, visible light, and EH, integrated frequency bands transceiver, communication on large integrated surface, integrated terrestrial, airborne and satellite networks, and edge artificial intelligence will be provided concrete roadmap for 6G radio services (Saad et al.; Chataut, & Akl, 2020). In 6G radio system, wireless device needs more power as compare to existing radio system for data services, video conversations, gamming, and provides an integrated service of Internet-of-Things (IoT) etc.. Thereby, it causes to battery draining problem. To mitigate this, the EH is a latest-and-promising technology that provides implicit power supply to energy constraint wireless devices.

The EH components and power management circuits have been addressed (Vullers et al., 2009), where an energy in form of electrical power is harvested from energy sources such as solar/light, thermoelectric, mechanical motion/vibration and electromagnetic radiation (EMR). In practical applications such as cellular systems, multisensory devices, robotics and autonomous systems, brain computer interactions, IoT, and blockchain, and distributed ledger technologies which is shown in Fig. 1, the EH technique can be used to enhance data rates, energy efficiency and lifetime of the 6G radio networks.

Figure 1. Proposed diagram of the SWIPT-enabled cooperative communication for 6G radio systems

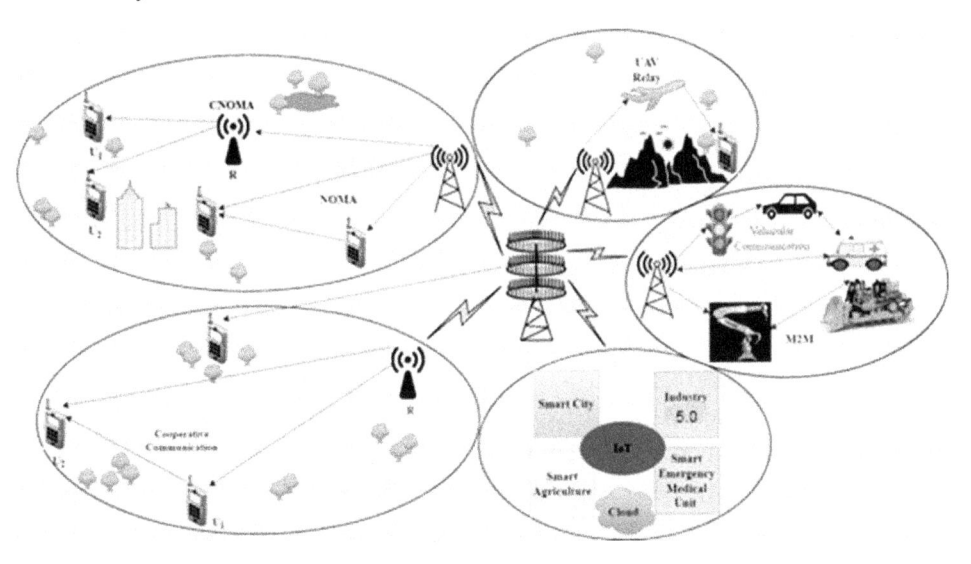

The EMR based energy source is state-of-the-art in wireless communications that exploits concept of the EH technique and provides sufficient energy to energy constraint nodes. The spatial multiplexing (SM) is implemented with assistance of the antenna diversity that provides strong links from the transmitter and the receiver to combat the effect of deep fading. On the basis of equipped antennas at the transmitter and the receiver, the SM can be classified into four categories. In the first approach, Varshney, L. R. (2008) has considered conventional SM system such as single-input-single-output (SISO) to exploit an ideal receiver to tradeoff the energy-and-information simultaneously with the help of the capacity-energy function. In the second approach, multiple-input-multiple-output (MIMO) broadcast system has been addressed (Zhang, & Ho, 2013) by exploiting optimal transmission strategy to exchange information and energy sufficiently, where a common transmitter sends energy signal to EH receiver and information bearing signal to information decoding receiver. In such a scenario, single receiver circuit cannot recover information and energy signal directly. Further, they have resolved this issue by designing practical receivers such as power splitting (PS) receiver and time-switching (TS) receiver. These two receivers are defined as follows: the PS receiver divides received signal into two separate signals with different power levels, one signal is used for EH and another for information decoding (ID) while the TS receiver divides block of time into two or more different time-slots to receive the energy and information from transmitted RF signal. In the third approach, the SWIPT enabled single-input-multiple-output (SIMO) system (Liu et al., 2013) has been considered by exploring the concept of the uniform PS transceiver to achieve tradeoff of the rate and energy. The fourth approach is multiple-input-single-output (MISO) multicasting for 5G radio system to meet the demand of high data rate. The MISO SWIPT system has been presented to investigate high data rate by exploiting PS based relaying (PSR) protocol (Khandaker et al., 2014) and TS based relaying (TSR) protocol (Nasir A.A. et al., 2017).

Cooperative relay communication (CRC) is one of the promising approach to enhance spectral efficiency, energy efficiency, and lifetime of wireless devices and networks as the wireless devices are deployed at the distant with diverse locations. One dimension half-duplex relay system (Nasir et al., 2013; Chen et al.; Nasir et al., Gu, & Aïssa, Liu et al., Huang et al., 2015; Kashef et al., Huang et al., 2016; Ding et al., 2017) has been assumed to improve transmission rate and capability of device for the EH by exploiting PS relaying (PSR) and TS relaying (TSR) protocols. Further, two-way HD-relay/relays system (Tutuncuoglu et al., 2015; Zeng et al., Liu et al., Salem & Hamdi, 2016; Alsharoa et al., 2017) has been considered to examine performance gains such as throughput, and energy efficiency by employing multi-access and broadcast (MABC) and time-division broadcast (TDBC) techniques under various relaying strategies such as decode-and-forward (DF), amplify-and-forward (AF), compress-and-forward (CF) and compute-and-forward (CaF). The problem with the HD relay is that it segments the time in two-slots and reduce the spectral efficiency by pre-log factor one-half. To overcome the limitation of the HD technique, the relevant transmission policy such as full-duplex (FD) has been implemented at any node to transmit and receive signals simultaneously over same frequency band. It has two prime keys to enhance spectral efficiency and reduce latency that meet the requirement of the 6G mobile radio networks. Two-hop FD-relay systems have been studied (Li & Murch, Zhong et al., 2014; Zeng et al., 2015; Liu et al., 2016, Wang et al., 2017; Le et al. 2018; Rabie et al., 2019; Guo et al., 2020) to improve spectral efficiency by removing pre-log factor one-half, and investigate outage probability. Two-way FD PS relay system (Wang et al., Chen et al., 2017) have been taken into account to boost-up spectral efficiency by exploiting relay selection policy. Further, two-way AF relay system (Kumar, & Hossain, 2020) has been presented by introducing the FD PS and TS protocols to enhance spectral efficiency and harvest efficient energy from RF-band signal and self-energy recycling (SER). Furthermore, wireless small base station communications system (Pérez et al., 2022) has been assumed to achieve outage probability by exploiting antenna dynamic FD relaying policy. Moreover, the FD-based SWIPT enabled TWRN has been considered (Kumar et al., 2023) to investigate the PS protocol to derive the closed-form outage probability, throughput, and energy efficiency for Nakagami-m fading channels, where different types of nonlinear power amplifier models are used at the relay. In the wireless communication, spectrum scarcity and energy constraint are challenging issues. To overcome this, the cognitive radio (CR) network is employed to enhance both spectral- and energy- efficiency. The Kumar et al. (2023) have presented the SWIPT enabled FD cooperative CR network to investigate performance gains by exploiting PS protocol over $\alpha-\mu$ channel. Recently, a survey on FD wireless for 6G networks has been studied briefly (Smida et al., 2023) and provided possible research direction towards substantial transformation of the poised wireless networks.

In modern era, the IoT-based 6G communication play important role to mitigate lots of problems related to human daily life. For this purpose, we need the IoT network that consists of millions of small wireless devices that require more power to handle the large amount of produced data and prolong the lifetime of the system. To overcome this challenges, the non-orthogonal multiple-access (NOMA) is a promising technique that support massive connectivity of the IoT device in the network and provides flexibility to harvest the energy from natural resources. The key idea of the NOMA is multi access in power-domain. It is different from orthogonal multiple access (OMA) (in terms of time/frequency/code-division multiple access) transmission policies. In the NOMA technique, the high power is allocated to bad channel users while low power is assigned to good channel users. The SWIPT enabled cooperative

NOMA (CNOMA) has been studied by employing PS scheme to analyze the outage probability and throughput (Liu et al., 20216). In the wireless communication, the power allocation (PA) policy authenticates and strengthen the wireless links with sufficient optimal power from source to destination via relay/without relay to improve the spectral and energy efficiency of wireless system. Further, the PA policies have been implemented (Yang et al., 2017) by exploiting SWIPT to enhance the performance of the HD based CNOMA networks. Furthermore, the beamforming vector and PS ratio have been investigated and jointly optimized (Xu et al., 2017). The FD-based SWIPT is needed to improve spectral efficiency significantly and provide flexibility for receiving and transmitting the signals simultaneously from multiple users. Nowadays, an integration of the FD and cooperative NOMA is achieved a milestone and provided flexibility to the IoT devices to perform SWIPT in wireless system. The SWIPT enabled FD-based NOMA network has been designed optimally by exploiting the PS receiver under consideration of the full- and partial CSI. (Liu et al., 2020). It is noticed that the performance of the FD-based NOMA network is degraded due to residual self-interference (RSI) at the FD relay. Further, Kurup and Babu (2020) have considered the SWIPT enabled FD-NOMA system to investigate adaptive PA (APA) and fixed PA (FPA) by exploiting TS protocol to enhance the performance gains in terms of outage probability and throughput for imperfect CSI (ICSI). They have also found that the APA scheme overcomes the RSI effect with high power source (base station). Furthermore, it has been noticed that the outage probability of the users is decreased under consideration of the APA scheme as compared to FPA scheme.

The unmanned aerial vehicle (UAV) is state-of-art in terrestrial wireless communication that provides flexibility in deployment, high speed mobility, and low cost wireless infrastructure. These factors encourage and lead to confirm the practical applications named as disaster management, public safety, reuse operation, and military operations etc. Recently, the UAV acts a relay that is introduced to enhance the coverage area significantly and performance gains as the source and destination terminals are placed at the long distances and obstructed by building, foliage and hills. The FD-UAV relay aided cooperative communication has been studied (Jayakody et al., 2020) to achieve performance gains such as outage probability and throughput by exploring the concept of the TS-based unified EH scheme. Instead of traditional wireless communication, the UAV-ground communication is more prone to overhearing of signals. Since, it broadcasts essential nature of the line-of-sight wireless channels. Thereby, the wireless system suffers from different security threats. A mobile UAV-relay system has been considered (Wang et al., 2020) to secure communication and achieved maximum secrecy rate in presence of the imperfect eavesdropper's locations. However, the FD jamming under the policy of the delay transmissions has not been studied still. The Ma et al. (2020) have presented the

SWIPT enabled FD jamming assisted cooperative communication system to analyse the secrecy throughput under the consideration of delay tolerant transmission and delay constrained transmission by employing PS- and TS- protocols with linear and non-linear EH models.

In ultramodern, most of the people uses smartphone that produces large amount of unpredicted data that leads to congestion on the wireless cellular system. As per the forecast of the CISCO (Tariq et al., 2020), the mobile data traffic has been appraised to increase with rate of around 55% from 2020 to 2030 and it will reach upto 607 exabytes (EB) in 2025 and 5016 EB in 2030. To overcome this challenges, the researcher are getting motivated to introduce the concept of the device-to-device (D2D) communication. It is emerging technique that is used to enhance the spectral efficiency, channel capacity, and extend the coverage area of the cellular networks. In the Budhiraja et al. (2021), a hexagonal cell in the cellular network with a centralized base station (BS), cell user equipment (CUE), D2D user equipment (DUE), and FD-based D2D transmitters (DDTs) have been presented to support massive connectivity and low latency by employing NOMA at the BS, where the DDTs and the CUEs harvest energy from the BS. Moreover, closed-form expression of the ergodic capacity of the DDTs and the DUEs, and outage probabilities have been derived for ICSI. In the 4G and 5G communication, the multiple access techniques such as a space division multiple access (SDMA) and a NOMA technologies have been exploited to reduce the overhead of data traffics and meet the requirements of the high data transmission rate. The SDMA technique employs to combat the effect of the deep fade and provide strong links with parallel transmission of the signals from transmitter to receiver. Apart from this, the NOMA technique is implemented to manage multi-user interference by forcing at least one user and successfully decode information with removal of interference of other users. Recently, the researchers are focused on 6G communications that will support the IoT networks, higher data rate, ultra-reliability, heterogeneous quality of service, enormous connectivity, sensing, localization, computing, and controlling (Tataria, et al., 2021). Thus, the 6G communication will be needed a more efficient use of the wireless resources and manage interference at the physical layer. The rate-splitting multiple access (RSMA) is one of the promising technique that is capable to fulfil the gap between the SDMA and NOMA. In the RSMA technique, user messages split into common data streams and private data streams, and support partially decoding interference and partially treating interference as noise. An advanced survey on the RSMA, and potential future research direction have been done by Mao et al. (2022). The FD based cooperative rate splitting scheme for DL multigroup multicast network has been studied (Li et al., 2022) to reduce the power consumption by maintaining the target rate for ICSI. The heterogeneous networks suffers from the cross-layer interference and worst channel conditions. The Kong & Lu (2023) have considered two-layer in the heterogeneous cellular network to investigate resource allocation of the cooperative RSMA scheme to achieve maximum sum rate of the small cell users, where in this policy, high quality channel user acts as FD relay to transmit comman signal to user with weak signals. Recently, the cognitive radio system has been presented (Gao et al., 2024) to obtain the outage probability and ergodic rate by exploiting RSMA scheme under the consideration of imperfect successive interference cancellation and channel estimation errors.

Figure 2. Bidirectional full-duplex relay system with high power station H, fixed road side relay R and mobile user u

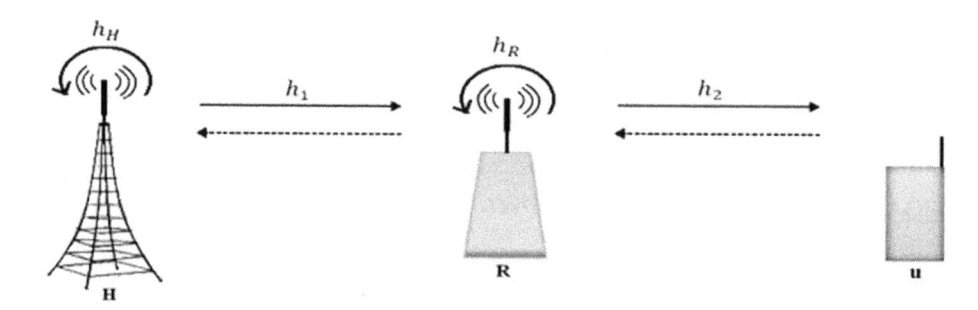

Motivated by advancement of wireless communication, the EH is state-of-art in 6G communication. The prime aim of this chapter is to explore the concept of simultaneously wireless power and information processing from RF signal with assistance of cooperative FD relay. For this purpose, we present two-way FD relay system and therein the relay and user harvest energy from high power station, **H** and information transfers directed to user from the **H** via relay **R**, where the relay also harvests energy from self-energy recycling (SER). The FD-based PS and TS protocols are developed to investigate closed-form expressions of the end-to-end (E2E) ergodic capacity for PS and TS protocols. Furthermore, the PS ratio and TS ratio expressions have also been obtained in closed-form. In addition, we evaluated closed-form expressions of the EH by all energy constrained devices for proposed protocols. It is notice that our proposed FD-based EH protocols superior to the HD-based EH protocols.

The remaining of this chapter is organized as follows. In the next Section, two-way FD relaying system model and its preliminaries are presented. Then, we propose the FD-based PS and TS protocols and derive the ergodic capacity for Rayleigh fading channel. After that, for better understanding, the proposed FD-based EH schemes are compared with HD-base EH policies. Finally, this chapter concludes and provides possible future research directions.

Figure 3. Communication structure of the power-splitting protocol for bidirectional FD relay system

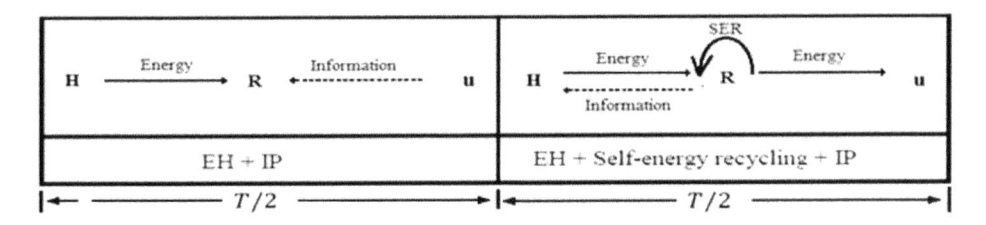

Energy → R ← Information u	Energy → R (SER) → Energy u
	Information
EH + IP	EH + Self-energy recycling + IP
← —— T/2 —— →	← —— T/2 —— →

PERFORMANCE OF SWIPT-ENABLED COOPERATIVE RELAYING COMMUNICATIONS

This Section presents cooperative relaying communication and its performance gains together with parallel processing the information and transfer the power.

COOPERATIVE RELAYING SYSTEM AND ITS PRELIMINARIES

We present simple bidirectional FD relay system which is shown in Fig. 2, where high power station **H** and mobile user **u** transfer power respectively to **u** and information to **H** via road side relay **R**. Due to dynamic motion of mobile user and deep fading channel, the direct link between **H** and **u** is not considered. The FD operation at the **H** and **R** is executed with the help of dual antenna, where one antenna is used for transmission and another for reception. Further, it is assumed that the **R** follows AF relaying protocol for transmission and reception of the signal. It is also, considered that the channel coefficient from **H** to **R** and from **u** to **R** are h_1 and h_2 respectively. Furthermore, it is assumed that the channels are reciprocal in each slot i.e., channel from **H** to **R** and channel from **u** to **R** are identical to channel from **R** to **H** and channel from **R** to **u** respectively. The h_H and the h_R are the residual self-interference channel response of **H** and **R** respectively. The **R** and the **u** are energy constrained wireless nodes, but both nodes have sufficient amount of energy for signal transmission and reception. These are powered by **H** with the assistance of RF energy signal.

END-TO-END ERGODIC CAPACITY BASED ON PROPOSED ENERGY HARVESTING PROTOCOLS

This Section elaborates the power-splitting and time-switching protocols and then formulate the E2E ergodic capacity for SWIPT enabled two-way FD relay system.

E2E ERGODIC CAPACITY BASED ON POWER-SPLITTING PROTOCOL FOR BIDIRECTIONAL FULL-DUPLEX AMPLIFY-AND-FORWARD RELAY SYSTEM

In this subsection, the concept of the PS protocol is introduced to investigate closed-form expressions of the ergodic capacity for bidirectional relay system. The Fig. 3 shows the analysis of the FD-based PS protocol. The PS protocol is framed into equal phases of the communication time block (CTB) T. In the first phase of the CTB $t \in [0, T/2]$, the **H** and **u** transfer RF power signal, $x_H[k]$ and information $x_u[k] \sim \mathcal{CN}(0,1)$ to **R** respectively. Thus, the received signal at the **R** is expressed as,

$$y_R[k] = h_1 x_H[k] + h_2 x_u[k] + n_R[k] \tag{1}$$

where k denotes symbol index and $n_R[k] \sim \mathcal{CN}(0, \sigma_R^2)$ denotes noise added at the **R**. The received signal, $y_R[k]$ at the **R** is splitted for EH and IP. Thereby, **R** can receive the energy signal from the **H** is written as,

$$y_{R,E}[k] = \sqrt{\alpha}\, y_R[k] \tag{2}$$

where α denotes PS ratio with $0 \le \alpha \le 1$. The R can receive an information from the u is expressed as,

$$y_{R,IF}[k] = \sqrt{(1-\alpha)}\, y_R[k] \tag{3}$$

It is notified that the term $\sqrt{(1-\alpha)}\, n_R[k]$ does not contain any information so that it is ignored. The equation (3) is carried out in the remaining phase of duration $t \in [T/2, T]$ which is amplified at the relay and forwards energy as well as information signal simultaneously to **u** and **H** respectively. Thereby, amplified signal at the **R** is to be given by,

$$x_R[k] = \sqrt{\beta_{PS}}\left(y_{R,IF}[k-1] + n_{R,P}[k-1]\right) \quad (4)$$

where $n_{R,P}[k]$ represents noise during IP which is produced by RF-to-baseband conversion and β_{PS} denotes amplification coefficient for PS protocol (Kumar and Hossain, 2020). The R collects energy signal from H and RSI due to antennas coupling at the R which is written as,

$$y'_{R,E}[k] = h_1 x_H[k] + h_R x_{R,N}[k] + n_R[k] \quad (5)$$

where $x_{R,N}[k]$ represents the normalized relayed signal at the **R** and h_R. From equation (2) and (5), total energy harvested in CBT T is given by

$$E_R^{PS} = \xi\left[P_H(1+\alpha)|h_1|^2 + P_R|h_R|^2\right]\frac{T}{2} \quad (6)$$

Where ξ denotes the energy conversion efficiency for EH with $0\leq\xi\leq 1$ for EH, the P_H represents the transmit power of H, and the h_R denotes the fading coefficient of self-interference channel. Note that proof is available in (Kumar and Hossain, 2020). It is assumed that the energy constrained relay consumes energy in second phase which is less than or, equal to total amount of energy harvested by the R during block of time, T i.e., $\frac{T}{2}P_R \leq E_R^{PS}$. By exploiting this condition, the relay transmit power P_R is evaluated as,

$$P_R \leq \frac{\xi P_H(1+\alpha)|h_1|^2}{1-\xi|h_R|^2} \quad (7)$$

In the remaining phase, the **H** receives information from the **R** and RF power signal of the **H** that can be given by

$$y_H[k] = h_1 x_R[k] + h_H x_H[k] + n_H[k] \quad (8)$$

In (8), second term does not contain useful information that can be removed by using notch filter. By substituting $x_R[k]$ from (4) into (8), we get

$$\tilde{y}_H[k] = \underbrace{\sqrt{\beta_{PS}}\sqrt{(1-\alpha)}h_1h_2x_u[k-1]}_{Information Signal} + \underbrace{\sqrt{\beta_{PS}}h_1n_{R,d}[k-1] + n_H[k]}_{Noise Signal} \quad (9)$$

Using (9), the signal-to-interference-plus-noise ratio (SINR) from mobile user **u** to the **H** is formulated as,

$$\gamma_{uH}^{PS} = \frac{P_u^{PS}(1-\alpha)|h_1|^2|h_2|^2}{|h_1|^2\sigma_R^2 + \frac{\sigma_H^2}{\beta_{PS}}} \quad (10)$$

Notice that the EH at the user is obtained with the help of the received signal so that the signal received at **u** in the remaining phase that can be expressed as,

$$y_{u,E}[k] = h_2 x_R[k] + n_u[k] \quad (11)$$

From (4) and (11), the user, **u** harvests the energy that can be obtained as,

$$E_u^{PS} = \xi|h_2|^2 \mathbb{E}\left[|x_R[k]|^2\right]\frac{T}{2} = \xi^2|h_2|^2\left[P_H(1+\alpha)|h_1|^2 + P_R|h_R|^2\right]\frac{T}{2} \quad (12)$$

From (12), the user transmit power, P_u^{PS} is given as,

$$P_u^{PS} = \frac{E_u^{PS}}{\frac{T}{2}} = \xi^2|h_2|^2\left[P_H(1+\alpha)|h_1|^2 + P_R|h_R|^2\right] \quad (13)$$

From (7), (10) and (13), the SINR is expressed as

$$\gamma_{uH}^{PS}(\alpha) = \frac{\xi^2 P_H|h_1|^4|h_2|^4}{1-\xi|h_R|^2}\gamma(\alpha) \quad (14)$$

Where

The SWIPT-Enabled Cooperative Full-Duplex Relaying Communication

$$\gamma(\alpha) = \cfrac{1}{\cfrac{|h_1|^2 \sigma_R^2}{1-\alpha^2} + \cfrac{\xi |h_2|^4 \sigma_H^2}{1+\alpha} + \cfrac{\left(1-\xi|h_R|^2\right)\sigma_R^2 \sigma_H^2}{\xi P_H (1-\alpha^2)(1+\alpha)|h_1|^2}} \tag{15}$$

Using (15), the E2E capacity performance gain from **u**-to-**H** for PS protocol (bits/sec/Hz) is written as,

$$C_{uH}^{PS} = \frac{1}{2} \log_2 \left(1 + \gamma_{uH}^{PS}(\alpha)\right) \tag{16}$$

From the (16), it is observed that the channel capacity obtained in (16) should be maximized to improve the performance gain of the relay system. The FD-based PS ratio, α is evaluated as,

$$\alpha = \underset{0 \leq \alpha \leq 1}{\text{Maximum}}\, C_{uH}^{PS}(\alpha) = 1 - \frac{1}{2}\left[\sqrt{a(a+4)} - a\right]; \text{ where } a = \frac{2|h_1|^2 \sigma_R^2}{\xi |h_2|^4 \sigma_H^2} \tag{17}$$

Notice that the Proof of the (17) is discussed in Appendix A (Kumar and Hossain, 2020). From (17), it is observed that the PS ratio does not depend on RSI channel gain. Finally, the ergodic capacity from **u**-to-**H** by exploiting PS protocol for same model is given by,

$$C_{E,uH}^{PS} = \mathbb{E}\left[C_{uH}^{PS}\right] = \frac{1}{2}\mathbb{E}\left[\log_2\left(1 + \gamma_{uH}^{PS}\right)\right] \tag{18}$$

where the expectation is realized by channels gain and RSI channel gain. For Rayleigh fading channel, the ergodic capacity with FD-based EH from **u**-to-**H** is formulated as,

$$C_{E,uH}^{PS} = \frac{1}{\mu_1 \mu_R \ln 2} \iint \left[\psi_2\left(|h_1|^2, |h_R|^2\right) - \psi_1\left(|h_1|^2, |h_R|^2\right)\right] e^{-\frac{|h_1|^2}{\mu_1}} e^{-\frac{|h_R|^2}{\mu_R}} d|h_1|^2 d|h_R|^2 \tag{19}$$

where,

$$\psi_l\left(|h_1|^2,|h_R|^2\right) = ci\left(\Gamma_l\left(|h_1|^2,|h_R|^2\right)\right)\cos\left(\Gamma_l\left(|h_1|^2,|h_R|^2\right)\right) + si\left(\Gamma_l\left(|h_1|^2,|h_R|^2\right)\right)\sin\left(\Gamma_l\left(|h_1|^2,|h_R|^2\right)\right);$$

for $l \in (1,2)$

$$\Gamma_1\left(|h_1|^2,|h_R|^2\right) = \frac{\sqrt{1-b_2|h_R|^2}}{\mu_2}\sqrt{\frac{b_3|h_1|^4 + b_5\left(1-b_2|h_R|^2\right)}{|h_1|^2\left(b_1|h_1|^4 + b_4\left(1-b_2|h_R|^2\right)\right)}}; \; b_1 = P_H\xi^2; \; b_2 = \xi; \; b_3 = \frac{\sigma_R^2}{1-\alpha^2};$$

$$\Gamma_2\left(|h_1|^2,|h_R|^2\right) = \frac{1}{\mu_2\sqrt{b_4}}\sqrt{\frac{b_3|h_1|^4 + b_5\left(1-b_2|h_R|^2\right)}{|h_1|^2}}; \; b_4 = \frac{\xi\sigma_H^2}{1+\alpha}$$

and $b_5 = \dfrac{\sigma_R^2\sigma_H^2}{\xi P_H\left(1-\alpha^2\right)(1+\alpha)}$ \hfill (20)

Mathematically, closed-form solution of the ergodic capacity in (20) is very difficult to track. Thus, it is investigated for same and also, fulfil the EH criteria of smart road side wireless node. For this, it is assumed that the high transmit power access point, **H** and road side relay **R** are fixed at a position and user, **u** is mobile. Note that $ci(\omega) = -\int_1^\infty \dfrac{\cos(\omega x)}{x}dx$ and $si(\omega) = -\int_1^\infty \dfrac{\sin(\omega x)}{x}dx$. If channel gain from **H**-to-**R**, $|h_1|^2$ and the RSI channel gain, $|h_R|^2$ are stationary (i.e., fixed) and channel gain from **u**-to-**R**, $|h_2|^2$ is exponentially distributed and characterized by Rayleigh fading, then E2E ergodic capacity (i.e., from **u**-to-**H**) for FD-based PS policy is formulated in closed-form as,

Notify that the proof of the (21) is available in Appendix B (Kumar and Hossain, 2020).

$$C_{E,uH}^{PS} = \frac{1}{\ln 2}\left[\sum_{j=1}^{2}(-1)^j\left\{ci\left(\Gamma_j\left(|h_1|^2,|h_R|^2\right)\right)\cos\left(\Gamma_j\left(|h_1|^2,|h_R|^2\right)\right)\right.\right.$$
$$\left.\left.+si\left(\Gamma_j\left(|h_1|^2,|h_R|^2\right)\right)\sin\left(\Gamma_j\left(|h_1|^2,|h_R|^2\right)\right)\right\}\right] \quad (21)$$

The SWIPT-Enabled Cooperative Full-Duplex Relaying Communication

Figure 4. Communication structure of the time-switching protocol for bidirectional FD relaying system

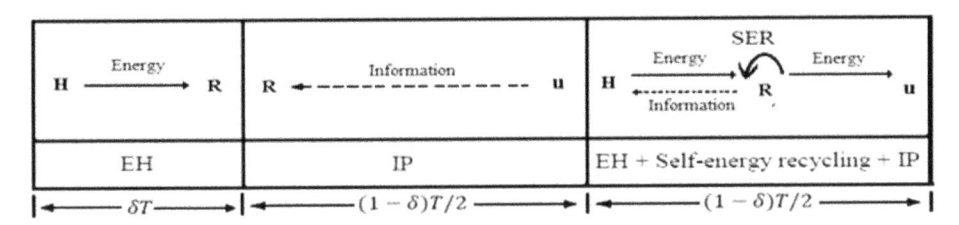

E2E ERGODIC CAPACITY BASED ON TIME-SWITCHING PROTOCOL FOR BIDIRECTIONAL FULL-DUPLEX AMPLIFY-AND-FORWARD RELAY SYSTEM

This section provides basic frame structure of the TS protocol for SWIPT enabled two-way FD AF relay system and obtains closed-form expressions of the TS ratio and E2E ergodic capacity for Rayleigh fading which is depicted in Fig. 4. To fulfil goals for the same, time frame structure of the communication between the end terminals is categorized into three phases of the communication block time T. In the first phase of CBT, $t \in [0, \delta/t]$, the H transfers energy signal to the R for EH, where δ denotes TS ratio with $0 \leq \delta \leq 1$. Thus, the received energy signal at the R is given by

$$y_{R,E}^{TS}[k] = h_1 x_H[k] + n_R[k] \tag{22}$$

From (22), the EH at the energy constrained node, **R** in first phase is expressed as

$$E_{R,TS}^{I} = \xi \mathbb{E}\left[\left|y_{R,E}^{TS}[k]\right|^2\right]\delta T = \xi P_H |h_1|^2 \delta T \tag{23}$$

Note that the noise added at the **R** does not involved in EH. In the second phase of CBT $t \in \left[\delta T, \delta T + (1-\delta)T/2\right]$, the user sends information to the **R**. Thereby, decoded information signal at the **R** is defined as

$$y_{R,IP}^{TS}[k] = h_2 x_u[k] + n_R^{'}[k] \tag{24}$$

where $n_R^{'}[k] \sim \mathcal{CN}(0, \sigma_R^2)$ represents antenna noise at the **R**. In the last phase of communication time $t \in \left[\delta T + (1-\delta)T/2, T\right]$, the **R** sends amplified information received in second phase and energy signal in first phase to the **H** and the **u** respectively at a time. Parallel, it gains energy from the **H** and self-interference signal from receiver antenna. Thereby, amplified information at the **R** can be formulated as

$$x_R^{TS}[k] = \sqrt{\beta_{TS}}\left(y_{R,IP}^{TS}[k-1] + n_{R,P}'[k-1]\right) \tag{25}$$

where $n_{R,P}'[k] \sim \mathcal{CN}(0, \sigma_R^2)$ denotes noise due to RF-to-baseband conversion. The information received at the **H** is written as,

$$y_H^{TS}[k] = h_1 x_R^{TS}[k] + h_H x_H[k] + n_H[k] \tag{26}$$

The second term of right side of (26) is removed with the help of the notch filter. Since, it does not content any information. By substituting (24) and (25) into (26), the information at the **R** can be expressed as

$$\tilde{y}_H^{TS}[k] = \underbrace{\sqrt{\beta_{TS}} h_1 h_2 x_u[k-1]}_{Information Signal} + \underbrace{\sqrt{\beta_{TS}} h_1 h_{R,P}'[k-1] + n_H[k]}_{Noise Signal} \tag{27}$$

The relay received energy signal from **H** and RSI via receiver antenna is given as

$$\tilde{y}_{R,E}^{TS}[k] = h_1 x_H[k] + h_R x_{R,N}[k] + n_R'[k] \tag{28}$$

Thereby, the energy harvested at the **R** in last phase is expressed as

$$E_{R,TS}^{III} = \xi \mathbb{E}\left[\left|\tilde{y}_{R,E}^{TS}[k]\right|^2\right]\frac{(1-\delta)T}{2} = \xi\left(P_H|h_1|^2 + P_R|h_R|^2\right)(1-\delta)\frac{T}{2} \tag{29}$$

From (23) and (29), aggregated EH at the **R** for TS policy is written as

$$E_R^{TS} = \xi\left(P_H|h_1|^2(1+\delta) + P_R|h_R|^2(1-\delta)\right)\frac{T}{2} \tag{30}$$

Note that the amplification coefficient, β_{TS} is formulated in (Kumar and Ashraf, 2020). In addition, energy harvested at the R in the block of time, T is greater than

The SWIPT-Enabled Cooperative Full-Duplex Relaying Communication

or equal to energy consumed at the R in last phase i.e., $E_R^{TS} \geq P_R(1-\delta)T/2$. Thus, transmit power of the **R**, P_R^{TS} given by

$$P_R^{TS} \leq \frac{\xi P_H (1+\delta)|h_1|^2}{\left(1-\xi|h_R|^2\right)(1-\delta)} \tag{31}$$

With the assistance of (27), the E2E SINR from **u**-to-**H** for TS protocol is expressed as,

$$\gamma_{uH}^{TS} = \frac{P_u^{TS}|h_1|^2|h_2|^2}{|h_1|^2 \sigma_R^2 + \frac{\sigma_H^2}{\beta_{TS}}} \tag{32}$$

It is also, assumed that the **u** is energy constrained node that the energy signal in last phase is written as,

$$y_{u,E}^{TS}[k] = h_2 x_R^{TS}[k] + n_u[k] \tag{33}$$

The EH at the **u** is calculated from (33) as,

$$E_u^{TS} = \xi|h_2|^2 \mathbb{E}\left[|x_R^{TS}[k]|^2\right]\frac{(1-\delta)T}{2} = \xi^2|h_2|^2\left[P_H(1+\delta)|h_1|^2 + P_R|h_R|^2(1-\delta)\right]\frac{T}{2} \tag{34}$$

Thus, user harvested transmit power, P_u^{PS} from (34) is given as,

$$P_u^{TS} = \frac{E_u^{TS}}{\frac{(1-\delta)T}{2}} = \xi^2|h_2|^2\left[P_H|h_1|^2\frac{1+\delta}{1-\delta} + P_R|h_R|^2\right] \tag{35}$$

By substituting (31), and (35) into (32), the E2E SINR from **u**-to-**H** for TS protocol is evaluated as,

123

$$\gamma_{uH}^{TS}(\delta) = \frac{\frac{(1+\delta)}{(1-\delta)}}{f_1\left[f_2 + f_3\frac{1-\delta}{1+\delta}\right]};$$

where

$$f_1 = 1-\xi|h_R|^2; \; f_2 = \frac{1}{\xi P_H |h_1|^2}\left[\frac{\sigma_H^2}{|h_1|^2} + \frac{\sigma_R^2}{\xi|h_2|^4}\right]; \; f_3 = \frac{\sigma_R^2 \sigma_H^2 f_1}{\xi^3 P_H^2 |h_1|^6 |h_2|^4} \quad (36)$$

The E2E capacity from **u**-to-**H** for TS protocol in bits/sec/Hz is formulated as

$$C_{uH}^{TS} = \frac{1}{2}(1-\delta)\log_2\left(1+\gamma_{uH}^{TS}(\delta)\right) \quad (37)$$

The performance gain in term of ergodic capacity for TS protocol is expressed in closed-form as,

$$C_{E,uH}^{TS} = \mathbb{E}\left[C_{uH}^{TS}\right] = \frac{1}{2}(1-\delta)\mathbb{E}\left[\log_2\left(1+\gamma_{uH}^{TS}\right)\right] \quad (38)$$

where expectation represents characteristic of channels gain and RSI channel gain. The TS ratio δ for two-way FD AF relaying network is given by,

$$\delta = \underset{0\leq\delta\leq 1}{Maximum}\, C_{uH}^{TS}(\delta) = \frac{1-f_0(1+p)-p}{1-f_0(1+p)+p} \quad (39)$$

Where $f_0 = f_1 f_2$, $p = W(\varepsilon)$ with $W(.)$ denotes Lambert W function is formulated as $\lambda = W(\lambda)e^{W(\lambda)}$ and $\varepsilon = \left(\frac{1}{f_0}-1\right)e^{-1}$. If the channels are exponentially distributed and characterized by Rayleigh fading, then E2E ergodic capacity is formulated as,

$$C_{E,uH}^{TS} = \frac{(1-\delta)}{\mu_1\mu_R \ln 2}\iint\left[\varphi_2\left(|h_1|^2,|h_R|^2\right)-\varphi_1\left(|h_1|^2,|h_R|^2\right)\right]e^{-\frac{|h_1|^2}{\mu_1}}e^{-\frac{|h_R|^2}{\mu_R}}d|h_1|^2\,d|h_R|^2$$

$$(40)$$

The SWIPT-Enabled Cooperative Full-Duplex Relaying Communication

where,

$$\varphi_n\left(|h_1|^2,|h_R|^2\right) = ci\left(\mathcal{D}_n\left(|h_1|^2,|h_R|^2\right)\right)\cos\left(\mathcal{D}_n\left(|h_1|^2,|h_R|^2\right)\right) + si\left(\mathcal{D}_n\left(|h_1|^2,|h_R|^2\right)\right)\sin\left(\mathcal{D}_n\left(|h_1|^2,|h_R|^2\right)\right);$$

for $n \in (1,2)$

$$\mathcal{D}_1\left(|h_1|^2,|h_R|^2\right) = \frac{1}{\mu_2}\sqrt{\frac{k_1|h_1|^4 + k_3\left(1-\xi|h_R|^2\right)}{k_2|h_1|^2 + B_0\left(|h_1|^2,|h_R|^2\right)}};$$

$$B_0\left(|h_1|^2,|h_R|^2\right) = \frac{k_0|h_1|^6}{1-\xi|h_R|^2}; \quad k_0 = \frac{1+\delta}{1-\delta}; \quad k_1 = \frac{\sigma_R^2}{\xi^2 P_H};$$

$$\mathcal{D}_2\left(|h_1|^2,|h_R|^2\right) = \frac{1}{\mu_2\sqrt{k_2}}\sqrt{\frac{k_1|h_1|^4 + k_3\left(1-\xi|h_R|^2\right)}{|h_1|^2}}; \quad k_2 = \frac{\sigma_H^2}{\xi P_H};$$

and $k_3 = \dfrac{\sigma_R^2 \sigma_H^2}{\xi^3 P_H^2 k_0}$ \hfill (41)

$$C_{E,uH}^{TS} = \frac{(1-\delta)}{\ln 2}\left[\sum_{j=1}^{2}(-1)^j\left\{ci\left(\mathcal{D}_j\left(|h_1|^2,|h_R|^2\right)\right)\cos\left(\mathcal{D}_j\left(|h_1|^2,|h_R|^2\right)\right)\right.\right.$$
$$\left.\left.+si\left(\mathcal{D}_j\left(|h_1|^2,|h_R|^2\right)\right)\sin\left(\mathcal{D}_j\left(|h_1|^2,|h_R|^2\right)\right)\right\}\right] \quad (42)$$

Notice that the channel gain from **H**-to-**R**, $|h_1|^2$ and RSI channel gain, $|h_R|^2$ are stationary, but channel gain from **u**-to-**R**, $|h_2|^2$ is exponentially distributed with Rayleigh fading. Thereby, E2E ergodic capacity for FD-based TS protocol from **u**-to-**H** is derived as,

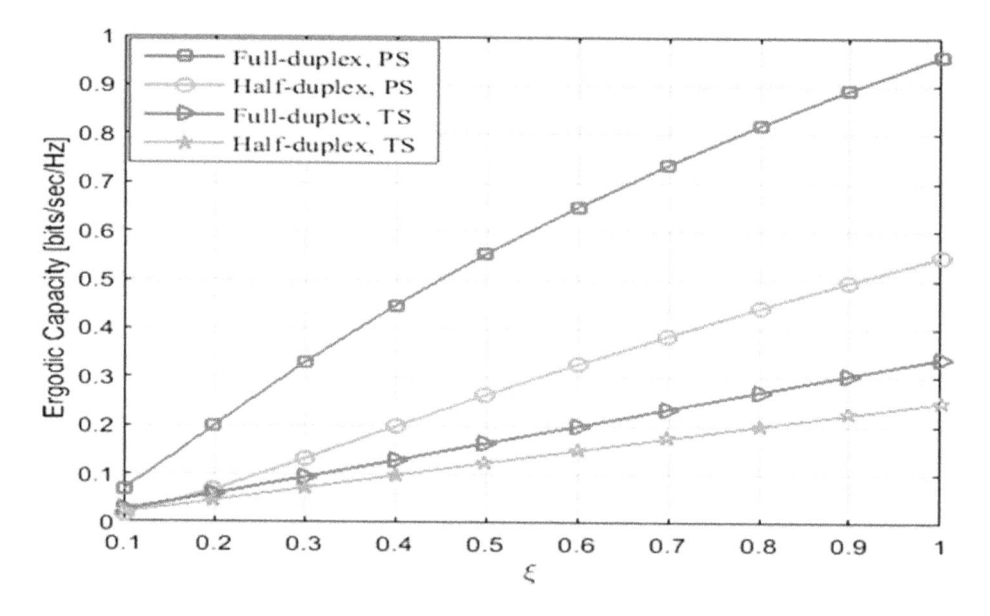

Figure 5. Plotted ergodic capacity from u-to-H verses energy conversion efficiency, ξ

NUMERICAL RESULTS AND DISCUSSIONS

This section provides performance gains of the considered system model for Rayleigh fading channel. For this purpose, it is assumed that the **H** and road side energy constrained relay, **R** are placed at fixed position while user, **u** is mobile. In the Figure 5, the E2E ergodic capacity from **u**-to-**H** is varied with ξ. As the energy conversion increases, performance gain is also increased for both the PSR-and-TSR protocols. It is observed that the PSR protocol performs better as compared to TSR protocol. Numerical results authenticate that ergodic capacity of the proposed FD-based PSR and TSR protocols are shown outstanding performance as compared to existing HD-based these protocols. It is also, found that the PSR protocol for both FD-and-HD relaying policies is always superior to the TSR protocol.

Further, the E2E ergodic capacity from **u**-to-**H** versus P_H is plotted which is shown in Figure 6. In the range of the 5dBm$\leq P_H \leq$14dBm, it is observed that performance gain such as HD-based TS protocol is poor as compared to other protocols. The performance of the proposed FD-based PS protocol is superior to other EH policies if the transmit power, P_H is reached beyond the 12 dBm. It is also, notice that if the P_H is greater than or equal to 23 dBm, the FD-based PS and TS protocol outperforms better as compared to existing HD-based PS and TS protocols. The EH at the road side energy constrained relay, **R** increases with transmit power of

The SWIPT-Enabled Cooperative Full-Duplex Relaying Communication

H, P_H which is shown in Fig. 7. The TS-based protocols show better performance gains as compared to PS-based EH if it is resumed transmit power between 35dBm$\leq P_H \leq$45dBm while if the transmit power, P_H is greater than or equal to 55 dBm then proposed FD-based EH policies shows superior performance behavior as compared to the HD-based EH policies. It is possible due to receive energy at the **R** from **H** in two phases while in the case of the HD relaying protocols, the **R** harvests energy from energy signal from the **H** in one phase. It is also, assumed that the mobile user **u** is energy constrained which is also harvests energy from the **H** via the **R** that is shown in Fig. 8. For this purpose, the EH at the **u** increases with the P_H. If it is compared the EH performance of user **u** with the EH at the **R**, the characteristics behavior of the EH is followed same. However, the magnitude of the EH at the **R** is greater than the magnitude of the EH at the **u**. Other way it can say that the fixed road side relay **R** is placed in the one hop distance from the **H** while the mobile user **u** is located in two-hop. Finally, it is analyzed the ergodic capacity performance with residual self-interference channel gain that is depicted in Fig. 9. The ergodic capacity for both FD-based PS and TS protocols displays constant behavior if the loopback interference is below the -45 dB. But performance gains for both protocols increases exponentially with RSI channel gain. It is assumed that the RSI channel gain must be nonzero for proposed FD-based protocol, because the performance gains of the system is improved with loopback interference channel gain. Furthermore, it is authenticated the FD-based PS protocol is highly desirable as compared to TS protocol.

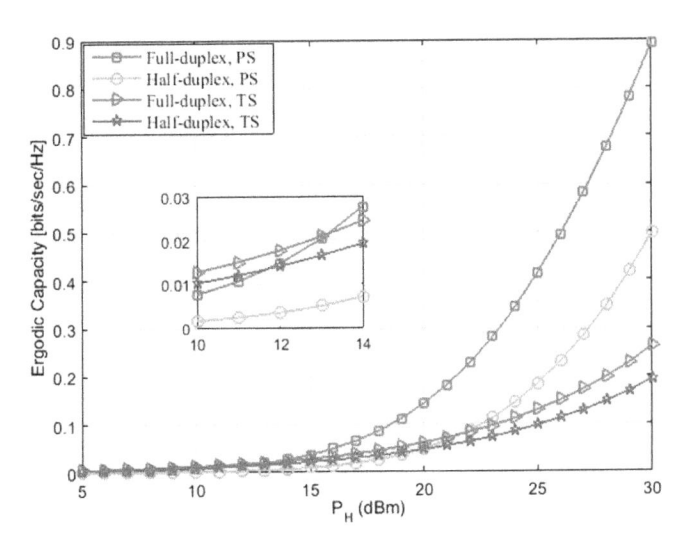

Figure 6. Ergodic capacity from u-to-H with transmit power of H, P_H

Figure 7. Variation of the energy-harvesting gain at R with transmit power, P_H

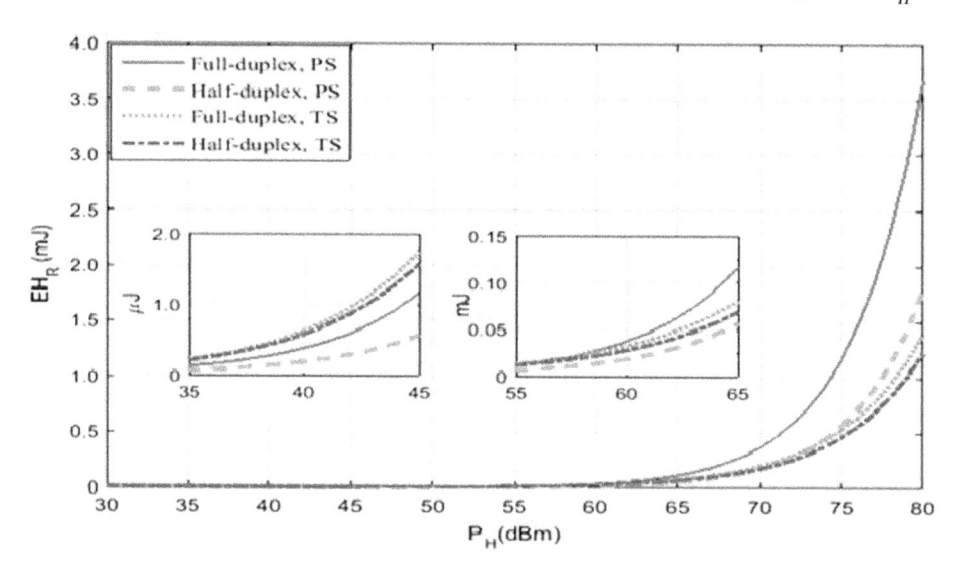

Figure 8. The EH at the mobile user u versus transmit power of H, P_H via the road side fixed relay R

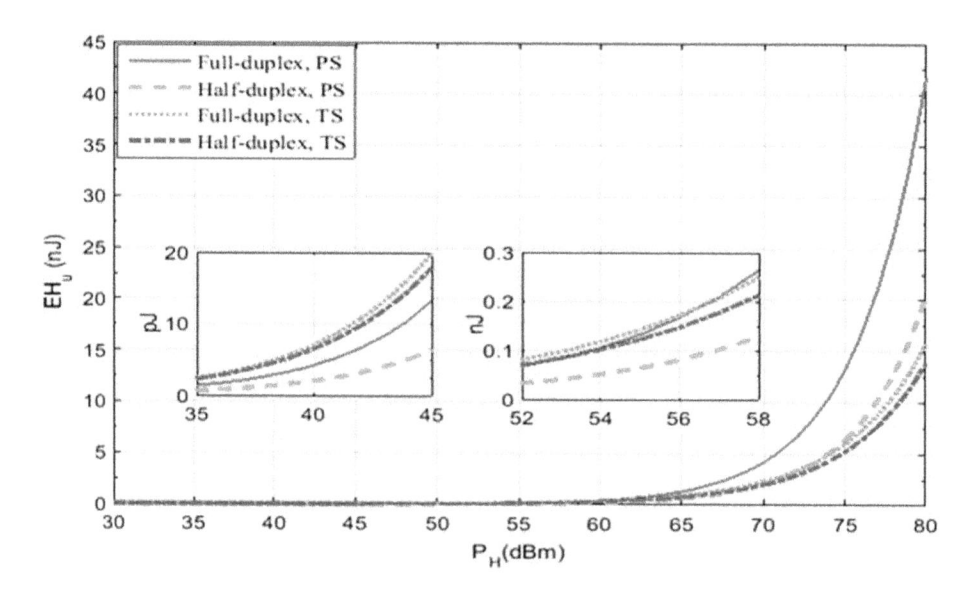

CONCLUSIONS AND FUTURE RESEARCH DIRECTIONS

In this chapter, two-way AF relay system is presented. It is assumed that energy-constrained road side relay **R** and the mobile user **u** harvest energy from the high transmit power **H**, where the **R** also collects energy from SER. It is observed that the proposed FD-based PS and TS protocols outperform as compared to HD-based PS and TS protocols. This chapter also provides closed-form expression of the end-to-end ergodic capacity of the proposed PS and TS protocols for Rayleigh fading channel. It is found that TS protocol is beneficial for lower value of transmit power of **H** while the PS protocol is suitable for higher value of transmit power of **H**. In addition, it is authenticated that proposed FD-based EH prolongs the lifetime of the energy constrained nodes. If the mobile user size is very small, the placement of the couple of the antenna for FD operation is very difficult and challenging issues, because antennas isolation method and analog-circuit-domain concepts are exploited to mitigate the self-interference by FD operation. Since, technique is suitable for the base station and the relay. Thus, possible research direction is needed to develop the smart antenna for FD operation. Some research opportunities also remain to enhance the capability of cooperative relay system for SWIPT with the help of queuing theory. In addition, a distributive and an adaptive transmission selection policy are possible research direction for cooperative buffer-aided relays system. Due to broadcast nature of the wireless communication, cooperative user relaying gets overhead with user information and forwards it to other users. To overcome, the SWIPT enabled cooperative NOMA technique has been used for perfect CSI. But it is still unknown for imperfect CSI. One of the major challenging problem of the UAV communication is deployment strategy and trajectory framework in terrestrial communication. These two problems have not been studied by exploring the concept of the RSMA scheme. Notice that the SWIPT enabled cooperative relaying system suffers from limited range of the energy receivers. It is resolved with the potential of the intelligent reflecting surface (IRS) technique. But, the RSMA in the SWIPT enabled network with the assistance of the IRS is not fully understood still. It is also noticed that the secure communication is more challenging issue as the energy receivers are located near to transmitter related to information receivers. Thus, the RSMA would be greatly employed to secure the communication in the SWIPT enabled wireless networks. The cognitive radio is one of the potential field in wireless communication that has potential to overcome the use of the radio resources. But an interference management between primary users (PUs) and secondary users (SUs) as the PUs are not know the existence of the SUs. In such a case, spectrum sharing is between SUs and PUs is very crucial and leads to security threats in the cognitive radio networks. In this scenario, the RSMA scheme is remains to overcome this prime challenges. It is also noticed that the SWIPT in the cognitive radio networks is not fully understood by exploiting RSMA scheme.

REFERENCES

Alsharoa, A., Ghazzai, H., Kamal, A. E., & Kadri, A. (2017). Optimization of a power splitting protocol for two-way multiple energy harvesting relay system. *IEEE Transactions on Green Communications and Networking*, *1*(4), 444–457. doi:10.1109/TGCN.2017.2724438

Budhiraja, I., Kumar, N., Tyagi, S., Tanwar, S., & Guizani, M. (2021). SWIPT-enabled D2D communication underlaying NOMA-based cellular networks in imperfect CSI. *IEEE Trans. on Vech. Tech*, *70*(1), 692–699.

Chataut, R., & Akl, R. (2020). Massive MIMO Systems for 5G and beyond Networks—Overview, Recent Trends, Challenges, and Future Research Direction. *Sensors (Basel)*, *20*(10), 2753. doi:10.3390/s20102753 PMID:32408531

Chen, G., Xiao, P., Kelly, J. R., Li, B., & Tafazolli, R. M. (2017). Full-duplex wireless-powered relay in two way cooperative networks. *IEEE Access : Practical Innovations, Open Solutions*, *5*, 1548–1558. doi:10.1109/ACCESS.2017.2661378

Chen, H., Li, Y., Rebelatto, J. L., Filho, B. F. U., & Vucetic, B. (2015). Harvest-then-cooperate: Wireless-powered cooperative communications. *IEEE Transactions on Signal Processing*, *63*(7), 1700–1711. doi:10.1109/TSP.2015.2396009

Ding, H., Wang, X., Costa, D. B. D., Chen, Y., & Gong, F. (2017). Adaptive time-switching based energy harvesting relaying protocols. *IEEE Transactions on Communications*, *65*(7), 2821–2837. doi:10.1109/TCOMM.2017.2693358

Gao, X., Li, X., Han, C., Zeng, M., Liu, H., Mumtaz, S., & Nallanathan, A. (2024). Rate-splitting multiple access-based cognitive radio network with ipSIC and CEEs. *IEEE Transactions on Vehicular Technology*, *73*(1), 1430–1434. doi:10.1109/TVT.2023.3305960

Guo, J., Zhang, S., Zhao, N., & Wang, X. (2020). Performance of SWIPT for full-duplex relay system with co-channel interference. *IEEE Transactions on Vehicular Technology*, *67*(2), 2311–2315. doi:10.1109/TVT.2019.2958626

Huang, G., Zhang, Q., & Qin, J. (2015). Joint time switching and power allocation for multicarrier decode-and-forward relay networks with SWIPT. *IEEE Signal Processing Letters*, *22*(12), 2284–2288. doi:10.1109/LSP.2015.2477424

Huang, X., & Ansari, N. (2016). Optimal cooperative power allocation for energy-harvesting-enabled relay networks. *IEEE Transactions on Vehicular Technology*, *65*(4), 2424–2434. doi:10.1109/TVT.2015.2424218

Jayakody, D. N. K., Perera, T. D. P., Ghrayeb, A., & Hasna, O. M. (2020). Self-energized UAV-assisted scheme for cooperative wireless relay networks. *IEEE Trans. on Vech. Tech*, *69*(1), 578–592.

Kashef, M., & Ephremides, A. (2016). Optimal partial relaying for energy-harvesting wireless networks. *IEEE/ACM Transactions on Networking*, *24*(1), 113–122. doi:10.1109/TNET.2014.2361683

Khandaker, M. R. A., & Kai-Kit Wong. (2014). SWIPT in MISO multicasting systems. *IEEE Wireless Communications Letters*, *3*(3), 277–280. doi:10.1109/WCL.2014.030514.140057

Kong, C., & Lu, H. (2023). Cooperative rate-splitting multiple access in heterogeneous networks. *IEEE Communications Letters*, *27*(10), 2807–2811. doi:10.1109/LCOMM.2023.3309818

Kumar, D., Singya, P. K., Krejcar, O., & Bhatia, V. (2023). On performance of a SWIPT enabled FD CRN with HIs and imperfect SIC over α–μ fading channel. *IEEE Transactions on Cognitive Communications and Networking*, *9*(1), 99–113. doi:10.1109/TCCN.2022.3220791

Kumar, D., Singya, P. K., Nebhen, J., & Bhatia, V. (2023). Performance of SWIPT-enabled FD TWR network with hardware impairments and imperfect CSI. *IEEE Systems Journal*, *17*(1), 1224–1234. doi:10.1109/JSYST.2022.3183501

Kumar, R., & Hossain, A. (2018). Experimental performance and study of low power strain gauge based wireless sensor node for structure health monitoring. *Wireless Personal Communications*, *101*(3), 1657–1669. doi:10.1007/s11277-018-5782-6

Kumar, R., & Hossain, A. (2020). Full-duplex wireless information and power transfer in two-way relaying networks with self-energy recycling. *Wireless Networks*, *26*(8), 6139–6154. doi:10.1007/s11276-020-02432-x

Kurup, R. R., & Babu, A. V. (2020). Power adaptation for improving the performance of time switching SWIPT-based full-duplex cooperative NOMA network. *IEEE Communications Letters*, *24*(12), 2956–2960. doi:10.1109/LCOMM.2020.3017624

Le, Q. N., Bao, V. N. Q., & An, B. (2018). Full-duplex distributed switch-and-stay energy harvesting selection relaying networks with imperfect CSI: Design and outage analysis. *Journal of Communications and Networks (Seoul)*, *20*(1), 29–46. doi:10.1109/JCN.2018.000004

Li, S., & Murch, R. D. (2014). An investigation into baseband techniques for single-channel full-duplex wireless communication systems. *IEEE Transactions on Wireless Communications*, *13*(9), 4794–4806. doi:10.1109/TWC.2014.2341569

Li, T., Zhang, H., Zhou, X., & Yuan, D. (2022). Full-duplex cooperative rate-splitting for multigroup multicast with SWIPT. *IEEE Transactions on Wireless Communications*, *21*(6), 4379–4393. doi:10.1109/TWC.2021.3129881

Liu, H., Kim, K. J., Kwak, K. S., & Vincent Poor, H. (2016). Power splitting-based SWIPT with decode-and-forward full-duplex relaying. *IEEE Transactions on Wireless Communications*, *5*(11), 7561–7577. doi:10.1109/TWC.2016.2604801

Liu, J., Xiong, K., Lu, Y., Fan, P., Zhong, Z., & Letaief, B. K. (2020). SWIPT-enabled full-duplex NOMA networks with full and partial CSI. *IEEE Transactions on Green Communications and Networking*, *4*(3), 804–818. doi:10.1109/TGCN.2020.2977611

Liu, P., Gazor, S., Kim, I. M., & Kim, D. I. (2015). Noncoherent relaying in energy harvesting communication systems. *IEEE Transactions on Wireless Communications*, *14*(12), 6940–6954. doi:10.1109/TWC.2015.2462838

Liu, Y., Ding, Z., Elkashlan, M., & Poor, H. V. (2016). Cooperative nonorthogonal multiple access with simultaneous wireless information and power transfer. *IEEE Journal on Selected Areas in Communications*, *34*(4), 938–953. doi:10.1109/JSAC.2016.2549378

Liu, Y., Wang, L., Elkashlan, M., Duong, T. Q., & Nallanathan, A. (2016). Two-way relay networks with wireless power transfer: Design and performance analysis. *IET Communications*, *10*(14), 1810–1819. doi:10.1049/iet-com.2015.0728

Ma, R., Wu, H., Ou, J., Yang, S., & Gao, Y. (2020). Power splitting-based SWIPT systems with full-duplex jamming. *IEEE Trans. on Vech. Tech*, *69*(9), 9822–9836.

Mao, Y., Dizdar, O., Clerckx, B., Schober, R., Popovski, P., & Poor, H. V. (2022). Rate-splitting multiple access: Fundamentals, survey, and future research trends. *IEEE Communications Surveys and Tutorials*, *24*(4), 2073–2126. doi:10.1109/COMST.2022.3191937

Nasir, A. A., Tuan, H. D., Ngo, D. T., Duong, T. Q., & Poor, H. V. (2017). Beamforming design for wireless information and power transfer systems: Receive power-splitting versus transmit time-switching. *IEEE Transactions on Communications*, *65*(2), 876–889. doi:10.1109/TCOMM.2016.2631465

Nasir, A. A., Zhou, X., Durrani, S., & Kennedy, R. A. (2013). Relaying protocols for wireless energy harvesting and information processing. *IEEE Transactions on Wireless Communications*, *12*(7), 3622–3636. doi:10.1109/TWC.2013.062413.122042

Nasir, A. A., Zhou, X., Durrani, S., & Kennedy, R. A. (2015). Wireless-powered relays in cooperative communications: Time-switching relaying protocols and throughput analysis. *IEEE Transactions on Communications*, *63*(5), 1607–1622. doi:10.1109/TCOMM.2015.2415480

Pérez, D. E., López, Q. L. A., Alves, H., & Latva-aho, M. (2022). Self-energy recycling for low-power reliable networks: Half-duplex or full-duplex? *IEEE Systems Journal*, *16*(3), 4780–4791. doi:10.1109/JSYST.2021.3127266

Rabie, K., Adebisi, B., Nauryzbayev, G., Badarneh, Q. S., Li, X., & Alouini, M. S. (2019). Full-duplex energy harvesting enabled relay networks in generalized fading channels. *IEEE Wireless Communications Letters*, *8*(2), 384–387. doi:10.1109/LWC.2018.2873360

Saad, W., Bennis, M., & Chen, M. (2020). A Vision of 6G Wireless Systems: Applications, Trends, Technologies, and Open Research Problems. *IEEE Network*, *34*(3), 134–142. doi:10.1109/MNET.001.1900287

Salem, A., & Hamdi, K. A. (2016). Wireless power transfer in multi-pair two-way AF relaying networks. *IEEE Transactions on Communications*, *64*(11), 4578–4591. doi:10.1109/TCOMM.2016.2607751

Smida, P., Sabharwal, A., Fodor, G., Alexandropoulos, G. C., Suraweera, H. A., & Chae, C. B. (2023). Full-duplex wireless for 6G: Progress brings new opportunities and challenges. *IEEE Journal on Selected Areas in Communications*, *41*(9), 2729–2750. doi:10.1109/JSAC.2023.3287612

Tariq, F., Khandaker, M. R. A., Wong, K. K., Imran, M. A., Bennis, M., & Debbah, M. (2020). A speculative study on 6G. *IEEE Wireless Communications*, *27*(4), 118–125. doi:10.1109/MWC.001.1900488

Tataria, H., Shafi, M., Molisch, A. F., Dohler, M., Sjoland, H., & Tufvesson, F. (2021). 6G wireless systems: Vision, requirements, challenges, insights, and opportunities. *Proceedings of the IEEE*, *109*(7), 1166–1199. doi:10.1109/JPROC.2021.3061701

Tutuncuoglu, K., Varan, B., & Yener, A. (2015). Throughput maximization for two-way relay channels with energy harvesting nodes: The impact of relaying strategies. *IEEE Transactions on Communications*, *63*(6), 2081–2093. doi:10.1109/TCOMM.2015.2427162

Varshney, L. R. (2008). Transporting information and energy simultaneously. *IEEE International Symposium on Information Theory*, (pp. 1612-1616). IEEE.

Vullers, R. J. M., Schaijk, R. V., Doms, I., Hoof, C. V., & Mertens, R. (2009). Micropower energy harvesting. *Elsevier Solid-State Circuits*, *53*(7), 684–693.

Wang, D., Zhang, R., Cheng, X., & Yang, L. (2017). Capacity-enhancing full-duplex relay networks based on power-splitting (PS) SWIPT. *IEEE Trans. on Vech. Tech*, *66*(6), 5446–5450.

Wang, D., Zhang, R., Cheng, X., Yang, L., & Chen, C. (2017). Relay selection in full-duplex energy harvesting two-way relay networks. *IEEE Transactions on Green Communications and Networking*, *1*(2), 182–191. doi:10.1109/TGCN.2017.2686325

Wang, W., Li, X., Zhang, M., Cumanan, K., Ng, D. W. K., Zhang, G., Tang, J., & Dobre, O. A. (2020). Energy-constrained UAV-assisted secure communications with position optimization and cooperative jamming. *IEEE Transactions on Communications*, *68*(7), 4476–4489. doi:10.1109/TCOMM.2020.2989462

Xiong, K., Fan, P., Zhang, C., & Letaief, K. B. (2015). Wireless information and energy transfer for two-hop non-regenerative MIMO-OFDM relay networks. *IEEE Journal on Selected Areas in Communications*, *33*(8), 1595–1611. doi:10.1109/JSAC.2015.2391931

Xu, Y., Shen, C., Ding, Z., Sun, X., Yan, S., Zhu, G., & Zhong, Z. (2017). Joint beamforming and power-splitting control in downlink cooperative SWIPT NOMA systems. *IEEE Transactions on Signal Processing*, *65*(18), 4874–4886. doi:10.1109/TSP.2017.2715008

Yang, Z., Ding, Z., Fan, P., & Dhahir, N. A. (2017). The impact of power allocation on cooperative non-orthogonal multiple access networks with SWIPT. *IEEE Transactions on Wireless Communications*, *16*(7), 4332–4343. doi:10.1109/TWC.2017.2697380

Zeng, Y., Chen, H., & Zhang, R. (2016). Bidirectional wireless information and power transfer with a helping relay. *IEEE Communications Letters*, *20*(5), 862–865. doi:10.1109/LCOMM.2016.2549515

Zeng, Y., & Zhang, R. (2015). Full-duplex wireless-powered relay with self-energy recycling. *IEEE Wireless Communications Letters*, *4*(2), 201–204. doi:10.1109/LWC.2015.2396516

Zhang, R., & Ho, C. K. (2013). MIMO broadcasting for simultaneous wireless information and power transfer. *IEEE Transactions on Wireless Communications*, *12*(5), 31989–32001. doi:10.1109/TWC.2013.031813.120224

Zhong, C., Suraweera, H., Zheng, G., Krikidis, I., & Zhang, Z. (2014). Wireless information and power transfer with full duplex relaying. *IEEE Transactions on Communications*, *62*(10), 3447–3461. doi:10.1109/TCOMM.2014.2357423

Chapter 5
Energy Harvesting and Smart Highways for Sustainable Transportation Infrastructure:
Revolutionizing Roads Using Nanotechnology

Mohanraj Gopal
School of Computer Science Engineering and Information Systems, Vellore Institute of Technology, Vellore, India

J. Lurdhumary
Department of Electronics and Communication Engineering, DMI College of Engineering, Chennai, India

S. Bathrinath
https://orcid.org/0000-0002-5502-6203
Department of Mechanical Engineering, Kalasalingam Academy of Research and Education, Krishnankoil, India

A. Parvathi Priya
Department of Chemistry, R.M.K. Engineering College, Chennai, India

Atul Sarojwal
Department of Electrical Engineering, MJP Rohilkhand University, Bareilly, India

S. Boopathi
https://orcid.org/0000-0002-2065-6539
Department of Mechanical Engineering, Muthayammal Engineering College, Namakkal, India

DOI: 10.4018/978-1-6684-9214-7.ch005

Copyright © 2024, IGI Global. Copying or distributing in print or electronic forms without written permission of IGI Global is prohibited.

ABSTRACT

The chapter explores the integration of nanotechnology, energy harvesting, and smart highways into global transportation infrastructure, aiming to create sustainable and efficient systems. Nanotechnology enhances road surface durability and functionality, offering increased strength, resilience, and self-healing properties. Energy harvesting techniques, such as piezoelectric and solar technologies, harness kinetic and solar energy from vehicular motion and sunlight, powering infrastructure, streetlights, and even the grid. Smart highways, enabled by interconnected sensors and communication systems, monitor traffic flow, adjust speed limits, provide real-time updates, and autonomously manage transportation systems. These innovations not only promise a sustainable transportation ecosystem but also catalyze economic growth, environmental preservation, and enhanced quality of life for communities worldwide.

INTRODUCTION

In the 21st century, the global community faces unprecedented challenges related to transportation, with burgeoning urbanization, population growth, and environmental concerns placing immense pressure on existing infrastructure. Traditional transportation systems heavily reliant on fossil fuels contribute significantly to air pollution, greenhouse gas emissions, and congestion, necessitating a paradigm shift towards sustainable alternatives. This has propelled the exploration and implementation of innovative technologies to transform transportation infrastructure into more efficient, eco-friendly, and resilient systems (Sultana et al., 2021).

As the world witness's rapid urbanization and population growth, cities become focal points for economic activities and human habitation. However, this surge in urban living also intensifies the demand for transportation, leading to increased traffic congestion, longer commute times, and environmental degradation. Sustainable transportation infrastructure is crucial to mitigate these challenges, ensuring that cities remain hubs of productivity without compromising the quality of life for their inhabitants (Ding & Liu, 2023).

The environmental toll of conventional transportation systems is undeniable, with vehicular emissions contributing significantly to air pollution and climate change. The need for sustainable transportation infrastructure arises from the imperative to reduce carbon footprints, embrace renewable energy sources, and foster eco-friendly practices. Addressing these environmental concerns is not only a responsibility but a necessity to ensure the health of the planet and its inhabitants. Traditional transportation systems often operate with inefficiencies, utilizing non-renewable energy sources and contributing to resource depletion. Sustainable transportation infrastructure seeks to optimize energy usage through innovative technologies like energy harvesting and nanotechnology. By harnessing energy from vehicular motion and sunlight, these systems not only reduce dependence on fossil fuels but also promote resource conservation, aligning with the principles of a circular and sustainable economy (Prus & Sikora, 2021).

The economic impact of inefficient transportation systems is profound, affecting productivity, commerce, and overall economic growth. Congestion and delays in transportation lead to financial losses for businesses and individuals alike. Sustainable transportation infrastructure presents an opportunity to boost economic development by improving the efficiency of logistics, reducing operational costs, and fostering innovation in emerging industries related to clean energy and smart technologies (Zhao et al., 2020). Beyond the environmental and economic aspects, the need for sustainable transportation infrastructure is deeply rooted in enhancing the quality of life for communities. Reduced air pollution, efficient traffic flow, and improved accessibility contribute to healthier, more livable urban environments. Smart highways, enabled by interconnected systems, provide real-time updates, reduce accidents, and offer a seamless travel experience. This, in turn, fosters community well-being, attracting businesses, residents, and investors to regions with forward-thinking, sustainable transportation solutions (Petru & Krivda, 2021).

Sustainable transportation infrastructure is crucial due to urbanization, environmental concerns, resource conservation, economic efficiency, and community well-being. Embracing innovative technologies and approaches is essential for building a sustainable and resilient future, meeting current demands while ensuring a resilient future in the complexities of the modern world.

Nanotechnology plays a pivotal role in revolutionizing transportation infrastructure by offering unprecedented improvements in road surface durability. Traditional road materials are susceptible to wear and degradation over time, leading to frequent maintenance and repair. Nanomaterials, with their unique properties at the nanoscale, enhance the strength and resilience of road surfaces. By incorporating nanoparticles into construction materials, roads become more resistant to abrasion, weathering, and structural damage. Moreover, nanotechnology introduces self-healing properties, allowing road surfaces to repair minor damages autonomously. This enhanced durability not only reduces maintenance costs but also extends the lifespan of transportation infrastructure, contributing to a more sustainable and cost-effective road network (Petru & Krivda, 2021; Pompigna & Mauro, 2022a).

Energy harvesting techniques, such as piezoelectric and solar technologies, address the growing demand for sustainable energy sources in transportation infrastructure. Piezoelectric materials embedded in roadways can convert the kinetic energy generated by vehicular motion into electrical power. This harvested energy can then be utilized to power streetlights, traffic management systems, and other infrastructure components. Similarly, solar technologies integrated into smart highways harness sunlight to generate renewable energy. This dual approach of harvesting energy from both vehicular motion and sunlight not only reduces the dependence on traditional power sources but also contributes to the overall sustainability of transportation infrastructure, paving the way for cleaner and more eco-friendly systems (Andriopoulou, 2012).

Smart highways represent a paradigm shift in transportation infrastructure, enabled by interconnected sensors and communication systems. These highways leverage real-time data and advanced analytics to monitor traffic flow, adjust speed limits, and provide dynamic route guidance to drivers. The importance of smart highways lies in their ability to optimize traffic patterns, reduce congestion, and enhance overall safety. Through autonomous traffic management, smart highways contribute to smoother traffic flow, reduced travel times, and fewer accidents. The integration of nanotechnology and energy harvesting further enhances the capabilities of smart highways, creating a holistic approach to sustainable and intelligent transportation infrastructure (Kima et al., 2018).

The integration of nanotechnology, energy harvesting, and smart highways not only addresses environmental concerns but also catalyzes economic growth. The adoption of innovative technologies in transportation infrastructure creates new industries and job opportunities. Nanotechnology research and development, manufacturing of energy harvesting devices, and the implementation of smart highway systems contribute to a growing sector that fosters economic resilience. Additionally, the efficiency gains and cost savings achieved through these advancements have a ripple effect on related industries, further stimulating economic growth on a regional and global scale (Ahmad et al., 2019; Zhang et al., n.d.-a).

Perhaps most crucially, the importance of these technologies lies in their potential to contribute to environmental preservation and enhance the quality of life for communities. The reduction in carbon emissions through sustainable energy practices, the extended lifespan of road infrastructure, and the optimization of traffic flow collectively contribute to a healthier and more sustainable urban environment. As communities experience improved air quality, reduced noise pollution, and enhanced accessibility, the overall well-being of residents is positively impacted. The integration of nanotechnology, energy harvesting, and smart highways represents a transformative approach to transportation infrastructure, aligning with the broader goals of sustainable development and a higher quality of life for people around the world (Pei et al., 2021a).

Gap Analysis: The summary discusses a comprehensive literature of global transportation infrastructure, focusing on weaknesses, inefficiencies, and areas for improvement. It also discusses the technological landscape, highlighting advancements in nanotechnology, energy harvesting, and smart highway technologies. The assessment also considers the environmental impact of these systems, including carbon emissions and ecological disruption. The analysis also includes an economic analysis of the costs and benefits of infrastructure maintenance and upgrades. The summary concludes with a social and community consideration.

Objectives

- Explore self-healing properties and their potential to reduce maintenance costs.
- Discuss applications in powering infrastructure, streetlights, and the grid.
- Explore benefits in traffic management, speed regulation, and dynamic updates.
- Explore how these technologies collectively enhance efficiency and sustainability.
- Evaluate economic implications such as job creation and industry development.

NANOTECHNOLOGY IN ROAD CONSTRUCTION

Enhanced Durability and Strength of Road Surfaces

Nanotechnology, operating at the scale of nanometers, has emerged as a revolutionary tool in road construction, promising to transform the fundamental characteristics of road surfaces. This section delves into the ways nanotechnology enhances the durability and strength of road infrastructure (A. K. Singh et al., 2020).

- *Nanomaterial Reinforcement:* Nanotechnology introduces a novel approach to road construction by incorporating nanomaterials, such as nanoparticles and nanofibers, into traditional construction materials like asphalt and concrete. These materials, when strategically integrated, significantly enhance the structural integrity of road surfaces. Nanoparticles act as reinforcements, creating a dense network within the material, thereby improving its load-bearing capacity and resistance to wear and tear. The result is a road surface that exhibits remarkable durability, capable of withstanding heavy traffic loads and adverse weather conditions.

- *Improved Resistance to Environmental Factors:* Traditional Road surfaces are susceptible to environmental factors like moisture, UV radiation, and temperature fluctuations, leading to cracks and deterioration over time. Nanotechnology addresses these vulnerabilities by imparting road surfaces with improved resistance. Nanomaterials can create a protective barrier, shielding the road against moisture ingress and UV damage. This enhanced resistance translates into longer lifespan and reduced maintenance requirements, contributing to a more sustainable and cost-effective road infrastructure.
- *Self-Healing Properties:* One of the most groundbreaking aspects of nanotechnology in road construction is its ability to confer self-healing properties to road surfaces. Nanomaterials with intrinsic healing capabilities can autonomously repair small cracks and damages that naturally occur during the lifespan of a road. The self-healing mechanism involves the nanomaterials filling in the gaps and restoring the structural integrity of the road. This not only reduces maintenance costs but also prolongs the life of the road, presenting a transformative solution for creating resilient and long-lasting transportation infrastructure.
- *Enhanced Friction and Skid Resistance:* Nanotechnology enables the modification of road surfaces at the nanoscale to improve friction and skid resistance. By introducing nanomaterials with unique surface characteristics, the road's texture can be fine-tuned to provide optimal grip for vehicle tires. This is particularly crucial for safety, as enhanced friction reduces the risk of accidents, especially in adverse weather conditions. The application of nanotechnology in this context contributes to creating safer roadways, aligning with the overarching goal of promoting road safety.
- *Sustainable Construction Practices:* Beyond the immediate benefits to road durability, the adoption of nanotechnology in road construction aligns with sustainability goals. The enhanced durability and longevity of road surfaces lead to a reduction in the frequency of repairs and the need for raw materials. This, in turn, minimizes the environmental impact associated with road construction and maintenance activities. Nanotechnology, therefore, contributes not only to the resilience of road infrastructure but also to the broader objectives of sustainable and eco-friendly construction practices.

The integration of nanotechnology in road construction is a significant advancement towards creating roads with enhanced durability, environmental resistance, self-healing capabilities, improved safety, and reduced environmental footprint, which could revolutionize the design, construction, and maintenance of road infrastructure for a sustainable and resilient transportation network (Boopathi, Umareddy, et al.,

2023; Boopathi & Davim, 2023; Vijayakumar et al., 2023). Nanotechnology, operating at the molecular and atomic scale, introduces a paradigm shift in road construction, ushering in transformative features that extend beyond traditional materials. This section focuses on two key aspects: the integration of self-healing properties for increased longevity and the enhancement of functionality and performance.

Self-healing Properties for Increased Longevity

Traditional road surfaces are prone to deterioration over time due to factors such as traffic-induced wear, weathering, and temperature fluctuations. Nanotechnology addresses this challenge by imparting self-healing properties to road materials. The incorporation of nanomaterials with intrinsic healing capabilities allows the road surface to autonomously repair micro-cracks and damages. This self-healing mechanism not only prolongs the life of the road but also reduces the frequency of maintenance interventions (Jordaan & Steyn, 2022).

Nanomaterials, such as shape-memory polymers or capsules containing healing agents, are embedded within the road structure. When cracks occur, these materials respond to the damage by triggering a self-repair process. The nanomaterials fill in the gaps, restoring the structural integrity of the road. This innovative approach significantly reduces the need for frequent repairs, leading to increased longevity of road surfaces. The application of nanotechnology in conferring self-healing properties represents a groundbreaking advancement in sustainable infrastructure, aligning with the global push for resilient and low-maintenance transportation networks.

Improved Functionality and Performance

Beyond longevity, nanotechnology enhances the functionality and performance of road surfaces. Nanomaterials, due to their unique properties at the nanoscale, can be engineered to provide specific functionalities that go beyond the capabilities of conventional materials. For instance, the incorporation of nanoparticles can lead to superior mechanical strength, allowing road surfaces to withstand heavier loads and increased traffic volumes (Rincón-Morantes et al., 2020).

Additionally, nanotechnology facilitates the modification of road surfaces to exhibit advanced characteristics, such as enhanced water resistance and reduced friction. Nano-engineered surfaces can repel water, preventing moisture-induced damage and enhancing the road's resilience against environmental factors. The manipulation of surface properties at the nanoscale also allows for the creation of smoother and more uniform road textures, contributing to improved driving comfort and reduced rolling resistance for vehicles.

The improved functionality and performance resulting from nanotechnology applications translate into a more efficient and reliable transportation infrastructure. Roads that resist damage, weathering, and offer enhanced performance not only reduce maintenance costs but also contribute to a safer and more comfortable driving experience. By pushing the boundaries of material science at the nanoscale, nanotechnology in road construction sets the stage for a new era of resilient and high-performance road networks that align with the demands of modern transportation systems (A. K. Singh et al., 2020). Nanotechnology is revolutionizing road construction by integrating self-healing properties and improved functionality, enhancing road surface longevity and creating efficient, durable, and responsive road networks that cater to the evolving needs of sustainable transportation infrastructure.

ENERGY HARVESTING TECHNIQUES

Energy harvesting techniques play a crucial role in the pursuit of sustainable transportation infrastructure. By harnessing energy from various sources, these technologies contribute to powering critical components of road networks, reducing dependency on traditional energy sources, and promoting eco-friendly practices. It delves into the various energy harvesting techniques and their applications in transportation infrastructure (Bai & Liu, 2021). Figure 1 depicts energy harvesting techniques for sustainable transportation infrastructure.

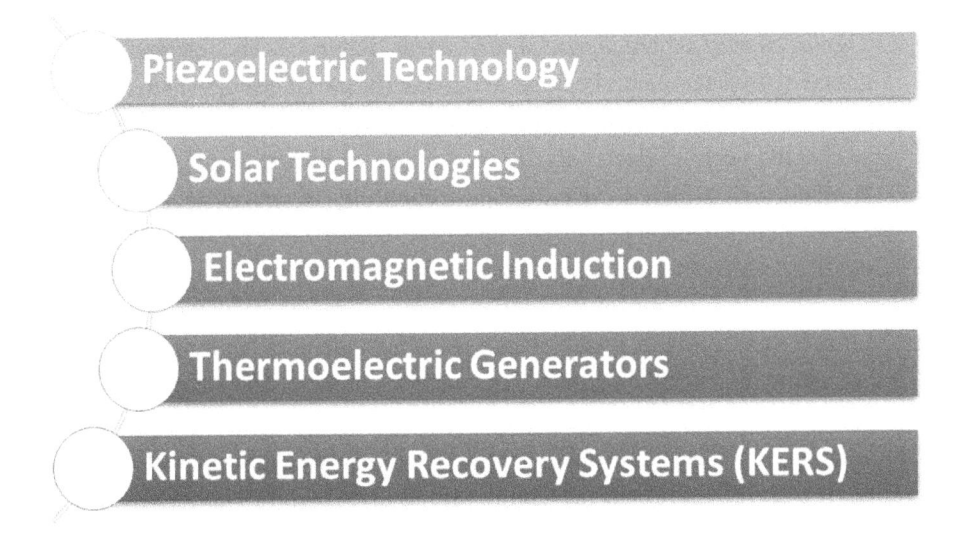

Figure 1. Energy harvesting techniques for sustainable transportation infrastructure

- Piezoelectric Technology
- Solar Technologies
- Electromagnetic Induction
- Thermoelectric Generators
- Kinetic Energy Recovery Systems (KERS)

Piezoelectric Technology

Piezoelectric materials generate electric charge in response to mechanical stress. In transportation, embedding piezoelectric elements in roadways allows the kinetic energy generated by moving vehicles to be converted into electrical energy. As vehicles pass over these piezoelectric sensors, mechanical pressure induces a charge separation, creating a potential difference that can be harvested for power. This technique holds promise for transforming the mechanical energy from vehicular motion into a sustainable source of electricity, contributing to the overall energy needs of transportation infrastructure(Elahi et al., 2020).

Piezoelectric technology stands at the forefront of innovative energy harvesting methods, offering a promising avenue for sustainable power generation within transportation infrastructure. This section delves into the principles of piezoelectricity, how it harnesses kinetic energy from vehicular motion, and its integration into roadways to generate power.

Principles of Piezoelectricity: At its core, piezoelectricity is a phenomenon where certain materials generate an electric charge in response to mechanical stress. In the context of transportation, this stress is provided by the mechanical vibrations produced when vehicles traverse road surfaces. Piezoelectric materials, commonly crystals or ceramics, are strategically embedded in the roadway, creating a network of sensors that can convert the kinetic energy from moving vehicles into electrical energy. This process enables the road itself to become an active participant in the generation of sustainable power.

Harnessing Kinetic Energy from Vehicular Motion: As vehicles travel over piezoelectric sensors embedded in the road, the weight and motion of the vehicles exert mechanical pressure on these materials. This mechanical stress induces a displacement of charge within the piezoelectric material, creating an electric potential difference. The resulting electrical charge is then captured and harnessed as usable electricity. By tapping into the kinetic energy generated by the motion of vehicles, piezoelectric technology offers a direct and efficient means of converting mechanical energy into a renewable power source without requiring additional infrastructure or devices on the vehicles themselves.

Integration into Roadways for Power Generation: The integration of piezoelectric technology into roadways involves a thoughtful design that maximizes energy harvesting potential. Piezoelectric sensors are embedded beneath the road surface, either directly within the pavement or in specialized modules within the road structure. The placement takes into consideration factors such as traffic density, road materials, and optimal locations for energy capture. This integration allows the road to continuously and passively collect energy from the natural movement of vehicles, making it a seamless and unobtrusive solution for sustainable power generation within transportation infrastructure.

Applications in Powering Infrastructure: The electricity generated through piezoelectric technology can be harnessed to power various components of transportation infrastructure. Streetlights, traffic signals, and electronic signage are prime candidates for this harvested energy. By utilizing the generated power locally, these applications reduce the dependence on external electricity sources and contribute to the overall sustainability of the transportation network. Additionally, excess energy can be stored for later use or fed back into the grid, enhancing the efficiency and reliability of the entire energy ecosystem.

Advantages and Considerations: The adoption of piezoelectric technology in roadways comes with notable advantages. It provides a continuous and renewable energy source, contributes to the reduction of carbon emissions, and minimizes the environmental impact associated with conventional power generation. However, considerations such as the cost of implementation, durability of materials, and the scalability of the technology on a large scale need to be carefully evaluated for widespread adoption.

In summary, piezoelectric technology's ability to harness kinetic energy from vehicular motion and its seamless integration into roadways offer a promising solution for sustainable power generation in transportation infrastructure. As advancements in materials and infrastructure design continue, piezoelectric systems have the potential to become integral components of smart and energy-efficient road networks, contributing to a more sustainable and resilient future in transportation.

Solar Technologies

Solar energy represents a clean and abundant source of power for transportation infrastructure. Integrating solar technologies into road surfaces and infrastructure components allows the capture of sunlight to generate electricity. Photovoltaic cells embedded in roadways or incorporated into noise barriers and other structures can convert sunlight into electrical energy. This harvested solar energy can be utilized to power streetlights, traffic signals, and various electronic systems, reducing the reliance on grid electricity and decreasing the carbon footprint of transportation networks (Bosso et al., 2021). Solar technologies have emerged as a cornerstone of sustainable energy solutions, offering the ability to harness the abundant power of sunlight. In the realm of transportation infrastructure, the integration of solar technologies presents a transformative approach to power generation. This section explores the principles of utilizing sunlight for renewable energy and the applications of solar technologies in powering infrastructure, streetlights, and the grid (Agrawal et al., 2024; Kumar B et al., 2024; Syamala et al., 2023).

Utilizing Sunlight for Renewable Energy: Solar technologies leverage the photovoltaic effect, where certain materials generate an electric current when exposed to sunlight. Photovoltaic cells, commonly made of silicon or other semiconductor materials, are arranged in solar panels to capture and convert sunlight into electricity. In transportation infrastructure, the large surface areas available, such as roadways, bridges, and noise barriers, provide ample space for the installation of solar panels. This allows for the efficient utilization of sunlight to generate renewable energy, reducing reliance on non-renewable energy sources and mitigating the environmental impact of conventional power generation (Nishanth et al., 2023).

Powering Infrastructure: Solar technologies contribute to powering various components of transportation infrastructure. Solar panels integrated into the structure of overpasses, bridges, and buildings associated with transportation networks can generate electricity to meet the energy demands of nearby infrastructure. This locally generated power reduces dependency on the grid, enhances the resilience of the transportation system, and promotes energy self-sufficiency. The decentralized nature of solar power in transportation infrastructure allows for increased flexibility and adaptability in meeting the diverse energy needs of different regions (Chandrika et al., 2023; Domakonda et al., 2023; Gnanaprakasam et al., 2023; Sundaramoorthy et al., 2023).

Illuminating Streetlights: Street lighting is a critical aspect of transportation infrastructure for safety and visibility. Solar technologies offer an eco-friendly solution for powering streetlights. Solar-powered Street lighting systems consist of photovoltaic panels, energy storage units (such as batteries), and efficient LED lighting. During the day, solar panels capture sunlight and convert it into electricity, which is stored for use during the night. This not only reduces the reliance on grid electricity but also ensures continuous illumination in areas where traditional power sources might be challenging to access or costly to implement.

Contributing to the Grid: Excess energy generated by solar technologies in transportation infrastructure can be fed back into the grid, contributing to the overall energy supply. Through grid-tied systems, surplus energy can be transmitted to the broader energy network, offsetting the demand for non-renewable sources and potentially providing financial incentives through net metering programs. This integration of solar-generated power into the grid aligns with broader efforts to transition towards a more sustainable and decentralized energy landscape (Domakonda et al., 2023).

Advantages and Considerations: The advantages of solar technologies in transportation infrastructure include reduced carbon emissions, lower operational costs, and increased energy resilience. However, considerations such as the initial installation cost, efficiency in different geographic locations, and maintenance requirements must be weighed against the long-term benefits. Ongoing advancements in solar technology efficiency and decreasing costs are making solar solutions increasingly viable for widespread adoption.

Electromagnetic Induction

Electromagnetic induction involves the generation of electrical current through the relative motion between a magnetic field and a conductor. In transportation, this principle can be applied to harvest energy from the movement of vehicles. Roadway surfaces equipped with electromagnetic generators can capture the kinetic energy produced by passing vehicles, converting it into electrical power. This method offers another avenue for sustainable energy harvesting, particularly in regions with high traffic density or along highways where continuous vehicular motion is prevalent (Bosso et al., 2021; Sanislav et al., 2021).

Thermoelectric Generators

Thermoelectric generators exploit temperature differentials to produce electricity. In transportation infrastructure, thermoelectric materials can be integrated into road surfaces to capture the heat generated by sunlight or the friction between tires and the road. This temperature gradient induces a flow of electric current, which can be harnessed for various applications. While still in early stages of development, thermoelectric generators hold potential for efficiently converting waste heat from transportation activities into usable electrical energy (Pan et al., 2021).

Kinetic Energy Recovery Systems (KERS)

Kinetic Energy Recovery Systems are employed in vehicles to capture and store energy during deceleration or braking. In transportation infrastructure, similar principles can be applied to capture the kinetic energy generated by vehicles during specific road conditions. This harvested energy can then be stored and used to power streetlights, electronic signage, or other energy-demanding components of the transportation network. KERS represents a dynamic approach to energy harvesting, responding directly to the motion patterns of vehicles within the infrastructure (Bai & Liu, 2021).

The use of energy harvesting techniques in transportation infrastructure is a significant step towards sustainability, as it utilizes various sources like vehicle motion, sunlight, and temperature differentials, thereby fostering self-sufficiency and eco-friendliness, thereby reducing environmental impact.

SMART HIGHWAYS AND INTERCONNECTED SYSTEMS

Smart highways represent a transformative paradigm in transportation infrastructure, integrating advanced technologies to enhance efficiency, safety, and overall functionality. The smart highway concept leverages cutting-edge innovations, including the Internet of Things (IoT), sensors, and communication systems, to create an intelligent and responsive roadway network (Gnanaprakasam et al., 2023; Sundaramoorthy et al., 2023). These highways go beyond traditional infrastructure by incorporating digital elements that enable real-time data collection, analysis, and decision-making. The overarching goal is to create a dynamic and adaptive transportation system capable of addressing the evolving challenges of modern urban mobility (Fantin Irudaya Raj & Appadurai, 2022).

Interconnected Sensors for Traffic Monitoring

A cornerstone of smart highways is the deployment of interconnected sensors that facilitate comprehensive traffic monitoring. These sensors, strategically placed along the roadway, collect real-time data on traffic flow, vehicle speed, and road conditions. Various sensor technologies, including cameras, radar, LiDAR, and inductive loops, work collaboratively to capture a detailed picture of the highway environment. The data collected by these sensors are then processed and analyzed, providing transportation authorities with valuable insights into traffic patterns, congestion points, and potential safety hazards. The seamless integration of interconnected sensors enables a holistic view of the highway's dynamics, empowering authorities to make informed decisions for optimizing traffic flow (Mahmood, 2021).

Real-time Updates and Autonomous Traffic Management

Smart highways excel in providing real-time updates to both transportation authorities and drivers. The data collected by interconnected sensors are processed in real-time to generate accurate and up-to-date information about traffic conditions (Hanumanthakari et al., 2023; Pramila et al., 2023; Ramudu et al., 2023). This information is then disseminated to drivers through dynamic signage, mobile applications, and other communication channels, allowing them to make informed decisions about route planning and travel times. Simultaneously, transportation authorities can use this real-time data to implement autonomous traffic management systems. These systems may include dynamic speed limits, variable message signs, and automated traffic signal adjustments to optimize traffic flow, reduce congestion, and enhance overall safety. The ability to dynamically manage traffic based on real-time conditions is a key feature of smart highways, contributing to a more responsive and adaptive transportation network (R. Singh et al., 2021).

Smart highways are a significant advancement in transportation infrastructure, integrating interconnected sensors, real-time data updates, and autonomous traffic management systems. They aim to improve efficiency, safety, and intelligence of road networks. By leveraging advanced technologies, smart highways can revolutionize urban mobility and provide a model for future, dynamic, adaptive transportation systems (Pompigna & Mauro, 2022b).

INTEGRATION OF NANOTECHNOLOGY AND ENERGY HARVESTING IN SMART HIGHWAYS

The integration of nanotechnology and energy harvesting represents a groundbreaking synergy in the development of smart highways, promising enhanced sustainability, efficiency, and functionality. This section explores how the convergence of nanotechnology and energy harvesting technologies can revolutionize smart highways (De Fazio et al., 2023).

Strengthening Road Surfaces with Nanomaterials: Nanotechnology plays a pivotal role in fortifying road surfaces within smart highways. The integration of nanomaterials, such as nanoparticles and nanofibers, enhances the structural integrity of the road, making it more resilient to wear and environmental factors. The addition of nanomaterials strengthens the road, reducing the frequency of maintenance and repairs. This enhanced durability is essential for accommodating the infrastructure demands of energy harvesting technologies, ensuring a stable platform for their integration into the roadway (Zhang et al., n.d.-b). The figure 2 depicts the convergence of nanotechnology technologies in smart highways.

Piezoelectric Energy Harvesting from Vehicular Motion: The marriage of nanotechnology and energy harvesting is particularly evident in the utilization of piezoelectric materials within the road surface. Nanoscale piezoelectric elements embedded in the road can convert the kinetic energy generated by moving vehicles into electrical energy. As vehicles traverse the road, the nanomaterials respond to the mechanical stress, producing a voltage potential that is harvested for power generation. This approach not only harnesses the benefits of nanotechnology in road durability but also taps into the sustainable energy potential of vehicular motion.

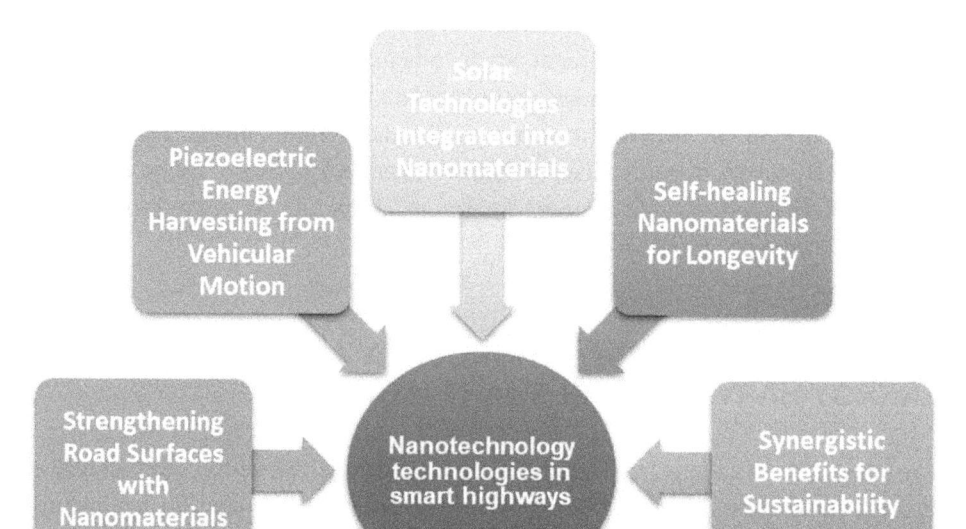

Figure 2. Convergence of nanotechnology technologies in smart highways

Solar Technologies Integrated into Nanomaterials: Nanotechnology also facilitates the integration of solar technologies into the road surface. Nanomaterials with photovoltaic properties can be embedded within the road, allowing the surface to capture solar energy and convert it into electricity. This dual-purpose application benefits from the strength and resilience provided by nanotechnology, ensuring that the road can withstand the mechanical stresses associated with both vehicular traffic and environmental conditions (Domakonda et al., 2023; Hema et al., 2023). The integration of solar technologies into nanomaterials enhances the overall energy harvesting capacity of smart highways, providing a renewable and continuous power source.

Self-healing Nanomaterials for Longevity: Smart highways benefit from the longevity provided by self-healing nanomaterials. Nanotechnology enables the incorporation of self-healing mechanisms within the road surface, allowing it to autonomously repair minor damages caused by wear or external factors. This self-healing capability not only extends the lifespan of the road but also ensures the continuous functionality of embedded energy harvesting technologies. The combination of self-healing properties with energy harvesting creates a resilient and low-maintenance infrastructure, further contributing to the sustainability of smart highways.

Synergistic Benefits for Sustainability: The integration of nanotechnology and energy harvesting in smart highways offers synergistic benefits that amplify the sustainability of transportation infrastructure. The strengthened road surfaces provided by nanotechnology support the efficient operation of energy harvesting technologies, contributing to a self-sustaining and eco-friendly ecosystem. This integration fosters a holistic approach to infrastructure development, where the durability and energy harvesting capabilities work in tandem to create a transportation network that is not only resilient but also environmentally conscious.

In conclusion, the convergence of nanotechnology and energy harvesting in smart highways represents a transformative approach to sustainable infrastructure. By reinforcing road surfaces with nanomaterials and harnessing energy through piezoelectric and solar technologies, smart highways exemplify the potential for innovative solutions at the intersection of nanotechnology and energy harvesting. This integration not only addresses the current challenges in transportation infrastructure but also paves the way for a greener and more resilient future.

Synergies between Nanotechnology and Smart Highway Technologies

The integration of nanotechnology and smart highway technologies creates a powerful synergy that goes beyond individual capabilities, offering combined benefits that significantly enhance the sustainability and efficiency of transportation infrastructure. This section explores the synergies between nanotechnology and smart highway technologies, highlighting the collective advantages they bring to the forefront (Du et al., 2023; Pei et al., 2021b).

Enhanced Durability and Self-healing Surfaces: Nanotechnology contributes to the development of smart highway surfaces with unparalleled durability. By incorporating nanomaterials into road construction, the resulting surfaces exhibit increased strength, resistance to wear, and improved lifespan. This durability is complemented by self-healing properties inherent to some nanomaterials. Smart highways that integrate self-healing nanotechnology can autonomously repair minor damages caused by traffic or environmental factors, ensuring prolonged functionality and reducing the need for frequent maintenance.

Strengthening Energy Harvesting Capacities: The integration of nanotechnology into road surfaces enhances the capabilities of energy harvesting technologies within smart highways. Nanomaterials, such as piezoelectric elements, not only provide structural reinforcement but also serve as efficient converters of kinetic energy from vehicular motion. The combined strength and energy harvesting capacity of nanotechnology create a symbiotic relationship, as the robust road surfaces support the effective operation of energy harvesting technologies, contributing to a more sustainable and self-sufficient energy ecosystem.

Dual-purpose Nanomaterials for Solar Integration: Nanotechnology enables the integration of solar technologies into road surfaces, paving the way for dual-purpose nanomaterials. These materials not only reinforce the road but also possess photovoltaic properties, allowing them to capture solar energy and convert it into electricity. The synergy between nanotechnology and solar integration results in road surfaces that are not only durable and resilient but also capable of generating renewable energy. This dual functionality contributes to the overall sustainability of smart highways, promoting eco-friendly practices and reducing dependence on traditional power sources.

Real-time Monitoring and Data-driven Decision Making: Smart highways leverage interconnected sensors, a key component of smart technologies, to collect real-time data on traffic conditions and road status. The integration of nanotechnology enhances the efficiency of these sensors by providing a stable and reliable platform. Nanomaterials contribute to the development of sensor-friendly road surfaces, ensuring optimal placement and longevity. The synergy between nanotechnology and real-time monitoring facilitates data-driven decision-making, allowing transportation authorities to respond promptly to traffic patterns, mitigate congestion, and enhance overall highway efficiency.

Holistic Approach to Sustainability: The synergies between nanotechnology and smart highway technologies foster a holistic approach to sustainability in transportation infrastructure. The combined benefits of enhanced durability, self-healing surfaces, energy harvesting, and real-time monitoring create a smart highway ecosystem that is not only resilient but also adaptive and environmentally conscious. This holistic approach aligns with broader goals of sustainable development, contributing to reduced maintenance costs, minimized environmental impact, and the creation of transportation networks that meet the evolving needs of society.

The integration of nanotechnology and smart highway technologies is a transformative move towards sustainability and efficiency in transportation infrastructure. This combination of nanotechnology's strengths and smart highway technologies' intelligence promises a resilient, eco-friendly, and technologically advanced future for transportation systems.

ENVIRONMENTAL AND ECONOMIC IMPACTS

The integration of nanotechnology, energy harvesting, and smart highway technologies into transportation infrastructure carries profound implications for both the environment and the economy. This section explores the dual impact, highlighting the potential reductions in carbon footprint and the economic growth, job creation, and industry development stimulated by these innovative solutions (Pompigna & Mauro, 2022c; Trubia et al., 2020). The integration of nanotechnology and smart highways is expected to have significant environmental and economic impacts as shown in Figure 3.

Figure 3. Environmental and economic impacts by integration of nanotechnology and smart highway

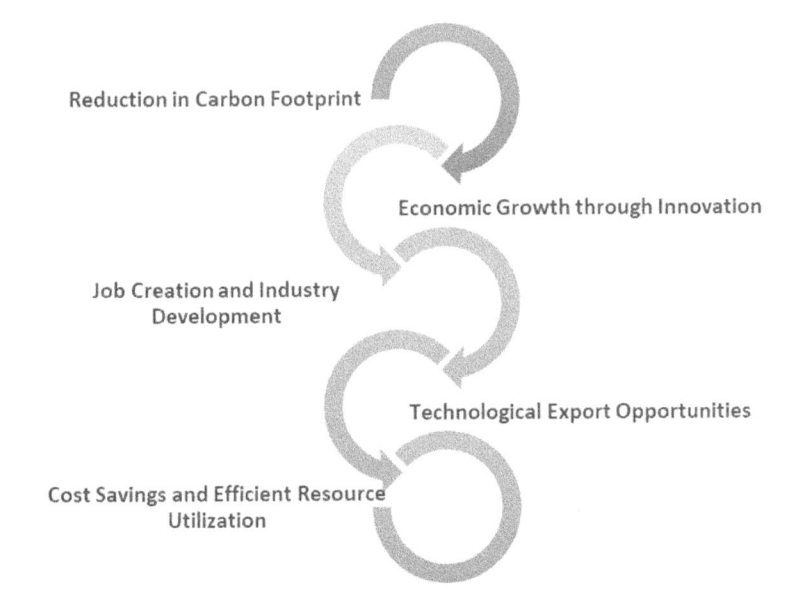

Reduction in Carbon Footprint: One of the primary environmental benefits of adopting nanotechnology, energy harvesting, and smart highways lies in the significant reduction of carbon footprint. The enhanced durability of road surfaces through nanotechnology minimizes the need for frequent repairs and maintenance, leading to reduced energy consumption and material usage associated with conventional construction practices. Energy harvesting technologies, such as piezoelectric systems and solar integration, contribute to the generation of clean, renewable energy, reducing dependency on fossil fuels. Smart highways, with real-time monitoring and traffic management, facilitate smoother traffic flow, preventing congestion and the associated emissions (Hussain et al., 2023). Collectively, these technologies contribute to a more sustainable transportation ecosystem, mitigating the environmental impact and aligning with global efforts to combat climate change.

Economic Growth through Innovation: The adoption of nanotechnology, energy harvesting, and smart highways fosters economic growth by driving innovation and technological advancement. The development and implementation of these cutting-edge technologies create new avenues for economic activity, attracting investment and stimulating growth in the technology and infrastructure sectors. Companies involved in the research, development, and deployment of nanomaterials, energy harvesting devices, and smart highway systems contribute to a thriving innovation ecosystem. This innovation-driven economic growth not only positions countries and regions at the forefront of technological leadership but also generates economic value through the creation of novel solutions for sustainable transportation infrastructure (Mohanty et al., 2023; Ravisankar et al., 2023).

Job Creation and Industry Development: The implementation of advanced technologies in transportation infrastructure results in job creation and the development of new industries. The construction, maintenance, and operation of smart highways require skilled professionals in areas such as engineering, information technology, and data analytics. Additionally, the manufacturing and deployment of nanomaterials, energy harvesting devices, and smart highway components create employment opportunities in emerging industries. As these technologies become more widespread, the demand for a skilled workforce to design, implement, and maintain smart transportation systems will continue to grow. The resulting job creation contributes to economic resilience and provides communities with opportunities for employment and skill development.

Technological Export Opportunities: Countries at the forefront of developing and implementing nanotechnology, energy harvesting, and smart highway solutions have the potential to become leaders in exporting these technologies to the global market. The expertise gained in designing and deploying innovative transportation infrastructure can be leveraged for international collaborations and partnerships. As other nations seek sustainable and efficient solutions for their transportation challenges, the technological expertise developed domestically can be exported, fostering economic ties and contributing to the global dissemination of sustainable transportation practices.

Cost Savings and Efficient Resource Utilization: From an economic perspective, the integration of these technologies into transportation infrastructure can lead to long-term cost savings. The enhanced durability of road surfaces through nanotechnology reduces the need for frequent repairs and maintenance, saving both time and resources. Energy harvesting technologies contribute to energy cost savings by utilizing renewable sources, and smart highways with real-time monitoring enable more efficient resource utilization, preventing unnecessary congestion and optimizing traffic flow. These cost-saving measures contribute to the overall economic efficiency of transportation systems, promoting financial sustainability in the long run.

The integration of nanotechnology, energy harvesting, and smart highways into transportation infrastructure has both environmental and economic benefits. It reduces carbon footprint, stimulates economic growth, creates jobs, and supports industry development, ultimately creating a sustainable and resilient transportation ecosystem.

Figure 4. Global implementation of nanotechnology, energy harvesting, and smart highway systems

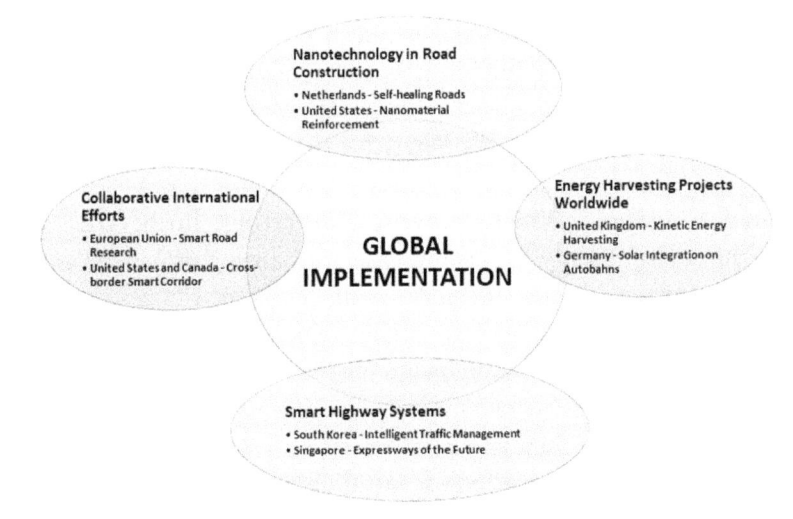

GLOBAL IMPLEMENTATION

The implementation of nanotechnology, energy harvesting, and smart highway systems has witnessed notable success stories worldwide, showcasing the transformative impact of these innovations on transportation infrastructure (Andriopoulou, 2012; Pei et al., 2021a; Zhao et al., 2020). The global implementation of nanotechnology, energy harvesting, and smart highway systems is depicted in Figure 4.

Successful Implementations of Nanotechnology in Road Construction

- **Netherlands - Self-healing Roads:** The Netherlands has been at the forefront of implementing nanotechnology in road construction. The country has pioneered the development of self-healing roads using nanomaterials. These roads are equipped with materials that possess self-healing properties, automatically repairing minor cracks and damages over time. This innovation has led to a significant reduction in maintenance costs and extended the lifespan of road surfaces, contributing to sustainable infrastructure development.
- **United States - Nanomaterial Reinforcement:** Several states in the U.S., including California and Texas, have embraced nanotechnology to reinforce road surfaces. Nanomaterials, such as nanoparticles and nanofibers, are being incorporated into asphalt and concrete to enhance the strength and durability of roads. These advancements have been particularly beneficial in regions with high traffic density and harsh weather conditions.

Energy Harvesting Projects Worldwide

- **United Kingdom - Kinetic Energy Harvesting:** In the UK, there have been successful energy harvesting projects focusing on kinetic energy from vehicular motion. Roads embedded with piezoelectric sensors capture the mechanical energy generated by passing vehicles, converting it into electrical energy. Pilot projects on highways and urban roads have demonstrated the feasibility of this technology, showcasing its potential to contribute to sustainable energy solutions.
- **Germany - Solar Integration on Autobahns:** Germany has been a pioneer in integrating solar technologies into its transportation infrastructure. Solar panels installed on noise barriers and overpasses along autobahns generate renewable energy. This dual-purpose application not only contributes to energy harvesting but also provides shading and noise reduction benefits.

Smart Highway Systems

- **South Korea - Intelligent Traffic Management:** South Korea has implemented smart highway systems that utilize interconnected sensors and real-time data analysis for intelligent traffic management. The smart highway systems in South Korea dynamically adjust speed limits, provide real-time traffic updates to drivers, and optimize traffic flow to reduce congestion. These systems contribute to enhanced road safety and efficiency.
- **Singapore - Expressways of the Future:** Singapore has been investing in smart highway initiatives as part of its vision for the "Expressways of the Future" program. This includes the deployment of smart sensors and communication systems to monitor and manage traffic conditions. The integration of technology allows for real-time updates, adaptive traffic control, and improved overall mobility.

Collaborative International Efforts

- **European Union - Smart Road Research:** The European Union has been fostering collaborative research initiatives focusing on smart roads. Projects like the "Smart Roads of the Future" involve multiple countries working together to develop and test innovative technologies, including nanotechnology applications and energy harvesting, with the aim of creating intelligent and sustainable road networks.
- **United States and Canada - Cross-border Smart Corridor:** Cross-border initiatives between the U.S. and Canada have seen the development of smart highway corridors. These projects involve the integration of advanced technologies for traffic monitoring, communication, and energy harvesting, emphasizing the potential for collaborative efforts to create seamless and intelligent transportation networks.

Global advancements in nanotechnology, energy harvesting, and smart highway systems showcase their versatility, fostering a global movement towards sustainable, efficient, and intelligent transportation infrastructures.

CHALLENGES AND FUTURE PROSPECTS

The integration of nanotechnology, energy harvesting, and smart highway technologies in transportation infrastructure presents both opportunities and challenges, necessitating a focus on addressing these challenges and preparing for future sustainable transportation advancements (Ahmad et al., 2019; Bosso et al., 2021; Pan et al., 2021; Sultana et al., 2021).

Challenges

- **Cost and Implementation:** One of the primary challenges is the initial cost associated with implementing these advanced technologies. Nanomaterials and sophisticated energy harvesting systems can be expensive, posing financial challenges for widespread adoption. Overcoming these cost barriers requires ongoing research to develop cost-effective solutions and incentivizing investments through government initiatives (Ahmad et al., 2019).
- **Public Acceptance and Perception:** The acceptance of innovative technologies by the public is essential for successful implementation. There may be resistance or skepticism regarding changes in infrastructure and concerns about the safety and reliability of these new technologies. Public awareness campaigns and effective communication strategies are needed to address these concerns and build trust in the reliability and benefits of these advancements.
- **Standardization and Regulations:** The lack of standardized practices and regulations poses a challenge to the integration of nanotechnology and energy harvesting in transportation infrastructure. Establishing clear standards and regulations is crucial to ensure consistency, safety, and interoperability. Governments and international organizations play a vital role in developing and enforcing these standards.
- **Long-Term Durability:** While nanotechnology enhances the durability of road surfaces, long-term performance and the effects of environmental factors need continuous evaluation. Understanding the longevity of nanomaterials and addressing potential degradation over time is crucial for maintaining the sustainability and effectiveness of these innovations.
- **Interdisciplinary Collaboration:** The successful integration of nanotechnology, energy harvesting, and smart highway technologies requires collaboration across multiple disciplines, including materials science, engineering, IT, and urban planning. Bridging the gap between these fields and fostering interdisciplinary collaboration is essential for holistic and effective solutions.

Future Prospects

- **Advancements in Nanomaterials:** Future developments in nanotechnology will likely bring forth advanced nanomaterials with enhanced properties.

Researchers are exploring materials with superior strength, self-healing capabilities, and improved cost-effectiveness. These advancements will contribute to more resilient and sustainable road surfaces (Ding & Liu, 2023; Prus & Sikora, 2021; Zhao et al., 2020).
- **Innovations in Energy Harvesting:** Ongoing research in energy harvesting technologies aims to improve efficiency and expand the range of sources that can be tapped for energy. Advancements in piezoelectric materials, solar technologies, and other innovative approaches may lead to more efficient and scalable energy harvesting solutions.
- **Integration of Artificial Intelligence (AI):** The incorporation of AI into smart highway systems holds significant promise. AI can optimize traffic management, predict maintenance needs, and enhance the adaptability of transportation infrastructure (Boopathi, Sureshkumar, et al., 2023; Boopathi & Kanike, 2023). Machine learning algorithms can analyze vast amounts of data from sensors, enabling more informed decision-making for traffic flow and road maintenance.
- **Hybrid and Multi-Modal Transportation:** The future of sustainable transportation involves a shift toward hybrid and multi-modal systems. Integrating nanotechnology and smart technologies with public transportation, electric vehicles, and alternative modes of transportation will contribute to creating more comprehensive and eco-friendly mobility solutions.
- **Global Collaboration and Knowledge Sharing:** Future prospects include increased global collaboration and knowledge-sharing initiatives. Countries and regions can learn from each other's successes and challenges, fostering a collaborative approach to sustainable transportation. International standards and best practices will facilitate the seamless integration of advanced technologies on a global scale.

The future of transportation will be shaped by a collaborative effort from researchers, policymakers, industry stakeholders, and the public, focusing on continuous innovation, interdisciplinary collaboration, and sustainability to create smarter, greener, and more resilient infrastructure.

CONCLUSION

The integration of nanotechnology, energy harvesting, and smart highway technologies is a significant shift in transportation infrastructure, promising sustainable, efficient, and resilient road networks. Exploration of these technologies has yielded key insights and conclusions.

The integration of nanotechnology, energy harvesting, and smart highway systems enhances the sustainability of transportation infrastructure by providing enhanced durability, self-healing properties, and renewable energy harvesting. Successful global implementations in countries like the Netherlands, the US, the UK, South Korea, and Singapore have demonstrated the adaptability and effectiveness of these innovations in diverse geographical and operational contexts.

The adoption of advanced technologies like self-healing roads, energy harvesting, and solar integration significantly reduces the carbon footprint associated with transportation infrastructure. These technologies also stimulate economic growth and job creation, fostering the development of new industries and putting nations at the forefront of technological leadership. This sustainable and environmentally conscious transportation ecosystem contributes to a more sustainable and prosperous future.

Challenges in the integration of nanomaterials, energy harvesting technologies, and artificial intelligence require strategic approaches like cost-effective solutions, public awareness campaigns, and clear standards. However, advancements in nanomaterials, energy harvesting, and AI integration offer promising prospects for superior nanomaterials, more efficient energy harvesting, and the evolution of smart systems, contributing to a more intelligent and adaptable transportation infrastructure.

The integration of nanotechnology, energy harvesting, and smart highway technologies is a transformative shift in transportation infrastructure management. As these innovations mature and gain widespread adoption, the future of transportation is expected to be more sustainable, efficient, and responsive to global society's needs. This ongoing journey towards smart and sustainable transportation promises a brighter, greener, and more connected future.

ABBREVIATIONS

UV: Ultraviolet
LED: Light Emitting Diode
KERS: Kinetic Energy Recovery System
IoT: Internet of Things
LiDAR: Light Detection and Ranging
AI: Artificial Intelligence
US: United States
UK: United Kingdom

REFERENCES

Agrawal, A. V., Shashibhushan, G., Pradeep, S., Padhi, S. N., Sugumar, D., & Boopathi, S. (2024). Synergizing Artificial Intelligence, 5G, and Cloud Computing for Efficient Energy Conversion Using Agricultural Waste. In B. K. Mishra (Ed.), Practice, Progress, and Proficiency in Sustainability. IGI Global. doi:10.4018/979-8-3693-1186-8.ch026

Ahmad, S., Abdul Mujeebu, M., & Farooqi, M. A. (2019). Energy harvesting from pavements and roadways: A comprehensive review of technologies, materials, and challenges. *International Journal of Energy Research*, *43*(6), 1974–2015. doi:10.1002/er.4350

Andriopoulou, S. (2012). *A review on energy harvesting from roads*. Academic Press.

Bai, S., & Liu, C. (2021). Overview of energy harvesting and emission reduction technologies in hybrid electric vehicles. *Renewable & Sustainable Energy Reviews*, *147*, 111188. doi:10.1016/j.rser.2021.111188

Boopathi, S., & Davim, J. P. (2023). Applications of Nanoparticles in Various Manufacturing Processes. In S. Boopathi & J. P. Davim (Eds.), Advances in Chemical and Materials Engineering. IGI Global. doi:10.4018/978-1-6684-9135-5.ch001

Boopathi, S., & Kanike, U. K. (2023). Applications of Artificial Intelligent and Machine Learning Techniques in Image Processing. In B. K. Pandey, D. Pandey, R. Anand, D. S. Mane, & V. K. Nassa (Eds.), Advances in Computational Intelligence and Robotics. IGI Global. doi:10.4018/978-1-6684-8618-4.ch010

Boopathi, S., Sureshkumar, M., & Sathiskumar, S. (2023). Parametric Optimization of LPG Refrigeration System Using Artificial Bee Colony Algorithm. In S. Tripathy, S. Samantaray, J. Ramkumar, & S. S. Mahapatra (Eds.), *Recent Advances in Mechanical Engineering* (pp. 97–105). Springer Nature Singapore. doi:10.1007/978-981-19-9493-7_10

Boopathi, S., Umareddy, M., & Elangovan, M. (2023). Applications of Nano-Cutting Fluids in Advanced Machining Processes. In S. Boopathi & J. P. Davim (Eds.), Advances in Chemical and Materials Engineering. IGI Global. doi:10.4018/978-1-6684-9135-5.ch009

Bosso, N., Magelli, M., & Zampieri, N. (2021). Application of low-power energy harvesting solutions in the railway field: A review. *Vehicle System Dynamics*, *59*(6), 841–871. doi:10.1080/00423114.2020.1726973

Chandrika, V. S., Sivakumar, A., Krishnan, T. S., Pradeep, J., Manikandan, S., & Boopathi, S. (2023). Theoretical Study on Power Distribution Systems for Electric Vehicles. In B. K. Mishra (Ed.), Advances in Civil and Industrial Engineering. IGI Global. doi:10.4018/979-8-3693-0044-2.ch001

De Fazio, R., De Giorgi, M., Cafagna, D., Del-Valle-Soto, C., & Visconti, P. (2023). Energy Harvesting Technologies and Devices from Vehicular Transit and Natural Sources on Roads for a Sustainable Transport: State-of-the-Art Analysis and Commercial Solutions. *Energies*, *16*(7), 3016. doi:10.3390/en16073016

Ding, X., & Liu, X. (2023). Renewable energy development and transportation infrastructure matters for green economic growth? Empirical evidence from China. *Economic Analysis and Policy*, *79*, 634–646. doi:10.1016/j.eap.2023.06.042

Domakonda, V. K., Farooq, S., Chinthamreddy, S., Puviarasi, R., Sudhakar, M., & Boopathi, S. (2023). Sustainable Developments of Hybrid Floating Solar Power Plants: Photovoltaic System. In P. Vasant, R. Rodríguez-Aguilar, I. Litvinchev, & J. A. Marmolejo-Saucedo (Eds.), Advances in Environmental Engineering and Green Technologies. IGI Global. doi:10.4018/978-1-6684-4118-3.ch008

Du, R., Xiao, J., Chang, S., Zhao, L.-C., Wei, K.-X., Zhang, W.-M., & Zou, H.-X. (2023). Mechanical energy harvesting in traffic environment and its application in smart transportation. *Journal of Physics. D, Applied Physics*, *56*(37), 373002. doi:10.1088/1361-6463/acdadb

Elahi, H., Munir, K., Eugeni, M., Atek, S., & Gaudenzi, P. (2020). Energy harvesting towards self-powered IoT devices. *Energies*, *13*(21), 5528. doi:10.3390/en13215528

Fantin Irudaya Raj, E., & Appadurai, M. (2022). Internet of things-based smart transportation system for smart cities. In Intelligent Systems for Social Good: Theory and Practice (pp. 39–50). Springer. doi:10.1007/978-981-19-0770-8_4

Gnanaprakasam, C., Vankara, J., Sastry, A. S., Prajval, V., Gireesh, N., & Boopathi, S. (2023). Long-Range and Low-Power Automated Soil Irrigation System Using Internet of Things: An Experimental Study. In G. S. Karthick (Ed.), Advances in Environmental Engineering and Green Technologies. IGI Global. doi:10.4018/978-1-6684-7879-0.ch005

Hanumanthakari, S., Gift, M. D. M., Kanimozhi, K. V., Bhavani, M. D., Bamane, K. D., & Boopathi, S. (2023). Biomining Method to Extract Metal Components Using Computer-Printed Circuit Board E-Waste. In P. Srivastava, D. Ramteke, A. K. Bedyal, M. Gupta, & J. K. Sandhu (Eds.), Practice, Progress, and Proficiency in Sustainability. IGI Global. doi:10.4018/978-1-6684-8117-2.ch010

Hema, N., Krishnamoorthy, N., Chavan, S. M., Kumar, N. M. G., Sabarimuthu, M., & Boopathi, S. (2023). A Study on an Internet of Things (IoT)-Enabled Smart Solar Grid System. In P. Swarnalatha & S. Prabu (Eds.), Advances in Computational Intelligence and Robotics. IGI Global. doi:10.4018/978-1-6684-8098-4.ch017

Hussain, Z., Babe, M., Saravanan, S., Srimathy, G., Roopa, H., & Boopathi, S. (2023). Optimizing Biomass-to-Biofuel Conversion: IoT and AI Integration for Enhanced Efficiency and Sustainability. In N. Cobîrzan, R. Muntean, & R.-A. Felseghi (Eds.), Advances in Finance, Accounting, and Economics. IGI Global. doi:10.4018/978-1-6684-8238-4.ch009

Jordaan, G. J., & Steyn, W. J. (2022). Practical Application of Nanotechnology Solutions in Pavement Engineering: Construction Practices Successfully Implemented on Roads (Highways to Local Access Roads) Using Marginal Granular Materials Stabilised with New-Age (Nano) Modified Emulsions (NME). *Applied Sciences (Basel, Switzerland)*, *12*(3), 1332. doi:10.3390/app12031332

Kima, S., Sternb, I., Shenc, J., Ahadd, M., & Baie, Y. (2018). Energy harvesting assessment using PZT sensors and roadway materials. *Int. J. of Thermal & Environmental Engineering*, *16*(1), 19–25. doi:10.5383/ijtee.16.01.003

Kumar, B. M., Kumar, K. K., Sasikala, P., Sampath, B., Gopi, B., & Sundaram, S. (2024). Sustainable Green Energy Generation From Waste Water: IoT and ML Integration. In B. K. Mishra (Ed.), Practice, Progress, and Proficiency in Sustainability. IGI Global. doi:10.4018/979-8-3693-1186-8.ch024

Mahmood, Z. (2021). Connected vehicles: A vital component of smart transportation in an intelligent city. In *Developing and Monitoring Smart Environments for Intelligent Cities* (pp. 198–215). IGI Global. doi:10.4018/978-1-7998-5062-5.ch008

Mohanty, A., Venkateswaran, N., Ranjit, P. S., Tripathi, M. A., & Boopathi, S. (2023). Innovative Strategy for Profitable Automobile Industries: Working Capital Management. In Y. Ramakrishna & S. N. Wahab (Eds.), Advances in Finance, Accounting, and Economics. IGI Global. doi:10.4018/978-1-6684-7664-2.ch020

Nishanth, J. R., Deshmukh, M. A., Kushwah, R., Kushwaha, K. K., Balaji, S., & Sampath, B. (2023). Particle Swarm Optimization of Hybrid Renewable Energy Systems. In B. K. Mishra (Ed.), Advances in Civil and Industrial Engineering. IGI Global. doi:10.4018/979-8-3693-0044-2.ch016

Pan, H., Qi, L., Zhang, Z., & Yan, J. (2021). Kinetic energy harvesting technologies for applications in land transportation: A comprehensive review. *Applied Energy*, *286*, 116518. doi:10.1016/j.apenergy.2021.116518

Pei, J., Guo, F., Zhang, J., Zhou, B., Bi, Y., & Li, R. (2021a). Review and analysis of energy harvesting technologies in roadway transportation. *Journal of Cleaner Production, 288*, 125338. doi:10.1016/j.jclepro.2020.125338

Pei, J., Guo, F., Zhang, J., Zhou, B., Bi, Y., & Li, R. (2021b). Review and analysis of energy harvesting technologies in roadway transportation. *Journal of Cleaner Production, 288*, 125338. doi:10.1016/j.jclepro.2020.125338

Petru, J., & Krivda, V. (2021). The transport of oversized cargoes from the perspective of sustainable transport infrastructure in cities. *Sustainability (Basel), 13*(10), 5524. doi:10.3390/su13105524

Pompigna, A., & Mauro, R. (2022). Smart roads: A state of the art of highways innovations in the Smart Age. *Engineering Science and Technology, an International Journal, 25*, 100986.

Pramila, P. V., Amudha, S., Saravanan, T. R., Sankar, S. R., Poongothai, E., & Boopathi, S. (2023). Design and Development of Robots for Medical Assistance: An Architectural Approach. In G. S. Karthick & S. Karupusamy (Eds.), Advances in Healthcare Information Systems and Administration. IGI Global. doi:10.4018/978-1-6684-8913-0.ch011

Prus, P., & Sikora, M. (2021). The impact of transport infrastructure on the sustainable development of the region—Case study. *Agriculture, 11*(4), 279. doi:10.3390/agriculture11040279

Ramudu, K., Mohan, V. M., Jyothirmai, D., Prasad, D. V. S. S. S. V., Agrawal, R., & Boopathi, S. (2023). Machine Learning and Artificial Intelligence in Disease Prediction: Applications, Challenges, Limitations, Case Studies, and Future Directions. In G. S. Karthick & S. Karupusamy (Eds.), Advances in Healthcare Information Systems and Administration. IGI Global. doi:10.4018/978-1-6684-8913-0.ch013

Ravisankar, A., Sampath, B., & Asif, M. M. (2023). Economic Studies on Automobile Management: Working Capital and Investment Analysis. In C. S. V. Negrão, I. G. P. Maia, & J. A. F. Brito (Eds.), Advances in Logistics, Operations, and Management Science. IGI Global. doi:10.4018/978-1-7998-9213-7.ch009

Rincón-Morantes, J. F., Reyes-Ortiz, O. J., & Ruge-Cardenas, J. C. (2020). Review of the use of nanomaterials in soils for construction of roads. *Respuestas, 25*(2), 213–223. doi:10.22463/0122820X.2959

Sanislav, T., Mois, G. D., Zeadally, S., & Folea, S. C. (2021). Energy harvesting techniques for internet of things (IoT). *IEEE Access : Practical Innovations, Open Solutions*, 9, 39530–39549. doi:10.1109/ACCESS.2021.3064066

Singh, A. K., Kulshreshtha, A., Banerjee, A., & Singh, B. R. (2020). Nanotechnology in road construction. *AIP Conference Proceedings*, 2224(1).

Singh, R., Sharma, R., Akram, S. V., Gehlot, A., Buddhi, D., Malik, P. K., & Arya, R. (2021). Highway 4.0: Digitalization of highways for vulnerable road safety development with intelligent IoT sensors and machine learning. *Safety Science*, 143, 105407. doi:10.1016/j.ssci.2021.105407

Sultana, S., Salon, D., & Kuby, M. (2021). Transportation sustainability in the urban context: A comprehensive review. *Geographic Perspectives on Urban Sustainability*, 13–42.

Sundaramoorthy, K., Singh, A., Sumathy, G., Maheshwari, A., Arunarani, A. R., & Boopathi, S. (2023). A Study on AI and Blockchain-Powered Smart Parking Models for Urban Mobility. In B. B. Gupta & F. Colace (Eds.), Advances in Computational Intelligence and Robotics. IGI Global. doi:10.4018/978-1-6684-9999-3.ch010

Syamala, M. C. R., K., Pramila, P. V., Dash, S., Meenakshi, S., & Boopathi, S. (2023). Machine Learning-Integrated IoT-Based Smart Home Energy Management System. In P. Swarnalatha & S. Prabu (Eds.), Advances in Computational Intelligence and Robotics (pp. 219–235). IGI Global. doi:10.4018/978-1-6684-8098-4.ch013

Trubia, S., Severino, A., Curto, S., Arena, F., & Pau, G. (2020). Smart roads: An overview of what future mobility will look like. *Infrastructures*, 5(12), 107. doi:10.3390/infrastructures5120107

Vijayakumar, G. N. S., Domakonda, V. K., Farooq, S., Kumar, B. S., Pradeep, N., & Boopathi, S. (2023). Sustainable Developments in Nano-Fluid Synthesis for Various Industrial Applications. In A. S. Etim (Ed.), Advances in Human and Social Aspects of Technology. IGI Global. doi:10.4018/978-1-6684-5347-6.ch003

Zhang M. Zhu L. Gao S. Yuan H. Liu T. (n.d.). Energy harvesting of novel smart concrete based on nanotechnology: Experimental and Numerical. *Available at* SSRN 4549309.

Zhao, X., Ke, Y., Zuo, J., Xiong, W., & Wu, P. (2020). Evaluation of sustainable transport research in 2000–2019. *Journal of Cleaner Production*, 256, 120404. doi:10.1016/j.jclepro.2020.120404

Chapter 6

AI and ML Adaptive Smart-Grid Energy Management Systems:
Exploring Advanced Innovations

S. Saravanan
https://orcid.org/0000-0001-8255-2623
Department of Electrical and Electronics Engineering, B.V. Raju Institute of Technology, Narsapur, India

Richa Khare
https://orcid.org/0000-0001-7975-2202
Amity School of Applied Sciences, Amity University, Lucknow, India

K. Umamaheswari
Department of Electrical and Electronics Engineering, Dr. Mahalingam College of Engineering and Technology, Pollachi, India

Smriti Khare
https://orcid.org/0000-0001-6701-1992
Amity School of Applied Sciences, Amity University, Lucknow, India

B. S. Krishne Gowda
Department of Commerce, Government College for Women, Chintamani, India

Sampath Boopathi
https://orcid.org/0000-0002-2065-6539
Department of Mechanical Engineering, Muthayammal Engineering College, Namakkal, India

DOI: 10.4018/978-1-6684-9214-7.ch006

Copyright © 2024, IGI Global. Copying or distributing in print or electronic forms without written permission of IGI Global is prohibited.

AI and ML Adaptive Smart-Grid Energy Management Systems

ABSTRACT

The chapter explores the transformative role of artificial intelligence (AI) and machine learning (ML) in shaping smart energy management systems (SEMS) and predicts innovations by 2030. It discusses AI principles in energy optimization, predictive analytics in smart grids, and renewable energy integration through AI-driven strategies. The chapter also addresses critical aspects like predictive maintenance, consumer-centric solutions, cybersecurity challenges, ethical considerations, and regulatory frameworks for responsible AI implementation. By examining challenges and prospects, it provides insights into the dynamic future of energy management driven by AI and ML advancements.

INTRODUCTION

Technological advancements and a shift towards sustainable practices have significantly transformed energy management systems. The shift from conventional methods to sophisticated, data-driven systems have reshaped energy production, distribution, and consumption. The energy sector is at the forefront of this transformation, with the integration of Artificial Intelligence (AI) paving the way for a new era of efficiency, optimization, and sustainability in energy management. The integration of AI into energy management systems has revolutionized the industry by processing vast datasets, enabling predictive analysis, and adaptive decision-making. AI algorithms, powered by Machine Learning models, extract valuable insights from energy grids, consumption patterns, weather fluctuations, and infrastructure conditions. This data enables proactive strategies, real-time optimization, and agile responses to changing energy demands (Ganesh & Xu, 2022a).

AI is revolutionizing energy management by transforming the core architecture of energy systems into intelligent, self-learning entities capable of foreseeing, adapting, and optimizing with precision. This includes predictive maintenance, demand forecasting, and grid optimization. AI also plays a crucial role in the seamless integration and harnessing of renewable energy sources, offering sustainable solutions to meet growing energy needs (Satav, Lamani, K. G., et al., 2024; Venkateswaran et al., 2023). This chapter explores the multifaceted realm of AI-driven energy management, navigating its applications, challenges, and ethical considerations. It envisions a future where AI not only optimizes energy but also paves the way towards a sustainable, interconnected, and resilient energy ecosystem. The journey of AI in energy management is not just about innovation but also a testament to the synergy between human ingenuity and technological advancement (Z. Liu et al., 2022a).

Energy management in the past was primarily manual and centralized, with limited adaptability to dynamic demands. However, sustainability and technological advancements have led to a paradigm shift with the emergence of smart grids. These grids use digital technologies for monitoring, control, and optimization of energy flow, with the true revolution emerging from the integration of Artificial Intelligence and Machine Learning (T. Liu, 2022).

AI and ML technologies are revolutionizing energy management by handling vast amounts of data, extracting actionable insights, and enabling predictive and proactive decision-making. AI algorithms, powered by ML models, analyze diverse datasets, including energy consumption patterns and weather forecasts, enabling real-time optimization and adaptive energy systems. AI-empowered systems have revolutionized energy management, from predictive maintenance to demand forecasting (Rashid et al., 2019). It also plays a pivotal role in integrating renewable energy sources into the grid, optimizing their utilization, and balancing supply and demand dynamics. This chapter explores the evolutionary journey of energy management systems, examining applications, challenges, and future possibilities of AI-enabled energy management, envisioning a sustainable and efficient energy future (Kumar B et al., 2024; Rahamathunnisa et al., 2023).

Gap Identified

Despite significant strides in integrating AI into energy management, critical gaps persist. One major challenge lies in data quality and availability, hindering the accuracy of AI algorithms for grid optimization and demand forecasting. Additionally, cybersecurity risks pose a threat to the secure functioning of AI-driven systems, raising concerns about potential attacks on critical energy infrastructure. Ethical considerations surrounding AI decision-making in resource allocation and the lack of standardized regulations further impede widespread adoption. Bridging these gaps necessitates concerted efforts in data infrastructure, cybersecurity protocols, ethical frameworks, and regulatory harmonization to unlock the full potential of AI in revolutionizing energy management.

Objectives

- To provide a detailed overview of how AI is revolutionizing various aspects of energy management, including grid optimization, demand forecasting, renewable energy integration, and predictive maintenance.
- To delve into the specific AI algorithms and machine learning techniques utilized in energy management systems. Explore their functionalities, strengths, and applications within the context of energy optimization.

- To assess the direct and indirect effects of AI integration on energy sustainability and efficiency. Examine how AI-driven systems contribute to reducing carbon footprints, optimizing resource utilization, and advancing sustainable energy practices.
- Highlight the challenges and limitations associated with AI adoption in energy management. Address issues such as data quality, cybersecurity risks, ethical considerations, and regulatory hurdles, providing strategies to overcome these obstacles.
- To Forecast the future trends and potential advancements in AI-driven energy management. Discuss emerging technologies, anticipated developments, and their implications for reshaping the energy landscape in the coming years.

FOUNDATIONS OF AI IN ENERGY MANAGEMENT

Artificial Intelligence and Machine Learning

Artificial Intelligence (AI) represents the simulation of human intelligence processes by machines, enabling them to learn, reason, and adapt. Within AI, Machine Learning (ML) is a subset that focuses on creating algorithms allowing systems to learn from data, improving their performance over time without explicit programming. In energy management, AI leverages historical consumption patterns, weather data, and infrastructure behavior to make predictions and optimize energy usage. ML algorithms, such as neural networks or decision trees, analyze vast datasets to identify trends, forecast demand, and optimize energy distribution (Vazquez-Canteli et al., 2020). ML's capability to handle complex, nonlinear relationships in energy systems proves invaluable. Supervised learning models predict energy demand based on historical data, while unsupervised learning techniques identify hidden patterns within energy consumption. Reinforcement learning algorithms optimize energy usage by constantly learning and adapting to changing conditions, mimicking human decision-making processes. The integration of AI and ML in energy management systems streamlines operations, enhances efficiency, and reduces costs by automating tasks like demand forecasting, load balancing, and predictive maintenance (Puri et al., 2023a). As these technologies evolve, they promise further advancements in energy conservation and sustainable practices.

Application of AI in Energy Optimization

AI's application in energy optimization spans various facets of energy management. Predictive analytics, a key AI-driven approach, enables utilities to forecast demand accurately. By analyzing historical data, weather patterns, and socio-economic factors, predictive models anticipate energy consumption, aiding in resource allocation and reducing waste. Moreover, AI augments grid stability through real-time monitoring and control. ML algorithms detect anomalies or potential failures in the grid, allowing for proactive interventions to prevent outages or disruptions (Murphey et al., 2012). AI-driven demand response mechanisms engage consumers by providing incentives to adjust their energy usage during peak times, reducing strain on the grid. AI-based energy optimization extends to predictive maintenance, where ML algorithms analyze equipment data to predict potential failures. By identifying maintenance needs beforehand, organizations minimize downtime and extend the lifespan of energy infrastructure, contributing to overall system reliability. AI is revolutionizing energy optimization by providing personalized energy solutions, analyzing individual behavior and preferences to offer energy-saving tips, promoting efficient usage patterns, and fostering smarter, more efficient systems that meet growing demands while promoting sustainability and cost-effectiveness (Fayyazi et al., 2023).

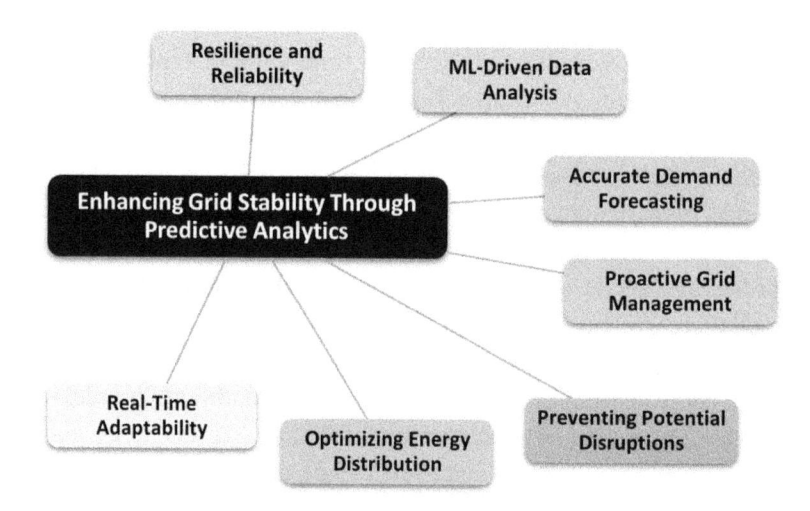

Figure 1. Enhancing grid stability through predictive analytics

ADVANCEMENTS IN SMART GRIDS

Smart grids represent a significant leap in energy infrastructure, integrating advanced technologies to enhance efficiency, reliability, and sustainability (Hema et al., 2023a; Kavitha et al., 2023a).

Enhancing Grid Stability Through Predictive Analytics: Smart grids leverage AI and ML to fortify stability. Predictive analytics, powered by ML algorithms, analyze vast data streams from sensors and meters to forecast demand and grid behavior accurately (Li et al., 2023; Puri et al., 2023b). This capability allows for proactive grid management, preventing potential disruptions and optimizing energy distribution as shown in Figure 1.
- **ML-Driven Data Analysis:** Smart grids harness Machine Learning (ML) algorithms to process and analyze extensive data streams collected from various sensors and meters within the grid infrastructure. These algorithms handle diverse data sets, including consumption patterns, weather conditions, and grid performance metrics.
- **Accurate Demand Forecasting:** Predictive analytics, powered by ML, enable precise forecasting of energy demand. By recognizing historical usage trends and considering external factors like weather and socio-economic patterns, these models anticipate future demand more accurately than traditional methods.
- **Proactive Grid Management:** The predictive capabilities of AI and ML empower grid operators with proactive management tools. By foreseeing potential fluctuations or spikes in energy demand, operators can take preventive measures to maintain stability and avoid disruptions before they occur.
- **Preventing Potential Disruptions:** ML algorithms identify potential stress points or vulnerabilities in the grid infrastructure. This early detection allows for preemptive actions to reinforce or redirect resources, averting potential disruptions and ensuring continuous energy supply.
- **Optimizing Energy Distribution:** The insights gleaned from predictive analytics optimize the distribution of energy resources. ML algorithms dynamically adjust energy flow, rerouting resources as needed to ensure efficient distribution, minimize losses, and maintain grid stability.
- **Real-Time Adaptability:** ML models continuously learn and adapt in real-time. This adaptability is crucial in handling sudden changes or unforeseen circumstances, allowing the grid to dynamically adjust operations to maintain stability.
- **Resilience and Reliability:** By leveraging predictive analytics, smart grids enhance resilience and reliability. Anticipating and addressing issues before they escalate bolsters the grid's ability to withstand challenges, reducing downtime and improving overall reliability.

AI-Enabled Demand Response Mechanisms: Another key advancement lies in demand response mechanisms. AI facilitates real-time monitoring and analysis of energy consumption patterns. By providing insights into peak usage times and consumer behavior, smart grids can implement demand response strategies, incentivizing consumers to modify their usage during high-demand periods (Ganesh & Xu, 2022b; Z. Liu et al., 2022b). This dynamic adjustment minimizes strain on the grid and reduces the likelihood of blackouts or brownouts as shown in Figure 2.

Figure 2. AI-enabled demand response mechanisms

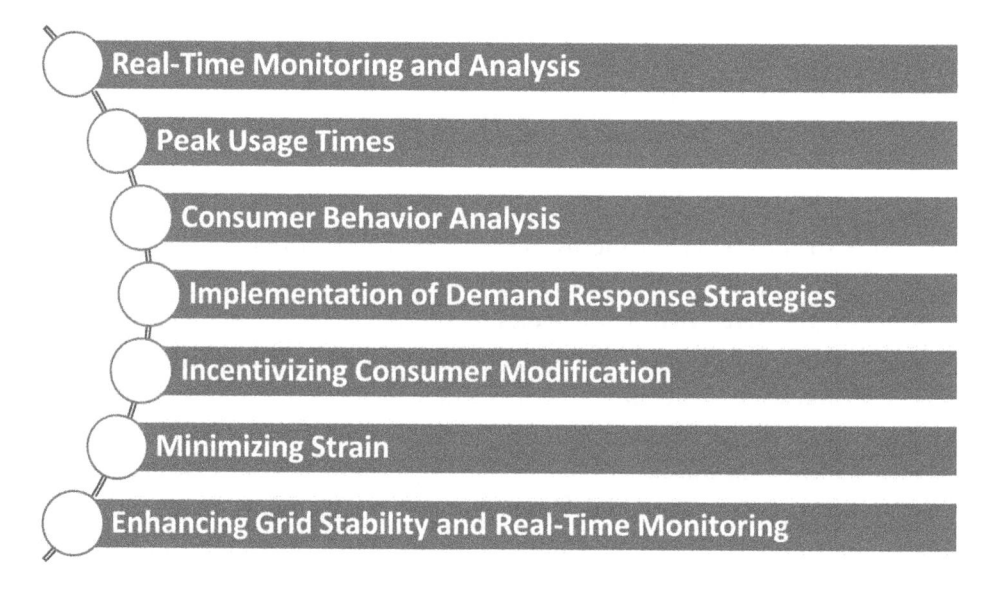

- Real-Time Monitoring and Analysis
- Peak Usage Times
- Consumer Behavior Analysis
- Implementation of Demand Response Strategies
- Incentivizing Consumer Modification
- Minimizing Strain
- Enhancing Grid Stability and Real-Time Monitoring

- **Real-Time Monitoring and Analysis:** AI technology enables continuous real-time monitoring and analysis of energy consumption patterns within smart grids. Through sophisticated algorithms, the system gathers data from various sources, such as smart meters and IoT devices, to capture real-time energy usage trends.
- **Insights into Peak Usage Times:** By leveraging AI's analytical capabilities, smart grids identify peak usage times and patterns in energy consumption. This insight allows for the identification of periods when the demand on the grid is at its highest, often coinciding with specific times of the day or certain events.
- **Consumer Behavior Analysis:** AI algorithms analyze consumer behavior patterns concerning energy usage. This analysis considers factors like historical usage, preferences, and responses to previous demand response programs.

- **Implementation of Demand Response Strategies:** Using the insights gleaned from real-time monitoring and consumer behavior analysis, smart grids implement demand response strategies. These strategies can include time-based pricing, incentives, or automated controls to encourage consumers to adjust their energy usage during peak demand periods.
- **Incentivizing Consumer Modification:** AI-driven demand response mechanisms incentivize consumers to modify their energy usage habits during high-demand periods. This could involve offering discounts, rebates, or rewards for reducing consumption or shifting usage to off-peak hours.
- **Minimizing Strain on the Grid:** By encouraging consumers to adjust their energy usage in response to demand signals, smart grids alleviate strain on the grid during peak periods. This proactive adjustment helps in balancing supply and demand, reducing the risk of grid overload and mitigating the likelihood of blackouts or brownouts.
- **Enhancing Grid Stability:** The dynamic adjustment of energy consumption through AI-enabled demand response contributes to the overall stability of the grid. By managing demand in real-time, smart grids maintain a more balanced and efficient energy distribution system.

Integration of Renewable Energy Sources: Smart grids play a pivotal role in integrating renewable energy sources like solar and wind power. AI algorithms forecast the intermittency of these sources based on weather patterns, enabling grid operators to balance supply and demand effectively. This integration not only promotes sustainability but also optimizes the utilization of renewable resources, making the grid more resilient and adaptable (Agrawal et al., 2024a; Kumar B et al., 2024; Rahamathunnisa et al., 2023).

Grid Modernization and Automation: Furthermore, smart grids embrace automation and modernization. AI-driven automation manages grid operations, dynamically rerouting energy flow to circumvent faults or congestion. This automated response enhances grid efficiency and resilience, minimizing human intervention and response time during contingencies.

Cyber-Physical Security Measures: However, these advancements also bring challenges, particularly in cybersecurity. Smart grids rely on interconnected systems, making them susceptible to cyber threats. AI-based cybersecurity measures are crucial for safeguarding against potential attacks, employing sophisticated algorithms to detect and respond to anomalies or malicious activities, ensuring the grid's integrity and reliability (Agrawal et al., 2023; Maguluri, Arularasan, et al., 2023a).

In summary, advancements in smart grids propelled by AI and ML technologies revolutionize the energy landscape, ensuring efficient, reliable, and secure energy distribution while accommodating the increasing integration of renewable sources.

PREDICTIVE MAINTENANCE IN ENERGY INFRASTRUCTURE

Machine Learning for Proactive Maintenance

Predictive maintenance employs advanced analytics, particularly Machine Learning (ML), to predict equipment failures before they occur. In the energy sector, predictive maintenance aims to anticipate and prevent breakdowns in critical infrastructure components like power plants, transformers, turbines, and distribution systems (Boopathi & Kanike, 2023a; Maheswari et al., 2023a; Ramudu et al., 2023a; Syamala et al., 2023).

- **Role of Machine Learning:** Machine Learning plays a pivotal role in predictive maintenance by analyzing historical data, sensor readings, and equipment performance metrics. ML algorithms detect patterns and anomalies in these datasets, enabling the identification of early indicators of potential failures (Boopathi & Kanike, 2023b; Maheswari et al., 2023b; Zekrifa et al., 2023).
- **Data-driven Predictive Models:** ML models, such as supervised learning algorithms (e.g., regression, decision trees) or unsupervised learning techniques (e.g., clustering, anomaly detection), process large volumes of data to create predictive models. These models learn from historical failure patterns and equipment behavior to forecast when maintenance is needed (Boopathi, Pandey, et al., 2023; Ingle et al., 2023; Sampath, C., et al., 2022).
- **Sensor Data and Predictive Analytics:** Sensors embedded within energy infrastructure continuously collect data regarding temperature, pressure, vibration, and other relevant parameters. ML algorithms analyze this sensor data, identifying deviations from normal operating conditions, which can signify impending issues.
- **Proactive Maintenance Strategies:** ML-driven predictive maintenance enables energy companies to adopt proactive maintenance strategies. By identifying potential equipment failures in advance, maintenance tasks can be scheduled optimally, reducing downtime, preventing costly repairs, and extending the lifespan of assets.
- **Cost and Resource Optimization:** Predictive maintenance powered by ML optimizes resource allocation. Rather than relying on fixed maintenance

schedules or reacting to unexpected failures, resources are allocated based on predictive insights, improving operational efficiency and reducing unnecessary maintenance.
- **Continuous Learning and Improvement:** ML models for predictive maintenance continually learn from new data. They adapt and improve over time, becoming more accurate in predicting equipment failures as they ingest more information, leading to enhanced reliability and performance of energy infrastructure (Nishanth et al., 2023; Rahamathunnisa et al., 2023).
- **Impact on Energy Sector:** Implementing machine learning-based predictive maintenance results in improved asset reliability, increased safety, reduced downtime, and enhanced overall performance of energy infrastructure, contributing to a more reliable and efficient energy supply.

Minimizing Downtime through Predictive Analytics

- **Early Detection of Equipment Issues:** Predictive analytics utilizes historical data, real-time monitoring, and machine learning algorithms to detect anomalies or deviations in equipment performance. By analyzing patterns, these systems identify potential issues before they escalate into significant problems (Koshariya et al., 2023; Maguluri, Ananth, et al., 2023).
- **Predictive Maintenance Planning:** Predictive analytics allows for the development of proactive maintenance schedules. By predicting when equipment is likely to fail, maintenance tasks can be scheduled during planned downtimes, minimizing disruptions to operations (P. R. Kumar et al., 2023a; Ramudu et al., 2023b).
- **Condition-Based Maintenance:** Rather than adhering to fixed schedules, predictive analytics enables condition-based maintenance. Sensors and monitoring systems collect data on equipment health, allowing for maintenance only when indicators suggest it's necessary, optimizing uptime.
- **Risk Mitigation and Prevention:** Predictive analytics assesses risk factors that could lead to downtime. By identifying vulnerabilities or failure patterns, proactive measures can be implemented to prevent potential issues before they impact operations (P. R. Kumar et al., 2023a).
- **Real-Time Monitoring and Alerts:** Continuous monitoring of equipment performance in real time allows for immediate detection of abnormalities. Alerts or notifications triggered by predictive analytics systems enable swift action to address emerging issues promptly.
- **Optimized Asset Performance:** Predictive analytics models optimize asset performance by providing insights into equipment efficiency and potential

failure points. This data-driven approach aids in making informed decisions to maximize equipment uptime.
- **Data-Driven Decision-Making:** Predictive analytics provides actionable insights based on data analysis. These insights guide decision-makers in allocating resources, scheduling maintenance, and investing in the most critical areas to minimize downtime.
- **Improved Reliability and Efficiency:** By leveraging predictive analytics, industries can enhance the reliability and efficiency of their operations. Anticipating and addressing potential problems before they occur leads to fewer unexpected disruptions and smoother operations overall.
- **Cost Savings and Increased Productivity:** Minimizing downtime through predictive analytics not only reduces maintenance costs but also increases productivity. Optimized maintenance schedules and reduced unexpected downtimes lead to improved output and revenue.

CONSUMER-CENTRIC ENERGY SOLUTIONS

Figure 3. AI-powered consumer-centric energy solutions

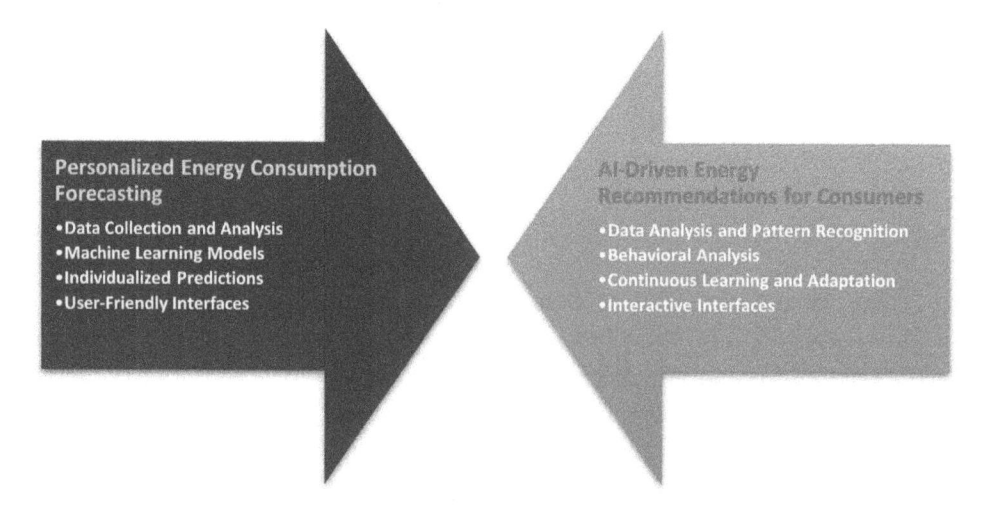

Personalized Energy Consumption Forecasting
- Data Collection and Analysis
- Machine Learning Models
- Individualized Predictions
- User-Friendly Interfaces

AI-Driven Energy Recommendations for Consumers
- Data Analysis and Pattern Recognition
- Behavioral Analysis
- Continuous Learning and Adaptation
- Interactive Interfaces

Personalized Energy Consumption Forecasting

Personalized energy consumption forecasting utilizes AI-powered algorithms to analyze individual or household energy usage patterns, preferences, and historical data (Boopathi, Kumar, et al., 2023; Domakonda et al., 2022; Kumara et al., 2023; Syamala et al., 2023).

- **Data Collection and Analysis:** AI algorithms collect and analyze data from smart meters, IoT devices, and historical usage patterns to understand individual consumption behavior. This data includes factors such as daily routines, weather conditions, appliance usage, and energy consumption peaks.
- **Machine Learning Models:** Machine learning models, such as neural networks or regression algorithms, process this data to create personalized consumption forecasts. These models continuously learn and adapt based on real-time data, refining predictions over time (Dhanya et al., 2023; Ravisankar et al., 2023).
- **Individualized Predictions:** The forecasts generated by these models provide consumers with personalized insights into their expected energy usage over specific periods. This information helps consumers plan and adjust their consumption habits to align with their preferences and optimize energy efficiency.
- **User-Friendly Interfaces:** User-friendly interfaces or mobile applications present these forecasts in accessible formats, offering users visual representations and insights into their predicted energy consumption. This empowers consumers to make informed decisions about their energy usage.

Personalized forecasts help consumers make informed decisions by providing insights into their energy usage patterns, enabling proactive consumption management and potential cost reduction. They also encourage energy efficiency by identifying areas for improvement and implementing energy-saving practices, promoting a sustainable lifestyle. Additionally, personalized forecasts help optimize energy budgets by planning and allocating resources based on predicted usage.

AI-Driven Energy Recommendations for Consumers

AI-driven energy recommendations utilize machine learning algorithms to provide personalized advice and suggestions to consumers regarding their energy usage (Agrawal et al., 2024b; B et al., 2024; Domakonda et al., 2022; Samikannu et al., 2022; Satav, Lamani, G, et al., 2024).

- **Data Analysis and Pattern Recognition:** AI algorithms analyze historical consumption data, behavior patterns, and external factors influencing energy usage to identify personalized recommendations for consumers.

- **Behavioral Analysis:** These algorithms assess individual behavior, preferences, and trends in energy consumption to create customized recommendations. For instance, they may suggest adjusting thermostat settings, optimizing appliance usage, or adopting energy-efficient practices based on specific usage patterns.
- **Continuous Learning and Adaptation:** The AI system continually learns from consumer feedback and real-time data, refining its recommendations to better suit individual preferences and evolving consumption patterns.
- **Interactive Interfaces:** User interfaces or applications deliver these recommendations to consumers in an easy-to-understand and actionable format. They might include personalized tips, reminders, or alerts aimed at improving energy efficiency.

AI-driven recommendations promote energy consciousness by raising consumer awareness about their energy usage habits, encouraging conscious consumption reduction. Tailored suggestions encourage sustainable practices, contributing to energy conservation efforts. Personalized recommendations empower consumer engagement by offering actionable steps to optimize energy usage, leading to more informed and efficient consumption.

INTEGRATION OF RENEWABLE ENERGY SOURCES

AI Strategies for Maximizing Renewable Energy Output

AI Strategies for Maximizing Renewable Energy Output (Syamala et al., 2023; Vennila et al., 2022) is shown Figure 4.

- **Forecasting Renewable Energy Generation:** AI algorithms leverage weather data, historical patterns, and real-time information to forecast renewable energy generation. Machine Learning models, like neural networks or regression algorithms, analyze these diverse data sets to predict solar irradiance, wind speeds, and other factors impacting renewable energy production.
- **Optimizing Generation Efficiency:** AI-driven systems optimize the efficiency of renewable energy generation. For instance, in solar photovoltaic systems, AI algorithms adjust panel angles or track the sun's position for maximum exposure. In wind farms, predictive analytics optimize turbine settings based on wind conditions, enhancing energy capture.

- **Demand-Side Management:** AI facilitates demand-side management by predicting energy demand patterns. By aligning renewable energy production with peak demand periods, AI helps in balancing supply and demand, reducing waste, and optimizing energy distribution.

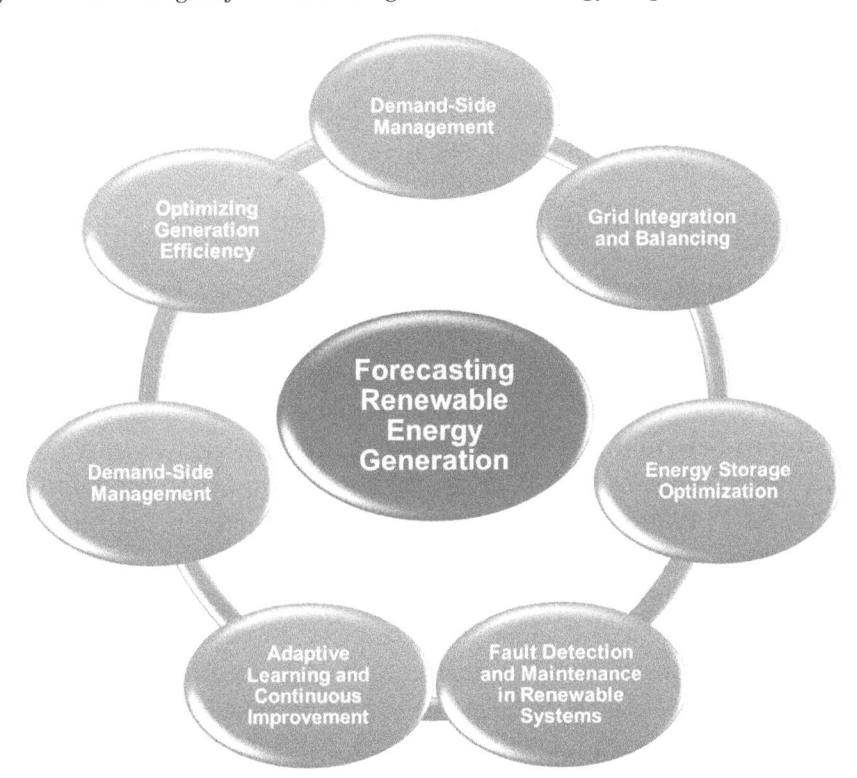

Figure 4. AI strategies for maximizing renewable energy output

- **Grid Integration and Balancing:** AI technologies enable effective integration of renewables into existing grids. Grid management systems use AI algorithms to balance fluctuating renewable energy inputs with the overall grid demand, preventing overloads and ensuring grid stability.
- **Energy Storage Optimization:** AI optimizes energy storage solutions by predicting when and how much energy to store based on production forecasts and demand patterns. This enables efficient use of storage systems, reducing energy waste and supporting continuous power supply.
- **Fault Detection and Maintenance in Renewable Systems:** AI algorithms monitor the health of renewable energy systems. Predictive maintenance

detects anomalies in equipment, reducing downtime by addressing issues before they lead to significant failures.
- **Adaptive Learning and Continuous Improvement:** AI systems continuously learn from new data, refining their models and strategies for maximizing renewable energy output. This continuous learning process enhances the accuracy of predictions and optimizations over time (Das et al., 2024).

AI strategies optimize renewable energy integration, grid stability, and cost efficiency, promoting renewable energy production and distribution, reducing operational costs and environmental impact.

ML-Driven Forecasting for Solar and Wind Power

- **Data Collection and Preprocessing:** Machine Learning (ML) models require diverse data sources, including historical weather patterns, solar irradiance, wind speeds, geographical characteristics, and past energy production data. Preprocessing involves cleaning and organizing this data for analysis.
- **Feature Selection and Model Training:** ML algorithms, such as neural networks, support vector machines, or regression models, are trained using this data. Features like time of day, weather conditions, and geographical factors are selected to create accurate predictive models.
- **Solar Power Forecasting:** For solar power, ML models predict energy output based on factors like sunlight intensity, cloud cover, and panel orientation. Neural networks, for instance, analyze historical solar irradiance and weather data to forecast energy production over specific time intervals (hours, days, or weeks) (Domakonda et al., 2023; Hema et al., 2023a).
- **Wind Power Forecasting:** ML models for wind power utilize historical wind speed, direction, and atmospheric pressure data. These models predict wind turbine performance and energy production by analyzing these factors, enabling forecasts for future wind power generation.
- **Ensemble Methods and Hybrid Models:** Ensemble methods, combining multiple ML algorithms or incorporating physical models with machine learning, enhance accuracy. Hybrid models merge meteorological data with ML techniques to improve predictions further.
- **Real-Time Adjustments and Continuous Learning:** ML models continuously learn and adapt. Real-time data updates enable adjustments in predictions, allowing for more accurate forecasts as new information becomes available (Boopathi, 2023).

- **Validation and Improvement:** ML-driven forecasting undergoes validation against actual production data to refine models. Continuous improvement involves recalibration based on discrepancies between predictions and real-world outcomes.
- **Integration into Energy Management Systems:** Forecasts generated by ML models are integrated into energy management systems. Grid operators use these predictions for demand-side management, optimal resource allocation, and grid stability.

Accurate forecasting aids in the integration of solar and wind power into the grid, providing reliable predictions for energy planners and operators. It enhances operational efficiency by optimizing resource allocation, reducing backup power reliance, and minimizing wastage. ML-driven forecasting also promotes cost savings and sustainability by maximizing renewable energy utilization.

CYBERSECURITY IN AI-DRIVEN ENERGY SYSTEMS

Addressing Security Challenges in Smart Grids

- **Cyber Threat Landscape:** Smart grids face diverse cyber threats, including data breaches, ransomware attacks, and unauthorized access. These threats pose risks to the confidentiality, integrity, and availability of energy infrastructure (Bikash Chandra Saha, 2022; Hema et al., 2023b; Kavitha et al., 2023b).
- **Vulnerabilities in Interconnected Systems:** Smart grids consist of interconnected devices, sensors, and communication networks, increasing the attack surface and susceptibility to cyber threats. Vulnerabilities in legacy systems and insufficient security protocols further compound these risks.
- **Risk Mitigation Strategies:** Addressing security challenges involves comprehensive strategies:
  - **Access Control and Authentication:** Implementing robust access controls, encryption, and multi-factor authentication to restrict unauthorized access to critical grid components.
  - **Network Segmentation:** Segmenting networks into distinct zones with restricted access limits the potential spread of cyber-attacks within the grid infrastructure.
  - **Regular Updates and Patch Management:** Timely installation of security patches and updates to address known vulnerabilities in software and hardware components.

- **AI-Driven Threat Detection:** Leveraging AI for threat detection involves (Agrawal et al., 2024a; Kumar Reddy R. et al., 2023):
    ◦ **Behavioral Analytics:** AI algorithms analyze network and device behavior, detecting anomalies that deviate from normal patterns, indicating potential security threats.
    ◦ **Anomaly Detection and Predictive Analytics:** ML models detect irregularities in data transmission or system behavior, predicting and preventing potential cyber-attacks before they occur.
- **Incident Response and Recovery:** Establishing robust incident response plans involves:
    ◦ **Real-Time Monitoring and Response:** Continuous monitoring of grid operations allows for immediate detection and response to security incidents.
    ◦ **Backup and Recovery Plans:** Regular data backups and well-defined recovery plans ensure minimal disruption in case of a successful cyber-attack.
- **Collaboration and Standards:** Industry-wide collaboration and adherence to cybersecurity standards, such as NIST (National Institute of Standards and Technology) guidelines or IEC (International Electrotechnical Commission) standards, are crucial in developing unified security frameworks for smart grids.
- **Training and Awareness:** Educating personnel about cybersecurity best practices, emphasizing the importance of adherence to security protocols, and fostering a culture of security-consciousness among employees is vital.
- **Regulatory Framework and Compliance:** Government and regulatory bodies play a significant role in establishing and enforcing cybersecurity regulations and standards for the energy sector, ensuring compliance across the industry.

Implementing comprehensive security measures strengthens the resilience of smart grids against cyber threats, ensuring continuous energy supply. It also protects critical infrastructure, maintaining system integrity and reliability. Robust cybersecurity measures boost consumer confidence in the grid infrastructure, fostering trust in the reliability and security of energy.

AI-Powered Threat Detection and Mitigation in Smart Grids

- **Advanced Analytics and Machine Learning:** AI algorithms, including machine learning and deep learning models, analyze vast volumes of data from smart grid components. They detect patterns, anomalies, and potential threats in real-time (Boopathi & Sivakumar, 2016; P. R. Kumar et al., 2023b; Ramudu et al., 2023a; Sampath, Pandian, et al., 2022).

- **Behavioral Analysis and Anomaly Detection:** AI systems establish normal behavior baselines for devices and network traffic. Any deviations from these established norms trigger alerts for potential security incidents. Anomaly detection algorithms continuously learn and adapt to identify new threats.
- **Predictive Analytics for Threat Forecasting:** AI-powered predictive models forecast potential security risks by analyzing historical data and identifying trends. These models anticipate possible threats, enabling proactive measures to prevent cyber-attacks or system compromises.
- **Dynamic Threat Response:** AI systems enable swift and automated responses to identified threats. They can isolate affected systems, apply security protocols, or initiate countermeasures in real-time to mitigate the impact of the threat.
- **Adaptive Learning and Continuous Improvement:** AI continuously learns from new data, adapting security measures to evolving threats. This adaptive capability enables the system to enhance its defense mechanisms and stay resilient against emerging cyber threats.
- **Automated Incident Handling:** AI streamlines incident handling by automating threat analysis, response decisions, and mitigative actions. This accelerates response times, reducing the window of vulnerability during cyber-attacks.
- **Integration with Security Operations Centers (SOCs):** AI augments the capabilities of Security Operations Centers by providing advanced threat intelligence, automating threat analysis, and aiding human operators in making informed decisions for threat response and mitigation.
- **Collaborative Threat Intelligence:** AI facilitates the sharing of threat intelligence and analysis across grids and cybersecurity communities. This collective approach enables a broader understanding of emerging threats and better preparation against cyber-attacks.

AI-powered systems enhance threat detection in smart grids, enabling proactive measures against potential attacks. Real-time response to security incidents minimizes downtime and reduces cyber-attack impact. AI's adaptive nature ensures security measures evolve to counter emerging threats, strengthening smart grid defenses against evolving cyber risks.

ETHICAL CONSIDERATIONS AND REGULATORY FRAMEWORK

Ensuring Responsible AI Implementation in Energy Management

- **Ethical Considerations:** Ethical frameworks must guide AI implementation, ensuring fairness, transparency, and accountability. Avoiding biases in data, algorithms, and decision-making processes is essential to ensure equitable outcomes for all stakeholders (Maguluri, Arularasan, et al., 2023b; Ugandar et al., 2023).
- **Transparency and Explain-ability:** AI systems should be transparent, providing clear explanations of how decisions are made. Explainable AI (XAI) ensures that AI-driven decisions are understandable and traceable, boosting trust and enabling scrutiny.
- **Data Privacy and Security:** Protecting sensitive consumer data is paramount. Compliance with data privacy regulations like GDPR (General Data Protection Regulation) ensures the lawful and ethical use of personal data in energy management systems.
- **Human Oversight and Control:** Retaining human oversight in decision-making processes involving AI is crucial. Human intervention ensures accountability and ethical judgment, especially in critical situations or when addressing complex ethical dilemmas.
- **Bias Mitigation and Fairness:** Proactively addressing biases in AI algorithms is essential to ensure fairness and prevent discrimination. Regular audits and testing for biases help in identifying and rectifying algorithmic biases.
- **Regulatory Compliance:** Adherence to industry-specific regulations and standards (such as ISO 27001 for information security or NIST frameworks for cybersecurity) ensures compliance and upholds ethical guidelines in AI implementation.
- **Risk Assessment and Mitigation:** Conducting thorough risk assessments helps in identifying potential ethical, social, or environmental risks associated with AI implementation. Mitigation strategies should be in place to address these risks proactively.
- **Continuous Monitoring and Evaluation:** Regular monitoring and evaluation of AI systems ensure ongoing compliance with ethical standards. Feedback loops enable refinement of algorithms and processes to align with evolving ethical considerations.
- **Stakeholder Engagement and Collaboration:** Involving stakeholders, including consumers, industry experts, and regulatory bodies, fosters a

collaborative approach to responsible AI implementation. Addressing concerns and soliciting feedback enhances transparency and trust.
- **Education and Training:** Providing education and training on AI ethics and responsible use ensures that personnel understand and adhere to ethical guidelines in AI implementation within the energy management domain.

Responsible AI implementation fosters trust and consumer confidence in energy management solutions. It reduces risks associated with biases, data breaches, and misuse, reducing legal and reputational liabilities. Ethically sound AI implementation also leads to sustainable, inclusive, and socially responsible solutions, promoting trust among consumers, stakeholders, and the public.

Policy and Governance in AI-Enabled Energy Systems

- **Regulatory Frameworks:** Establishing regulatory frameworks specific to AI applications in the energy sector is essential. These frameworks should address data privacy, security standards, transparency, and accountability in AI-driven energy systems (Sundaramoorthy et al., 2023).
- **Ethical Guidelines:** Developing and adhering to ethical guidelines for AI use in energy management ensures responsible deployment. This includes principles for fairness, transparency, accountability, and the avoidance of biases.
- **Data Governance and Privacy:** Policies should outline clear guidelines for data governance, emphasizing the secure collection, storage, and sharing of data in compliance with privacy regulations. Ensuring the ethical use of consumer data is paramount.
- **Standardization and Interoperability:** Standards for interoperability among AI systems in energy management facilitate seamless integration and collaboration between different platforms, ensuring compatibility and efficiency.
- **Transparency and Explain-ability:** Policies should mandate transparency and explain-ability in AI decision-making processes. Users should understand how AI algorithms work and the basis for the decisions they make.
- **Human Oversight and Accountability:** Establishing mechanisms for human oversight and accountability in AI systems ensures that human judgment is retained in critical decision-making processes. This includes defining roles and responsibilities for oversight.
- **Compliance and Auditing:** Mandating regular compliance checks and audits ensures adherence to policies and standards. Audits assess AI systems for biases, fairness, security, and adherence to ethical guidelines.

- **Education and Workforce Development:** Policies should support education and training programs to build AI expertise within the energy sector. This includes training on AI ethics, data handling, and AI system management.
- **Public Engagement and Participation:** Involving stakeholders and the public in policy formulation ensures that diverse perspectives and concerns are considered. Public engagement builds trust and supports informed decision-making.
- **International Collaboration:** Encouraging collaboration and information sharing among nations fosters the development of global standards and best practices for AI in energy systems, addressing cross-border implications.

Effective policies in energy systems ensure ethical and responsible use of AI, building trust and confidence among consumers, stakeholders, and the public. These policies also promote innovation by providing a clear roadmap and guidelines for AI adoption in energy, fostering advancements and improvements.

CHALLENGES AND FUTURE PERSPECTIVES

Challenges in AI Implementation in the Energy Sector

- **Data Quality and Availability:** Inconsistent data quality and limited access to comprehensive datasets pose challenges for AI implementation. Ensuring reliable, diverse, and accessible data is crucial for accurate AI-driven insights.
- **Legacy Infrastructure Integration:** Retrofitting AI into existing legacy infrastructure can be complex. Compatibility issues and the need for interoperability with older systems can hinder seamless integration.
- **Cost and Investment:** High upfront costs associated with AI implementation, including infrastructure upgrades, skilled personnel, and ongoing maintenance, may act as barriers, particularly for smaller energy companies.
- **Regulatory Compliance and Standards:** Adhering to evolving regulatory standards, privacy laws, and ethical guidelines while implementing AI in energy systems presents challenges. Keeping up with compliance requirements can be demanding.
- **Cybersecurity Risks:** AI systems are susceptible to cyber threats, making security a paramount concern. Safeguarding AI-powered systems against attacks and ensuring data privacy is crucial.
- **Skills Gap and Workforce Training:** The shortage of skilled professionals proficient in AI technologies is a challenge. Training and upskilling the workforce to manage and utilize AI effectively is necessary.

Strategies to Overcome Implementation Barriers

- **Data Management and Quality Improvement:** Prioritize data quality initiatives and invest in data collection, cleaning, and aggregation processes to ensure high-quality, standardized datasets for AI implementation.
- **Gradual Integration and Pilot Projects:** Implement AI in phases, starting with pilot projects to test functionalities and demonstrate ROI. Gradually scaling successful implementations reduces risk and facilitates smoother integration.
- **Collaboration and Partnerships:** Foster collaboration between energy companies, tech firms, and research institutions to share knowledge, resources, and best practices for AI implementation.
- **Regulatory Engagement and Compliance:** Engage with regulatory bodies to develop clear guidelines and frameworks that balance innovation with regulatory compliance, ensuring AI deployment aligns with ethical standards.
- **Cybersecurity Measures:** Prioritize cybersecurity protocols by employing robust encryption, authentication, and monitoring systems to protect AI-enabled energy infrastructure from cyber threats.
- **Workforce Development:** Invest in workforce training programs and partnerships with educational institutions to bridge the skills gap and develop a talent pool proficient in AI technologies.

Future Perspectives

- **AI-Driven Predictive Maintenance:** Further advancements in predictive maintenance using AI can optimize energy infrastructure, reducing downtime and maintenance costs.
- **Decentralized Energy Systems:** AI can facilitate the integration of decentralized energy sources by optimizing their management and ensuring grid stability.
- **Energy Market Forecasting:** AI-powered models can enhance energy market forecasting accuracy, enabling better decision-making in energy trading and pricing.
- **Continued Ethical Advancements:** Continued focus on ethical AI development, promoting fairness, transparency, and responsible AI use within the energy sector.

Anticipated Innovations in Energy by 2030 and Beyond

- **AI-Enabled Energy Optimization:** Advanced AI algorithms will continue to optimize energy consumption, leveraging real-time data for predictive analytics and personalized energy management. AI-driven smart grids will efficiently balance supply and demand, minimizing wastage (Agrawal et al., 2024a; Kumar B et al., 2024; Satav, Lamani, K. G., et al., 2024).
- **Renewable Energy Integration:** Further advancements in renewable energy technologies, like improved solar panels and wind turbines, alongside innovative energy storage solutions, will accelerate the integration of renewables into the grid.
- **Distributed Energy Systems:** Decentralized energy systems, enabled by AI and Internet of Things (IoT) technologies, will empower communities to produce and manage their energy, fostering local energy markets and resilience.
- **Grid Modernization and Resilience:** Enhanced grid infrastructure with smart sensors, predictive maintenance, and self-healing capabilities will bolster grid resilience. Microgrids and energy islands will provide backup during disruptions.
- **Electrification and Clean Transportation:** Increased adoption of electric vehicles (EVs) will drive innovations in charging infrastructure and smart grid solutions to manage their charging patterns and integration into the energy ecosystem.
- **Hybrid Energy Solutions:** Integration of multiple renewable sources, energy storage, and traditional energy forms will lead to hybrid energy systems. These systems will optimize efficiency by leveraging the strengths of various energy sources.
- **Energy-Positive Buildings:** Advancements in building design, incorporating AI for energy management, will enable the construction of energy-positive structures that generate more energy than they consume, contributing to a net-zero future.
- **AI-Driven Energy Market Platforms:** AI-powered energy marketplaces and platforms will facilitate peer-to-peer energy trading, enabling consumers to buy, sell, and exchange energy directly among themselves.
- **Quantum Computing in Energy:** Quantum computing applications will revolutionize energy modeling, material discovery for efficient solar cells, and optimization of energy systems, unlocking unprecedented computational power.

- **Circular Economy in Energy:** Embracing circular economy principles will drive innovations in recycling, repurposing, and reusing energy components, reducing waste and maximizing resource efficiency.

Impact and Transformation

- **Sustainability and Resilience:** Innovations will accelerate the transition towards cleaner, more sustainable energy systems, reducing carbon footprints and enhancing energy security and resilience (Boopathi, Kumar, et al., 2023; Domakonda et al., 2022; Samikannu et al., 2022).
- **Empowered Consumers:** Consumers will have greater control over their energy usage, fostering energy independence, cost savings, and environmental consciousness.
- **Technological Convergence:** The convergence of AI, IoT, and other emerging technologies will create interconnected, intelligent energy ecosystems, transforming the way energy is produced, managed, and consumed.
- **Economic Growth and Job Creation:** The energy sector's evolution will stimulate economic growth, fostering innovation, and generating new job opportunities in technology, engineering, and renewable energy sectors (Hussain et al., 2023; Ingle et al., 2023).

CONCLUSION

The integration of Artificial Intelligence (AI) in energy management has revolutionized the sector, enhancing efficiency, sustainability, and resilience. Key insights from this exploration highlight the transformative nature of AI's role in energy management.

AI plays a crucial role in energy optimization, enhancing operational efficiency and balancing supply and demand dynamics. Its integration has significantly contributed to sustainable energy practices, integrating renewable energy sources, optimizing resource utilization, and minimizing wastage. However, challenges such as data quality, cybersecurity vulnerabilities, ethical implications, and regulatory compliance persist. Future innovation in AI algorithms and emerging technologies promises further enhancements in energy optimization, decentralized energy systems, and consumer-centric solutions. Collaborative efforts from stakeholders across industries, policymakers, researchers, and technology innovators are needed to address challenges, promote ethical standards, and steer the energy sector towards a sustainable and AI-driven future. The realization of AI's full potential in energy management requires collaboration among stakeholders across industries, policymakers, researchers, and technology innovators.

AI's integration into energy management is a significant shift, combining technological innovation with sustainable practices. Despite challenges, it offers potential for optimized, efficient, and sustainable energy systems, highlighting its transformative role in the energy landscape.

REFERENCES

Agrawal, A. V., Magulur, L. P., Priya, S. G., Kaur, A., Singh, G., & Boopathi, S. (2023). Smart Precision Agriculture Using IoT and WSN. In Advances in Information Security, Privacy, and Ethics (pp. 524–541). IGI Global. doi:10.4018/978-1-6684-8145-5.ch026

Agrawal, A. V., Shashibhushan, G., Pradeep, S., Padhi, S. N., Sugumar, D., & Boopathi, S. (2024). Synergizing Artificial Intelligence, 5G, and Cloud Computing for Efficient Energy Conversion Using Agricultural Waste. In B. K. Mishra (Ed.), Practice, Progress, and Proficiency in Sustainability. IGI Global, doi:10.4018/979-8-3693-1186-8.ch026

B, M. K., K, K. K., Sasikala, P., Sampath, B., Gopi, B., & Sundaram, S. (2024). Sustainable Green Energy Generation From Waste Water. In *Practice, Progress, and Proficiency in Sustainability* (pp. 440–463). IGI Global. doi:10.4018/979-8-3693-1186-8.ch024

Bikash Chandra Saha, M. S., Deepa, R., Akila, A., & Sai Thrinath, B. V. (2022). IOT based smart energy meter for smart grid. Academic Press.

Boopathi, S. (2023). Internet of Things-Integrated Remote Patient Monitoring System: Healthcare Application. In A. Suresh Kumar, U. Kose, S. Sharma, & S. Jerald Nirmal Kumar (Eds.), Advances in Healthcare Information Systems and Administration. IGI Global., doi:10.4018/978-1-6684-6894-4.ch008

Boopathi, S., & Kanike, U. K. (2023a). Applications of Artificial Intelligent and Machine Learning Techniques in Image Processing. In *Handbook of Research on Thrust Technologies' Effect on Image Processing* (pp. 151–173). IGI Global. doi:10.4018/978-1-6684-8618-4.ch010

Boopathi, S., & Kanike, U. K. (2023b). Applications of Artificial Intelligent and Machine Learning Techniques in Image Processing. In B. K. Pandey, D. Pandey, R. Anand, D. S. Mane, & V. K. Nassa (Eds.), Advances in Computational Intelligence and Robotics. IGI Global. doi:10.4018/978-1-6684-8618-4.ch010

Boopathi, S., Kumar, P. K. S., Meena, R. S., Sudhakar, M., & Associates. (2023). Sustainable Developments of Modern Soil-Less Agro-Cultivation Systems: Aquaponic Culture. In Human Agro-Energy Optimization for Business and Industry (pp. 69–87). IGI Global.

Boopathi, S., Pandey, B. K., & Pandey, D. (2023). Advances in Artificial Intelligence for Image Processing: Techniques, Applications, and Optimization. In B. K. Pandey, D. Pandey, R. Anand, D. S. Mane, & V. K. Nassa (Eds.), Advances in Computational Intelligence and Robotics. IGI Global. doi:10.4018/978-1-6684-8618-4.ch006

Boopathi, S., & Sivakumar, K. (2016). Optimal parameter prediction of oxygen-mist near-dry wire-cut EDM. *Inderscience: International Journal of Manufacturing Technology and Management*, *30*(3–4), 164–178. doi:10.1504/IJMTM.2016.077812

Das, S., Lekhya, G., Shreya, K., Lydia Shekinah, K., Babu, K. K., & Boopathi, S. (2024). Fostering Sustainability Education Through Cross-Disciplinary Collaborations and Research Partnerships: Interdisciplinary Synergy. In P. Yu, J. Mulli, Z. A. S. Syed, & L. Umme (Eds.), Advances in Higher Education and Professional Development. IGI Global. doi:10.4018/979-8-3693-0487-7.ch003

Dhanya, D., Kumar, S. S., Thilagavathy, A., Prasad, D. V. S. S. S. V., & Boopathi, S. (2023). Data Analytics and Artificial Intelligence in the Circular Economy: Case Studies. In B. K. Mishra (Ed.), Advances in Civil and Industrial Engineering. IGI Global. doi:10.4018/979-8-3693-0044-2.ch003

Domakonda, V. K., Farooq, S., Chinthamreddy, S., Puviarasi, R., Sudhakar, M., & Boopathi, S. (2022). Sustainable Developments of Hybrid Floating Solar Power Plants: Photovoltaic System. In Human Agro-Energy Optimization for Business and Industry (pp. 148–167). IGI Global.

Domakonda, V. K., Farooq, S., Chinthamreddy, S., Puviarasi, R., Sudhakar, M., & Boopathi, S. (2023). Sustainable Developments of Hybrid Floating Solar Power Plants: Photovoltaic System. In P. Vasant, R. Rodríguez-Aguilar, I. Litvinchev, & J. A. Marmolejo-Saucedo (Eds.), Advances in Environmental Engineering and Green Technologies. IGI Global. doi:10.4018/978-1-6684-4118-3.ch008

Fayyazi, M., Sardar, P., Thomas, S. I., Daghigh, R., Jamali, A., Esch, T., Kemper, H., Langari, R., & Khayyam, H. (2023). Artificial Intelligence/Machine Learning in Energy Management Systems, Control, and Optimization of Hydrogen Fuel Cell Vehicles. *Sustainability (Basel)*, *15*(6), 5249. doi:10.3390/su15065249

Ganesh, A. H., & Xu, B. (2022). A review of reinforcement learning based energy management systems for electrified powertrains: Progress, challenge, and potential solution. *Renewable & Sustainable Energy Reviews, 154*, 111833. doi:10.1016/j.rser.2021.111833

Hema, N., Krishnamoorthy, N., Chavan, S. M., Kumar, N., Sabarimuthu, M., & Boopathi, S. (2023b). A Study on an Internet of Things (IoT)-Enabled Smart Solar Grid System. In *Handbook of Research on Deep Learning Techniques for Cloud-Based Industrial IoT* (pp. 290–308). IGI Global. doi:10.4018/978-1-6684-8098-4.ch017

Hema, N., Krishnamoorthy, N., Chavan, S. M., Kumar, N. M. G., Sabarimuthu, M., & Boopathi, S. (2023a). A Study on an Internet of Things (IoT)-Enabled Smart Solar Grid System. In P. Swarnalatha & S. Prabu (Eds.), Advances in Computational Intelligence and Robotics. IGI Global. doi:10.4018/978-1-6684-8098-4.ch017

Hussain, Z., Babe, M., Saravanan, S., Srimathy, G., Roopa, H., & Boopathi, S. (2023). Optimizing Biomass-to-Biofuel Conversion: IoT and AI Integration for Enhanced Efficiency and Sustainability. In N. Cobîrzan, R. Muntean, & R.-A. Felseghi (Eds.), Advances in Finance, Accounting, and Economics. IGI Global. doi:10.4018/978-1-6684-8238-4.ch009

Ingle, R. B., Swathi, S., Mahendran, G., Senthil, T. S., Muralidharan, N., & Boopathi, S. (2023). Sustainability and Optimization of Green and Lean Manufacturing Processes Using Machine Learning Techniques. In N. Cobîrzan, R. Muntean, & R.-A. Felseghi (Eds.), Advances in Finance, Accounting, and Economics. IGI Global. doi:10.4018/978-1-6684-8238-4.ch012

Kavitha, C. R., Varalatchoumy, M., Mithuna, H. R., Bharathi, K., Geethalakshmi, N. M., & Boopathi, S. (2023). Energy Monitoring and Control in the Smart Grid: Integrated Intelligent IoT and ANFIS. In M. Arshad (Ed.), Advances in Bioinformatics and Biomedical Engineering. IGI Global. doi:10.4018/978-1-6684-6577-6.ch014

Koshariya, A. K., Kalaiyarasi, D., Jovith, A. A., Sivakami, T., Hasan, D. S., & Boopathi, S. (2023). AI-Enabled IoT and WSN-Integrated Smart Agriculture System. In *Artificial Intelligence Tools and Technologies for Smart Farming and Agriculture Practices* (pp. 200–218). IGI Global. doi:10.4018/978-1-6684-8516-3.ch011

Kumar, B. M., Kumar, K. K., Sasikala, P., Sampath, B., Gopi, B., & Sundaram, S. (2024). Sustainable Green Energy Generation From Waste Water: IoT and ML Integration. In B. K. Mishra (Ed.), Practice, Progress, and Proficiency in Sustainability. IGI Global. doi:10.4018/979-8-3693-1186-8.ch024

Kumar, P. R., Meenakshi, S., Shalini, S., Devi, S. R., & Boopathi, S. (2023). Soil Quality Prediction in Context Learning Approaches Using Deep Learning and Blockchain for Smart Agriculture. In R. Kumar, A. B. Abdul Hamid, & N. I. Binti Ya'akub (Eds.), Advances in Computational Intelligence and Robotics. IGI Global. doi:10.4018/978-1-6684-9151-5.ch001

Kumar Reddy, R. V., Rahamathunnisa, U., Subhashini, P., Aancy, H. M., Meenakshi, S., & Boopathi, S. (2023). Solutions for Software Requirement Risks Using Artificial Intelligence Techniques. In T. Murugan & N. E. (Eds.), Advances in Information Security, Privacy, and Ethics (pp. 45–64). IGI Global. doi:10.4018/978-1-6684-8145-5.ch003

Kumara, V., Mohanaprakash, T., Fairooz, S., Jamal, K., Babu, T., & Sampath, B. (2023). Experimental Study on a Reliable Smart Hydroponics System. In *Human Agro-Energy Optimization for Business and Industry* (pp. 27–45). IGI Global. doi:10.4018/978-1-6684-4118-3.ch002

Li, J., Herdem, M. S., Nathwani, J., & Wen, J. Z. (2023). Methods and applications for Artificial Intelligence, Big Data, Internet of Things, and Blockchain in smart energy management. *Energy and AI*, *11*, 100208. doi:10.1016/j.egyai.2022.100208

Liu, T. (2022). *Reinforcement learning-enabled intelligent energy management for hybrid electric vehicles*. Springer Nature.

Liu, Z., Gao, Y., & Liu, B. (2022). An artificial intelligence-based electric multiple units using a smart power grid system. *Energy Reports*, *8*, 13376–13388. doi:10.1016/j.egyr.2022.09.138

Maguluri, L. P., Ananth, J., Hariram, S., Geetha, C., Bhaskar, A., & Boopathi, S. (2023). Smart Vehicle-Emissions Monitoring System Using Internet of Things (IoT). In Handbook of Research on Safe Disposal Methods of Municipal Solid Wastes for a Sustainable Environment (pp. 191–211). IGI Global.

Maguluri, L. P., Arularasan, A. N., & Boopathi, S. (2023). Assessing Security Concerns for AI-Based Drones in Smart Cities. In R. Kumar, A. B. Abdul Hamid, & N. I. Binti Ya'akub (Eds.), Advances in Computational Intelligence and Robotics. IGI Global. doi:10.4018/978-1-6684-9151-5.ch002

Maheswari, B. U., Imambi, S. S., Hasan, D., Meenakshi, S., Pratheep, V., & Boopathi, S. (2023a). Internet of Things and Machine Learning-Integrated Smart Robotics. In Global Perspectives on Robotics and Autonomous Systems: Development and Applications (pp. 240–258). IGI Global. doi:10.4018/978-1-6684-7791-5.ch010

Maheswari, B. U., Imambi, S. S., Hasan, D., Meenakshi, S., Pratheep, V. G., & Boopathi, S. (2023b). Internet of Things and Machine Learning-Integrated Smart Robotics. In M. K. Habib (Ed.), (pp. 240–258). Advances in Computational Intelligence and Robotics. IGI Global. doi:10.4018/978-1-6684-7791-5.ch010

Murphey, Y. L., Park, J., Kiliaris, L., Kuang, M. L., Masrur, M. A., Phillips, A. M., & Wang, Q. (2012). Intelligent hybrid vehicle power control—Part II: Online intelligent energy management. *IEEE Transactions on Vehicular Technology*, *62*(1), 69–79. doi:10.1109/TVT.2012.2217362

Nishanth, J. R., Deshmukh, M. A., Kushwah, R., Kushwaha, K. K., Balaji, S., & Sampath, B. (2023). Particle Swarm Optimization of Hybrid Renewable Energy Systems. In B. K. Mishra (Ed.), Advances in Civil and Industrial Engineering. IGI Global. doi:10.4018/979-8-3693-0044-2.ch016

Puri, V., Mondal, S., Das, S., & Vrana, V. G. (2023). Blockchain propels tourism industry—An attempt to explore topics and information in smart tourism management through text mining and machine learning. *Informatics (MDPI)*, *10*(1), 9. doi:10.3390/informatics10010009

Rahamathunnisa, U., Sudhakar, K., Padhi, S. N., Bhattacharya, S., Shashibhushan, G., & Boopathi, S. (2023). Sustainable Energy Generation From Waste Water: IoT Integrated Technologies. In A. S. Etim (Ed.), Advances in Human and Social Aspects of Technology. IGI Global. doi:10.4018/978-1-6684-5347-6.ch010

Ramudu, K., Mohan, V. M., Jyothirmai, D., Prasad, D., Agrawal, R., & Boopathi, S. (2023a). Machine Learning and Artificial Intelligence in Disease Prediction: Applications, Challenges, Limitations, Case Studies, and Future Directions. In Contemporary Applications of Data Fusion for Advanced Healthcare Informatics (pp. 297–318). IGI Global.

Ramudu, K., Mohan, V. M., Jyothirmai, D., Prasad, D. V. S. S. S. V., Agrawal, R., & Boopathi, S. (2023b). Machine Learning and Artificial Intelligence in Disease Prediction: Applications, Challenges, Limitations, Case Studies, and Future Directions. In G. S. Karthick & S. Karupusamy (Eds.), Advances in Healthcare Information Systems and Administration. IGI Global. doi:10.4018/978-1-6684-8913-0.ch013

Rashid, R. A., Chin, L., Sarijari, M. A., Sudirman, R., & Ide, T. (2019). Machine learning for smart energy monitoring of home appliances using IoT. *2019 Eleventh International Conference on Ubiquitous and Future Networks (ICUFN)*, 66–71. 10.1109/ICUFN.2019.8806026

Ravisankar, A., Sampath, B., & Asif, M. M. (2023). Economic Studies on Automobile Management: Working Capital and Investment Analysis. In C. S. V. Negrão, I. G. P. Maia, & J. A. F. Brito (Eds.), Advances in Logistics, Operations, and Management Science. IGI Global. doi:10.4018/978-1-7998-9213-7.ch009

Samikannu, R., Koshariya, A. K., Poornima, E., Ramesh, S., Kumar, A., & Boopathi, S. (2022). Sustainable Development in Modern Aquaponics Cultivation Systems Using IoT Technologies. In *Human Agro-Energy Optimization for Business and Industry* (pp. 105–127). IGI Global.

Sampath, B., Pandian, M., Deepa, D., & Subbiah, R. (2022). Operating parameters prediction of liquefied petroleum gas refrigerator using simulated annealing algorithm. *AIP Conference Proceedings*, *2460*(1), 070003. doi:10.1063/5.0095601

Sampath, B. C. S., & Myilsamy, S. (2022). Application of TOPSIS Optimization Technique in the Micro-Machining Process. In M. A. Mellal (Ed.), Advances in Mechatronics and Mechanical Engineering. IGI Global. doi:10.4018/978-1-6684-5887-7.ch009

Satav, S. D., & Lamani, D. G, H. K., Kumar, N. M. G., Manikandan, S., & Sampath, B. (2024). Energy and Battery Management in the Era of Cloud Computing. In Practice, Progress, and Proficiency in Sustainability (pp. 141–166). IGI Global. doi:10.4018/979-8-3693-1186-8.ch009

Sundaramoorthy, K., Singh, A., Sumathy, G., Maheshwari, A., Arunarani, A. R., & Boopathi, S. (2023). A Study on AI and Blockchain-Powered Smart Parking Models for Urban Mobility. In B. B. Gupta & F. Colace (Eds.), Advances in Computational Intelligence and Robotics. IGI Global. doi:10.4018/978-1-6684-9999-3.ch010

Syamala, M., Komala, C., Pramila, P., Dash, S., Meenakshi, S., & Boopathi, S. (2023). Machine Learning-Integrated IoT-Based Smart Home Energy Management System. In *Handbook of Research on Deep Learning Techniques for Cloud-Based Industrial IoT* (pp. 219–235). IGI Global. doi:10.4018/978-1-6684-8098-4.ch013

Ugandar, R. E., Rahamathunnisa, U., Sajithra, S., Christiana, M. B. V., Palai, B. K., & Boopathi, S. (2023). Hospital Waste Management Using Internet of Things and Deep Learning: Enhanced Efficiency and Sustainability. In M. Arshad (Ed.), Advances in Bioinformatics and Biomedical Engineering. IGI Global. doi:10.4018/978-1-6684-6577-6.ch015

Vazquez-Canteli, J. R., Dey, S., Henze, G., & Nagy, Z. (2020). CityLearn: Standardizing research in multi-agent reinforcement learning for demand response and urban energy management. *arXiv Preprint arXiv:2012.10504*.

Venkateswaran, N., Vidhya, K., Ayyannan, M., Chavan, S. M., Sekar, K., & Boopathi, S. (2023). A Study on Smart Energy Management Framework Using Cloud Computing. In P. Ordóñez De Pablos & X. Zhang (Eds.), Practice, Progress, and Proficiency in Sustainability. IGI Global. doi:10.4018/978-1-6684-8634-4.ch009

Vennila, T., Karuna, M., Srivastava, B. K., Venugopal, J., Surakasi, R., & Sampath, B. (2022). New Strategies in Treatment and Enzymatic Processes: Ethanol Production From Sugarcane Bagasse. In Human Agro-Energy Optimization for Business and Industry (pp. 219–240). IGI Global.

Zekrifa, D. M. S., Kulkarni, M., Bhagyalakshmi, A., Devireddy, N., Gupta, S., & Boopathi, S. (2023). Integrating Machine Learning and AI for Improved Hydrological Modeling and Water Resource Management. In V. Shikuku (Ed.), Advances in Environmental Engineering and Green Technologies. IGI Global. doi:10.4018/978-1-6684-6791-6.ch003

Chapter 7

Smart Highways–Based Piezoelectric Vehicle Speed Sensor

Luay Y. Taha
Pennsylvania State University, Altoona, USA

ABSTRACT

In this chapter, the authors proposed a piezoelectric based vehicle speed sensor. The sensors produce pulses when stressed by a vehicle's front and back wheels, passed over them. Due to the vehicle physical dimension, the sensor pulse, due to the back wheel stressing, has some delay time as compared with the front wheel sensing pulse, due to the front wheel stressing. The time delay between these pulses is estimated using microcontroller program. Then, the program computes the vehicle speed from the distance to time delay ratio.

BACKGROUND

In recent years, there is an increasing demand to monitor the vehicle speed in smart highways and roads. Detection speed violation is essential to control the traffic. Also, type of vehicle is required to make a complete study about vehicles passing on that road.

Different state-of-the art sensor technologies were implemented to sense and detect the vehicle movement. The sensing technologies include the use of Anisotropic Magneto-Resistive (AMR) with piezoelectric sensors (Markevicius et al., 2019, 2020), image sensor (Lu et al., 2020), three point-laser sensors (Keipour et al., 2022), distributed fiber-optic acoustic sensor (Liu et al., 2018, 2019), three-

DOI: 10.4018/978-1-6684-9214-7.ch007

Copyright © 2024, IGI Global. Copying or distributing in print or electronic forms without written permission of IGI Global is prohibited.

axis digital magnetic sensors (Markevicius et al., 2018), wireless magnetic sensor network (Zhang et al., 2017), micro-electromechanical system (MEMS) SmatRock sensors (Zhang et al., 2021), camera (Shim et al., 2021), and GPS and MEMS gyro (Juang et al., 2015).

Table 1 illustrates these sensor technologies. The table also shows the speed estimation algorithms (SEA) and system principle. It also highlights some merits and challenges. The SEA includes adaptive signature cropping (ASC), pulse extraction, feature extraction, frame difference method (FDM), Image motion detection (IMD), laser change of Hight triggering (LCOHT), visual servoing controller (VSC), wavelet-denoising algorithm (WDA), dual-threshold algorithm (DTA), cross correlation (CC), sum of absolute differences (SAD), circular convolution (CCON), signals centres of mass (SCOM), dynamic time warping (DTW), Normalized cross-correlation (NCC), smoothed coherence transform (SCOT), phase transform (PHT), playback speed (PBS), and Kalman filtering. These algorithms can be classified as a time domain approaches, frequency domain approaches, or a combination between them. Most of these algorithms detect the vehicle, estimate the speed, and classify the vehicle type.

The comparison shows wide variations of the sensing technologies and speed estimation algorithms. Although Table 1 addresses differences between these technologies and estimation algorithms, the decision about the optimum system is not available due to the lack of large database needed in testing and comparison. The speed estimation is open research and currently under development.

The aim of this chapter is to propose a novel piezoelectric based vehicle speed sensor system. The system detects the moving vehicle in highways using piezoelectric array sensors placed in all lanes. The layout of the proposed array sensors is shown in Figure 1. These sensors generate pulses according to the piezoelectric generator effect. Then the pulses are processed to estimate the speed using speed estimation algorithm. A hardware and software testbench are developed to validate the research result and can become a seed to create future courses related to the power and energy area. A simulation is also provided in this chapter using MATLAB and SIMULINK. Also, experimental testing is conducted using our design prototype. Results have shown a considerable accuracy in sensing the vehicle speed when varied according to the highways speed limits, and when the input signal is subjected to a disturbance noise signal.

The motivation and contribution of this chapter is as follows:

1. Demonstrate the design of the proposed piezoelectric speed sensor and how it works.
2. Explain the sensor signal conditioning circuit needed to produce the sensor output pulses.

3. Explain the proposed speed estimation algorithm and tested it using simulation and practical data.
4. Explain and discus the results from simulation and experiments.
5. Show chapter conclusion based on the results.

Table 1. Comparison between different speed sensor systems used in smart highways. The abbreviations used in this table are explained in Table 2

Reference	Sensor Technology	SEA	System Principle	Merits	Challenges
Markevicius, 2019, 2020	AMR sensing with piezoelectric (PVDF) sensor as reference	ASC	ASC based speed estimation	2.6% average MAPE	computation complexity issue
Rajab, 2014	Piezoelectric sensor	Pulse extraction, feature extraction	Speed estimation after feature extraction and vehicle classification	98.9% accuracy	Complex hardware
Lu, 2020	CMOS image sensor with the resolution of 720 9 288, at 25 frames per second	FDM, IMD	finding a projection histogram then a group of key bins are selected to estimate the speed	0.3 km/h average error	the error of range from - 2 km/h to 2 km/h
Keipour, 2022	Three point-laser sensors	LCOHT, VSC	Detect vehicle using LCOHT, then estimate speed using VSC	Fast UAV landing on moving platforms	sun reflection causes 13.6% failures
Liu, 2018, 2019	distributed fiber-optic acoustic sensor	WDA, DTA	extracts features then DTA is applied for speed estimation	error is less than 5%,	detection requires improvement
Markevicius, 2018	Three-axis digital magnetic sensors using LIS3MDL and STM32F401RBT6	CC, SAD, CCON, SCOM	Change in magnetic field magnitude observed in two sensor nodes	Four algorithms were introduced	not always reliable when the sample size is small
Zhang, 2017	wireless magnetic sensor network using SX1278, and HMC5983	DTW	calculate the similarity between the signatures at a pair of sensors using DTW	accuracy is better than 98%.	not suitable for time distorted waveforms
Zhang, 2021	MEMS SmartRock sensors	NCC, SCOT, PHT	Capture the triaxial acceleration, rotation, and stress data, then estimate the speed	Three CC algorithms, were introduced	Low accuracy, 84.7%, tested up to 50 km/h

Table 1 continued

Reference	Sensor Technology	SEA	System Principle	Merits	Challenges
Shim, 2021	Dashcams	PBS estimator	detection phase then estimation phase that calibrates and measures the speed	99.99% accuracy	Non-real time speed estimation, after accidents
Juang, 2015	GPS, MEMS gyro	Kalman filter	Sensor fusion approach	Simple implementation, tuning, cost effective.	Issues at low speed estimation

Table 2. Abbreviations used in Table 1

Abbreviation	Name	Abbreviation	Name
AMR	anisotropic magneto-resistive	CCON	circular convolution
ASC	adaptive signature cropping	SCOM	signals centers of mass
MAPE	mean absolute percentage error	UAV	Unmanned Aerial Vehicle
FDM	frame difference method	DTW	dynamic time warping
IMD	Image motion detection	MEMS	micro-electromechanical system
LCOHT	laser change of hight triggering	NCC	Normalized cross-correlation
VSC	visual servoing controller	SCOT	smoothed coherence transform
WDA	wavelet-denoising algorithm	PHAT	phase transform
DTA	dual-threshold algorithm	PBS	playback speed
CC	Cross correlation	GPS	Global position system
SAD	sum of absolute differences		

PIEZOELECTRIC SPEED SENSOR

The aim of this section is to show a brief explanation about piezoelectric effects, piezoelectric materials, and their properties. Then, present the proposed sensor mathematical model. Simulation of the model and the speed estimation principle is also provided in this section.

Figure 1. Layout of array sensors used in the proposed piezoelectric speed sensor system

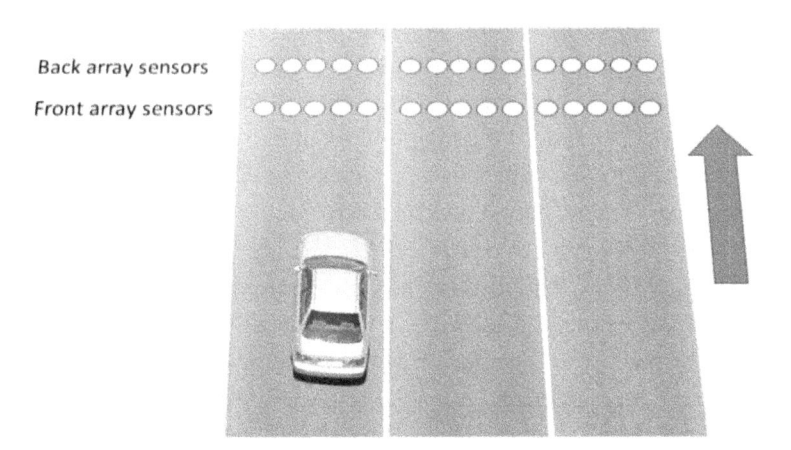

Piezoelectric Effects

The piezoelectric phenomenon occurs in certain materials with dipole moments resulting from non-centrosymmetric crystal structures (Taha, 2009). The direct piezoelectric effect occurs when a charge is generated due to a change in the dipole movement caused by the application of a mechanical stress to the crystal. The converse piezoelectric effect occurs when a strain is generated on the crystal by the application of an electric field (Piekarski, 2005). The root of the word "piezo" means "pressure" hence the original meaning of the word piezoelectricity implies "pressure electricity". The prefix "piezo" comes from the Greek work "piezein", meaning "to squeeze" or "to press" (Jaffe, 1958) and (Cady, 2018). As the piezoelectric materials have theses properties, they can be used as stress sensors, actuators, and energy converters.

Domains are the regions of local alignment of adjoining dipoles. The alignment gives a net dipole moment to the domain resulting in a net polarization. In an unpoled piezoelectric material, the polarizations among neighboring domain are randomly oriented and there is no resultant polarization. Thus, the material does not show piezoelectric effect. However, the domain can be aligned by a process called poling. Poling means the application of a strong electric field at a temperature slightly below the Curie point. After removing the electric field, most of the dipoles are frozen into a status of "near aligned". The material becomes anisotropic and so a polarization occurs (Ceramics, M. E., 2001). After poling, the material possesses piezoelectricity, as shown in the following examples.

Example 1: If an external force F produces compressive or tensile strain in a cylindrical piezoelectric material, the resulting change in dipole moment causes a voltage to appear between the electrodes. If the material is compressed so that it resumes its original form, i.e. before poling, the voltage will have the same polarity as the poling voltage, if it is stretched by a force F, the voltage across the electrodes will have opposite polarity to the poling voltage. This behaviour is denoted as *generator action*: the conversion of mechanical energy into electrical energy (Taha, 2009). Figure 2 illustrates the piezoelectric behaviour. All piezoelectric sensors use the generator action in their sensing strategies.

Example 2: Alternatively, the material has a reverse piezoelectric effect, called the *motor (or actuator) action*, when a mechanical strain is generated from an applied electric field. This action is not discussed in this chapter as we only focus on the sensor action.

Figure 2. Piezoelectric material behavior for different applied forces: (a) the material under no-load condition, (b) after applying a compressive force F, (c) after applying a stretched force F

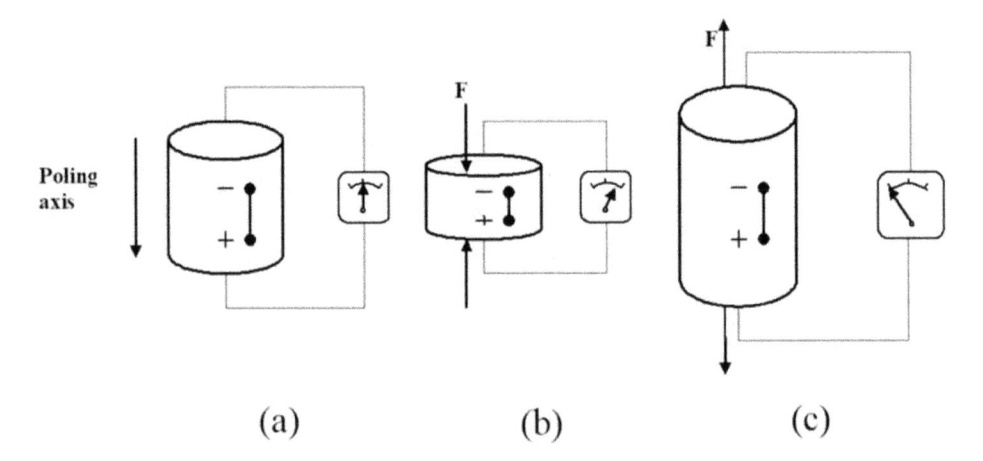

Piezoelectric Materials and Properties

Different types of piezoelectric materials are available by manufacturers, such as Lead Zirconate Titanate (PZT), Lead Lanthanum Zirconate Titanate (PLZT), Quartz (SiO_2), Polymers, Barium Titanate (BaTiO3), Lead Titanate (PbTiO3 and Zinc Oxide (ZNO). These piezoelectric materials are classified as piezopolymers (PVDF or PVF2), single crystal materials, piezoceramics, piezocomposites and piezofilms

(Taha, 2009). Several comparisons between piezoelectric materials are possible depending upon the behaviours and applications. Since this chapter is focused on the piezoelectric sensing, only the three important constants are used for comparisons: the charge (or strain) constant (d_{11}, d_{31}, d_{33}), the relative permittivity ($\varepsilon_{11}^T, \varepsilon_{31}^T, \varepsilon_{33}^T$) and the coupling factor (k_{11}, k_{31}, k_{33}). The relative permittivity is the ratio of absolute permittivity to the permittivity of free space (8.85 x 10^{-12} F/m). All these constants are generally given two subscript indices (such as 11, 31, 33) which refer to the direction of the two related quantities (e.g. stress and strain for elasticity, displacement, and electric field for permittivity). The reason for that is the piezoelectric materials are anisotropic, thus, their physical constants (such as elasticity, permittivity, etc.) are tensor quantities and relate to both direction of the applied stress, electric field, etc. and to directions perpendicular to these. In addition to the two subscript indices, a superscript index is also used to indicate a quantity that is kept constant (such as a constant stress T used in $\varepsilon_{11}^T, \varepsilon_{31}^T, \varepsilon_{33}^T$). The direction of positive polarization is usually chosen to coincide with Z-axis of a rectangular system of crystallographic axes X, Y, Z. If directions of X, Y, and Z are represented by 1, 2 and 3 respectively and the shear about these axes by 4, 5, and 6 respectively. For example:

1. μ_{11}^T refers to the relative permittivity for the dielectric displacement and electric field in direction 1 under conditions of constant stress T.
2. For generator (sensor), d_{33} refers to the induced polarization in direction 3 per unit stress applied in direction 3. Alternatively, for motors, it is the induced strain in direction 3 per unit electric field applied in direction 3.
3. For generator (sensor), d_{31} refers to the induced polarization in direction 3 per unit stress applied in direction 1. Alternatively, for motors, it is the induced strain in direction 1 per unit electric field applied in direction 3.

Table 3 illustrates a comparison between piezoelectric materials based on these factors (Lam, 2006). It shows that the PZT material scores the best properties due to its highest strain constant (d_{33}), highest relative permittivity ε, and highest coupling factor k. Thus, in this chapter, the PZT material is selected during the design, experiments, and simulations.

Piezoelectric Sensor Mathematical Model

Assuming that a force F is applied on a piezoelectric material in the 33 mode. To find the voltage on the top and bottom faces of the material, a transformer model can be used (Platt et al., 2005). Figure 3 illustrates the piezoelectric sensor transformer model.

Table 3. Comparison between different piezoelectric materials (Lam, 2006)

Material	Strain Coefficient (10⁻¹² C/N)	Relative Permittivity ε	Coupling Coefficient k²%
BaTiO$_3$	82 - 145 (d$_{33}$)	625 - 1350	39 - 46
LiNbO$_3$	19.2 (d$_{33}$)	44	17.2
LiTaO$_3$	8.0 (d$_{33}$)	41	4.7
PVDF	-12.0 (d$_{31}$)	13	0.18
SiO$_2$	2.3 (d$_{11}$)	4.5	0.11
PZT	240 - 550 (d$_{33}$)	1100 - 3200	55 - 73
ZnO	10 – 12 (d$_{33}$)	8.5	7.5

Table 4 illustrates all symbols used in Figure 3.

Figure 3. The piezoelectric generator (sensor) transformer model

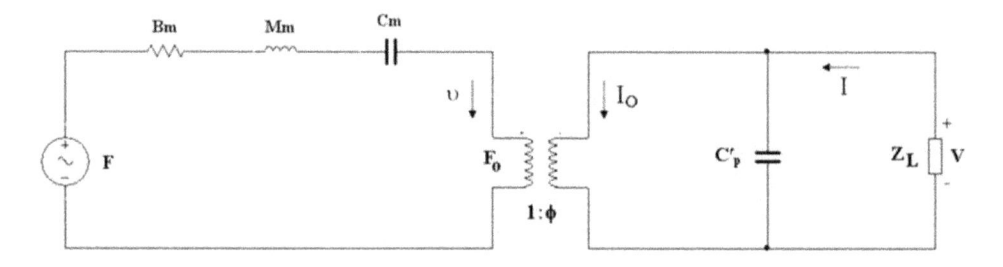

Table 4. Illustrations about symbols used in Figure 3

Symbol	Definition	Unit
F	input force	(N)
F$_o$	net force transferred into electrical voltage by piezoelectric effect, and is opposite to the applied force	(N)
C$_m$	piezoelectric short circuit compliance = $s_{33}^E \dfrac{t}{A}$	(m/N)
s_{33}^T	mechanical compliance	(m²/N)
C'$_p$	blocked capacitance = $C_p - \dfrac{d_{33}^2}{C_m}$	(C/V)

Smart Highways-Based Piezoelectric Vehicle Speed Sensor

Table 4 continued

Symbol	Definition	Unit
C_p	free capacitance = $\varepsilon_{33}^T \dfrac{A}{t}$	(C/V)
υ	velocity of the piezoelectric element in the z-axis direction = $\dfrac{dx}{dt}$	(m/s)
x	displacement of the moving piezoelectric element in the z-axis direction	(m)
Φ	mechanical to electrical conversion ratio: $\dfrac{V}{F_o} = \dfrac{1}{I_o} = \dfrac{C_m}{d_{33}}$	(V/N)
M_m	mass of piezoelectric element	(kg)
B_m	damping of the piezoelectric element	(N.s/m)
t	thickness of the piezoelectric element	(m)
A	cross sectional area of the piezoelectric element	(m²)
I_o	current passing through the piezoelectric terminals	(A)
I	output current passing through connected load impedance	(A)
Z_L	load impedance	(Ω)
V	generated output voltage = $\Phi\upsilon$	(V)

Figure 3 shows that the piezoelectric sensor output voltage depends upon the electro-mechanical parameters and the external load impedance Z_L. For resistive load, which is mostly used in applications, it can be proven that the sensor transfer function $\dfrac{V(s)}{F(s)}$ is given by (Taha, 2009):

$$\frac{V(s)}{F(s)} = \frac{sd_{33}R_L}{D(s)}, \tag{1}$$

where:

$$D(s) = b_3 s^3 + b_2 s^2 + b_1 s + b$$

$$b_3 = M_m R_L \left(C_p C_m - d_{33}^2\right)$$

$$b_2 = M_m C_m + B_m R_L \left(C_p C_m - d_{33}^2 \right)$$

$$b_1 = B_m C_m + R_L C_p$$

$$b_0 = 1$$

Piezoelectric Sensor Output Simulation

To visualize the sensor output voltage when a force is applied, a MATLAB simulation is applied to (1). The following example shows a typical sensor simulation.

Example 3: A pulse force F of 1kN max is applied to a disk piezoelectric material type PZT 5. The material has a thickness and diameter of 4.75 mm and 10 mm, respectively. The pulse width is equal to 9 ms with a rise time t_r and a fall time t_f equal to 1.125 ms. The PZT 5 constants are as follows: d_{33}= 240×10^{-12} C/N, $s_{33}^T = 16.2 \times 10^{-12}$ m²/N, $\epsilon_{33}^T = 1100$, B_m= 5×10^{-8} N.s/m. The material is connected to a 800kΩ load resistance (R_L). A MATLAB script is written to simulate the piezoelectric material based on (1) and the above parameters. Result from the simulation shows that the transfer function can be reduced to:

$$\frac{V(s)}{F(s)} = \frac{8.233 \times 10^{11} s}{s^3 + 1.223 \times 10^4 s^2 + 5.524 \times 10^{11} s + 4.288 \times 10^{15}} \tag{2}$$

Thus, the piezoelectric sensor can be modelled as a third order system if a resistive load is connected. However, the higher order terms (s^3 and s^2) have negligible effects on the transfer function. Thus, the sensor transfer function can be reduced to a first order system given by:

$$\frac{V(s)}{F(s)} = \frac{1.49s}{s + 7762} \tag{3}$$

Figure 4 illustrates the visualization of the applied force and the output voltage based on (2). Results show that a narrow pulse output voltage is generated during the rise and the fall time of the applied force. Results also shows that the peak voltage is high. This may not be suitable in sensor applications that requires low output voltage (such as +5V). Thus, a charge amplifier may be connected to reduce the output voltage (Bartolome, 2010). Alternatively, the load resistance may be adjusted

Smart Highways-Based Piezoelectric Vehicle Speed Sensor

to decrease the output voltage. Figure 5 illustrates the visualization of the applied force and the output voltage using the same parameters in example 3, except that the load resistance R_L is changed to 20kΩ. Results show an acceptable peak voltage.

Due to the change of R_L, the transfer function is changed to:

$$\frac{V(s)}{F(s)} = \frac{8.233 \times 10^{11} s}{s^3 + 4.89 \times 10^5 s^2 + 5.524 \times 10^{11} s + 1.715 \times 10^{17}} \tag{4}$$

Eliminating the higher order terms (s^3 and s^2), Equation (4) can be simplified to:

$$\frac{V(s)}{F(s)} = \frac{1.49s}{s + 3.105 \times 10^5} \tag{5}$$

From (1), (3) and (5), the general form of the transfer function can be written as:

$$\frac{V(s)}{F(s)} = \frac{Ks}{s + p} \tag{6}$$

where: $K = \dfrac{d_{33} R_L}{b_1} = \dfrac{d_{33} R_L}{B_m C_m + R_L C_p}$ and $p = \dfrac{1}{b_1} = \dfrac{1}{B_m C_m + R_L C_p}$

From (6), the output voltage ramp response $v(t)$, assuming $F(t) = \dfrac{F_{max}}{t_r} t, t \geq 0$, is given by:

$$v(t) = \frac{K F_{max}}{t_r p} \left(1 - e^{-at}\right) \tag{7}$$

In example 3, the value of the rise time t_r is equal to 1.125 ms. The maximum ramp input $F_{max} = 1kN$. Then, the maximum value of $v(t)$ is given by:

$$v(t)_{max} = \frac{K F_{max}}{t_r p} \tag{8}$$

Figure 4. Visualization of the applied force (a) and the output voltage (b) in example 3, using $R_L = 800$ kΩ

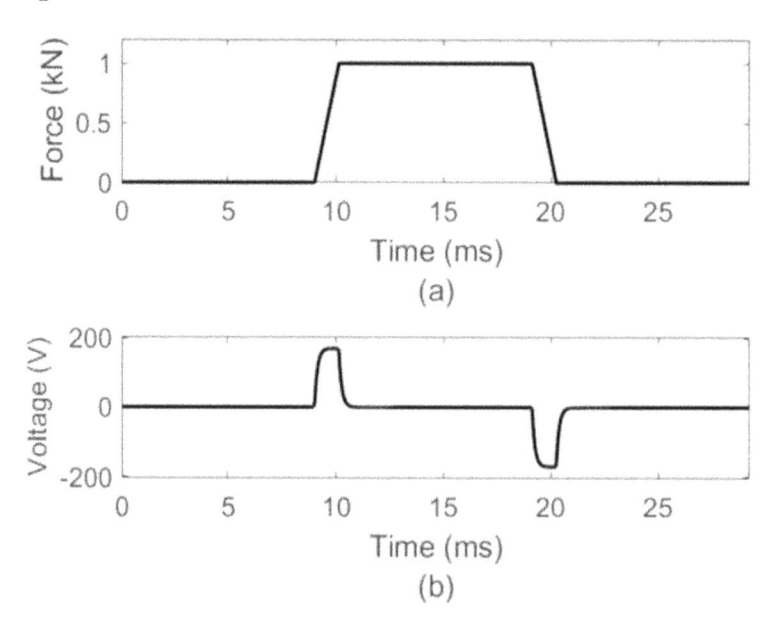

Figure 5. Visualization of the applied force (a) and the output voltage (b) in example 3, using $R_L = 20$ kΩ

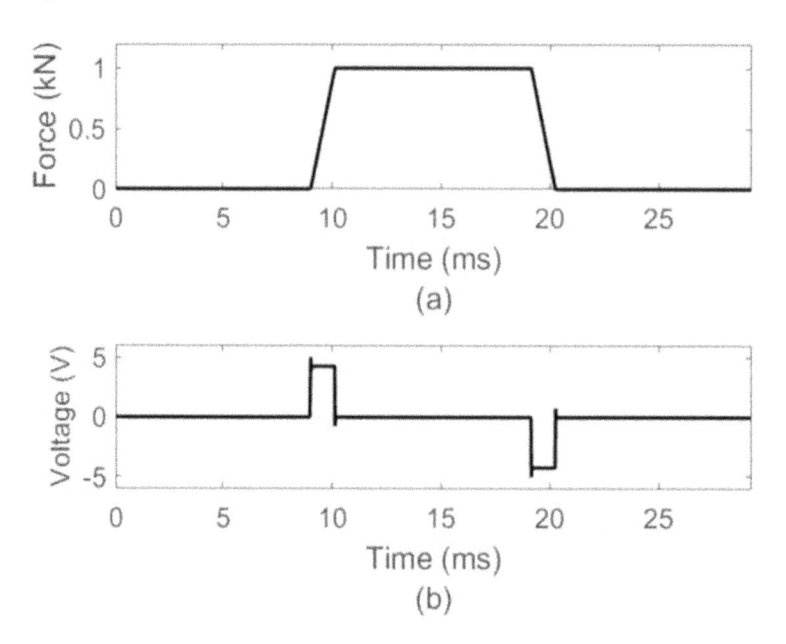

Smart Highways-Based Piezoelectric Vehicle Speed Sensor

Substituting the values of R_L in (8), the maximum value of $v(t)$ is equal to 170.64 V for $R_L = 800k\Omega$, and 4.26 V for $R_L = 20k\Omega$ These values are identical to the values in Figure 4 and Figure 5.

Using (7) and (8), the simulation is repeated by varying R_L and recording the maximum output voltage. Figure 6 illustrates the visualization of $v(t)_{max}$ when R_L is varied from 100Ω–$1M\Omega$. The value of R_L must not exceed $22.8k\Omega$ to get acceptable range of $v(t)_{max}$ below +5V. However, R_L may change if the sensor is used in applications that requires more than +5V.

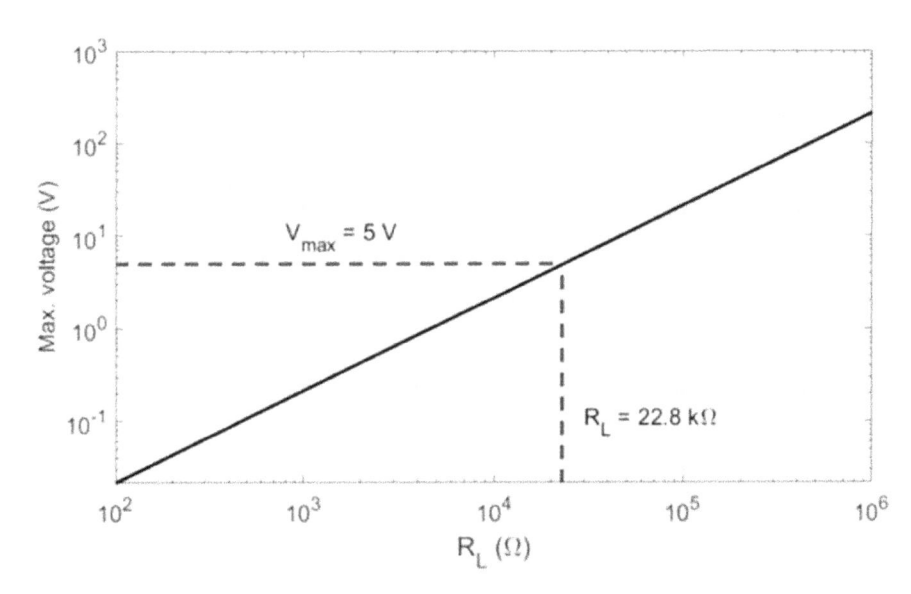

Figure 6. Visualization of $v(t)_{max}$ when R_L is varied from 100 Ω-1MΩ

Speed Estimation Principle

From the results in example 3, the piezoelectric sensor can be applied to estimate the vehicle speed from the knowledge of the time difference between the peaks of two consecutive positive pulses. The following example shows the speed sensing procedure.

Example 4: Figure 7 illustrates two disk *piezoelectric materials* (PEM1 and PEM2) embedded in the road asphalt. Type of PEM's is the same as in example 3. The distance L (m) between the two materials is fixed and is equal to 1 m. The figure also illustrates one vehicle wheel, moving from left to right at a speed of $u(m/s)$, and stressing the two PEM's. The wheel generates a pulse force F of 1 kN on the disk piezoelectric material type PZT 5.

According to example 3, two pulses (one positive and one negative) will be generated from PEM1, followed by another two pulses (one positive and one negative) from PEM2. Since the negative pulses are not important in sensing, it can be removed by adding a diode at the output of PEM's. Figure 8(a) shows the applied pulse force F. Figure 8(b) illustrates the positive and negative sensing pulses. Figure 8(c) illustrates the positive sensing pulses, after removing the negative pulses using diode rectifier. The vehicle speed is then estimated from the knowledge of the time T between the peaks of the two positive pulses, as follows (Underwood et al., 2023) and (Rall et al., 2023):

$$u = \frac{L}{T} \qquad (9)$$

In this section, we assume, for illustration purposes, that T is estimated from graph, However, the general case is to estimate T using time delay estimation algorithm. This will be discussed later in this chapter.

Figure 7. Visualization of one vehicle wheel stressing two piezoelectric materials (PEM's) embedded in the road asphalt

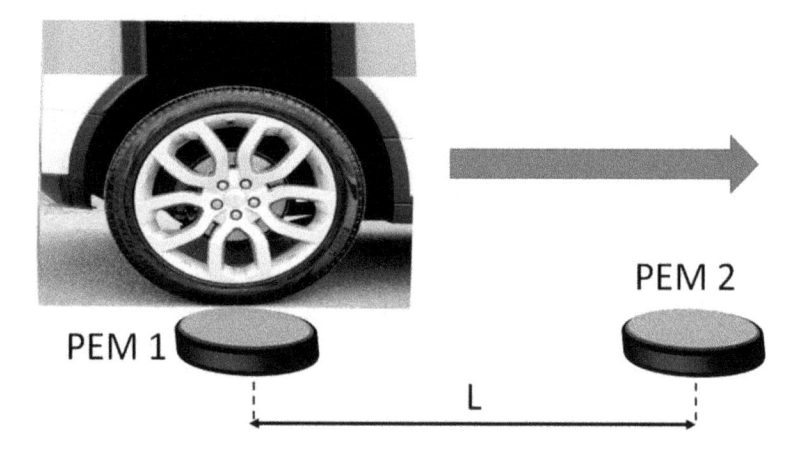

Smart Highways-Based Piezoelectric Vehicle Speed Sensor

A MATLAB script is written to estimate and plot the speed from the knowledge of T. Figure 9 illustrates the estimated speed versus the actual speed when varied from 20 km/h to 140 km/h. The estimation is under the assumption that the sensor output is noise free. The mean squared error (MSE) is equal to 0.053993. Repeating the simulating by adding a noise, so that the signal to noise ratio (SNR) becomes 10 dB, results in increasing the MSE to 0.0634, as shown in Figure 10. Results indicate an outstanding performance in estimating the speed with a negligible MSE error.

Figure 8. Typical sensing pulses in Example 4: (a) the applied force F, (b) the sensor output V1 with positive and negative pulses, (c) the sensor output V2 after removing the negative pulses

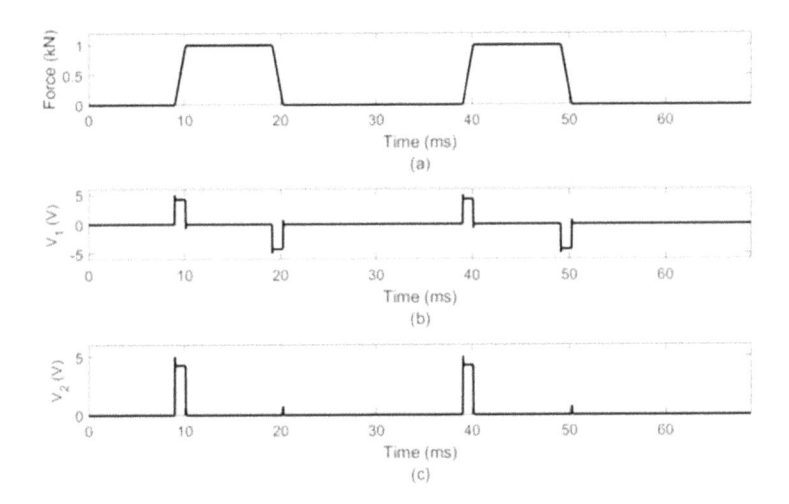

PIEZOELECTRIC SPEED SENSOR SYSTEM

In this section, the hardware and software of the piezoelectric sensor system is discussed and tested by simulation and practical experiment. The hardware part includes the sensor construction and the electrical connections. The software includes the method of estimating the vehicle speed using speed estimation algorithm.

Piezoelectric Speed Sensor Construction

The piezoelectric speed sensor system is constructed from three main parts: piezoelectric primary sensor (PPS), signal conditioning unit (SCU), and speed estimation unit (SEU), as shown in Figure 11.

Figure 9. Typical estimated speed vs. actual speed in Example 4, assuming a noise free sensing signal

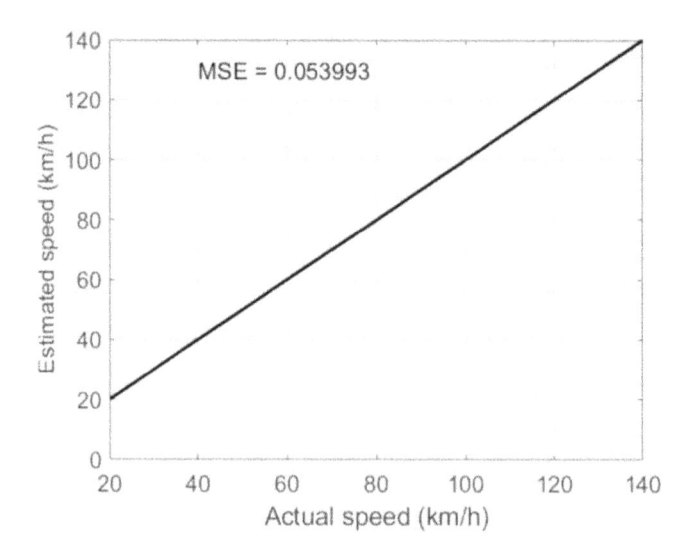

Figure 10. Typical estimated speed vs. actual speed in Example 4. assuming that the sensing signal contains noise of level 10 dB

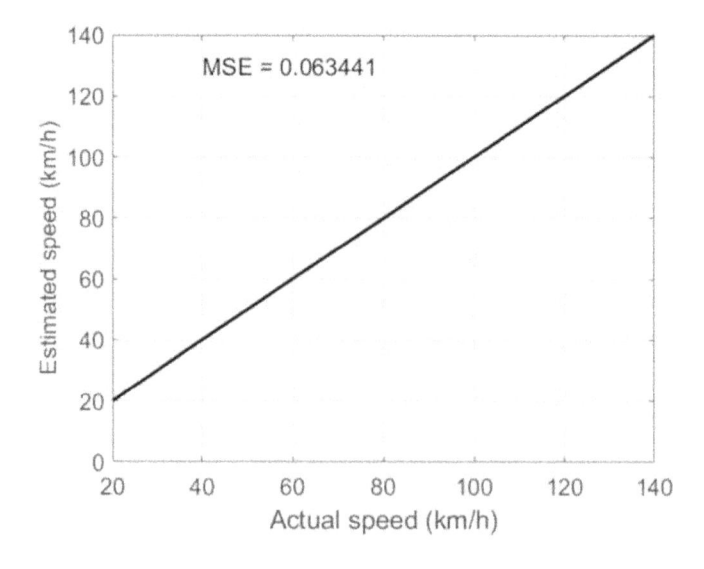

The PPS is constructed from PZT piezoelectric stacks, stacks housing unit, top and base plates, support plate, and spacer. Figure 12 illustrates the schematic diagram of the constructed PPS.

Smart Highways-Based Piezoelectric Vehicle Speed Sensor

Figure 11. Block diagram of the piezoelectric speed sensor system

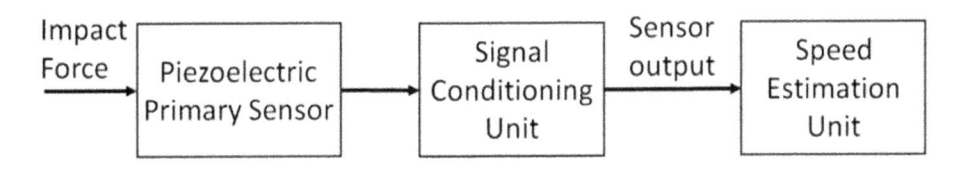

Figure 12. Schematic diagram of the constructed PPS

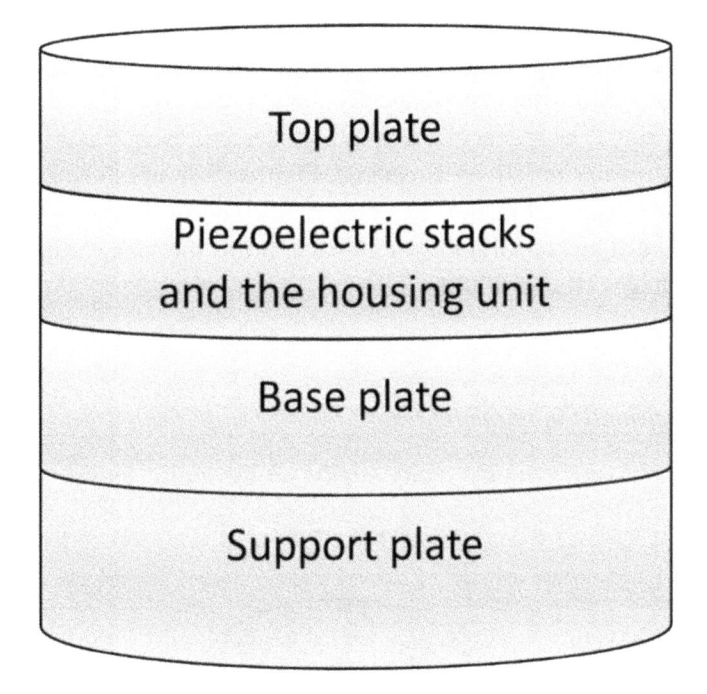

Each PZT stack is constructed from three layers of PZT-53HD materials (He-Shuai, 2018), electrically connected in parallel. Figure 13 illustrates the parallel connection of the three layers in the stack. This connection has the merit of producing low terminal voltage, as compared with higher voltages produced from series connection. Each stack has a diameter D of 10 mm and a thickness h of 14.25 mm (4.75 mm per layer). The stack is working in the 33 mode (z-axis). Applying a force on the top plate results in a change of the thickness by Δh. This change results in a voltage developed on the top and bottom face of the stack according to the piezoelectric direct effect (sensor) explained before. To improve the piezoelectric sensing, five of these stacks, electrically connected in parallel, as shown in Figure 14, are housed in a polycarbonate unit with a 1018 AISI low carbon steel base plate. The stacks themselves are held in place using a small 3D printed stack holder made of PLA plastic. This maximizes the stress distribution of the vehicle through the asphalt to the piezoelectric. The top and base Plates are both made from polycarbonate plastic. This material has many advantages such as high resistance to impact force, high electrical insulator capability, easy machinability, and high stiffness. Finally, the support plate is made from AISI 1018 low-carbon steel. This type of steel offers a good balance of toughness, strength, and ductility. The merits of the support plate are to add rigid support to the base plate and enclose the wires and fasteners. The support plate is also CNC machined, so machinability is crucial. The spacer is made of neoprene rubber and selected due to its high elasticity, low stiffness, high fatigue life, and water resistance. This is desired to allow the top plate to "float" and return to its original position without permanently absorbing energy. The spacer functions like a return spring. The spacer also acts as a gasket to prevent water from entering. Therefore, high-water resistivity is desired. Figure 15 shows the constructed PPS using SOLIDWORKS 3D model (Sherren, 2022). The stacks output terminals are connected to the signal conditioning unit that controls the sensor output voltage to avoid high voltages (Karki, 2000).

Figure 13. Construction of the one piezoelectric stack using three PZT-5H materials connected electrically in parallel

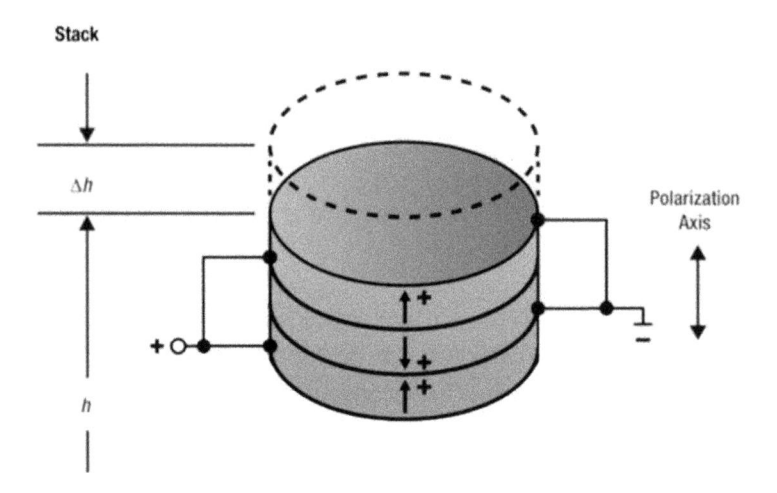

The SCU is connected to the PPS using two external wires as shown in Figure 16. The simplest SCU circuit is to use an external resistor to decrease the sensor terminal voltage. However, this may result is some electrical loading issues between the sensor and the connected controller. A better way is possible using charge amplifier.

Figure 14. Electrical connection of the five piezoelectric stacks. The R and C refer to the stack top and bottom faces, respectively.

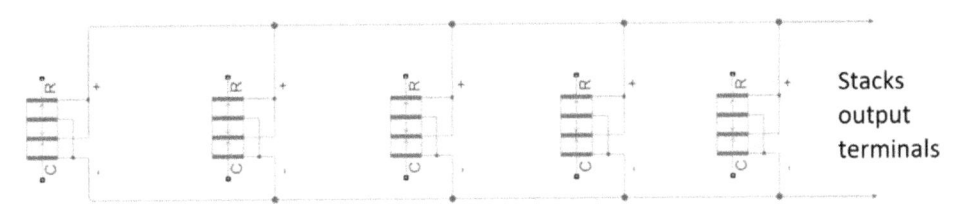

Figure 15. The constructed PPS using SOLIDWORKS 3D model. The top plate has a diameter of 10 cm.

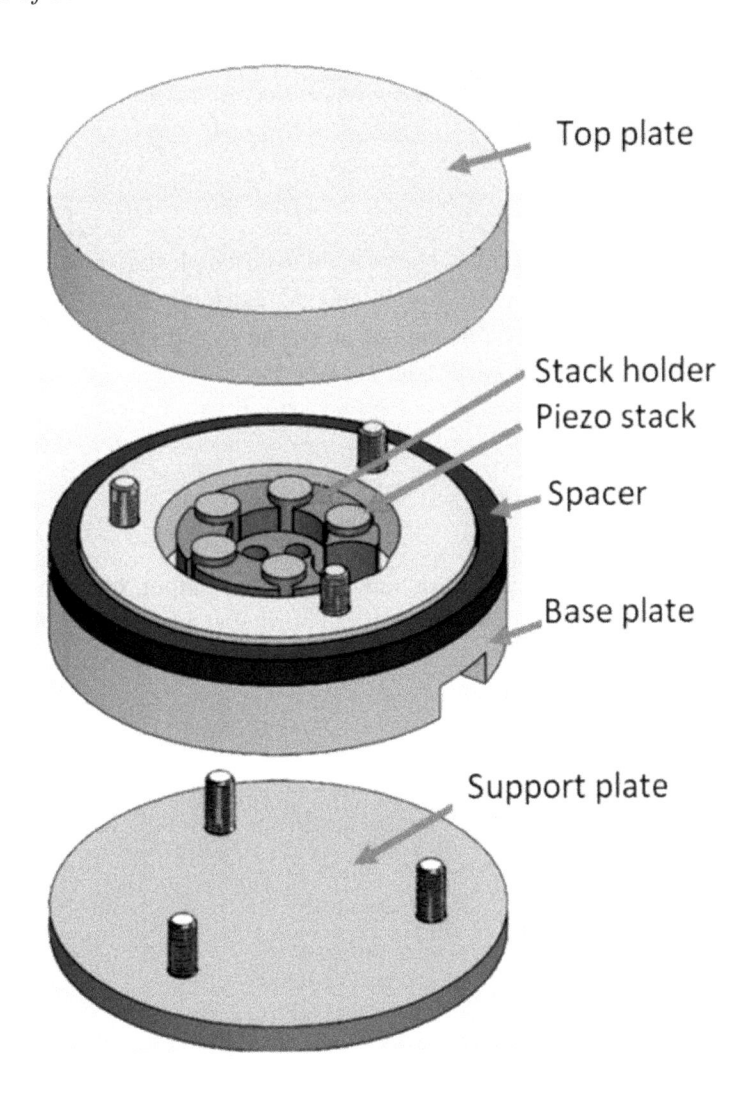

Figure 16. Connection the PPS to the SCU

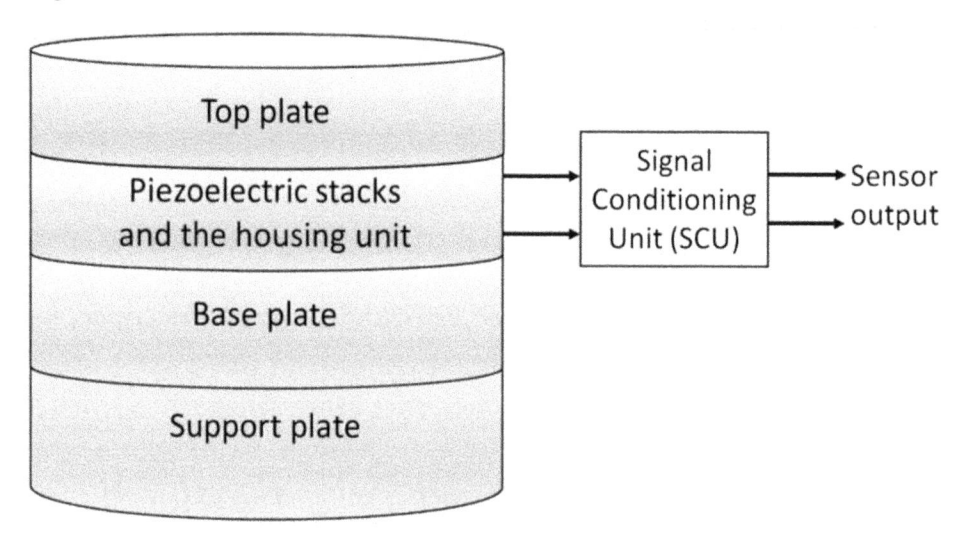

The charge amplifier circuit is constructed from a high slew-rate OP-AMP using two feedback branches: the resistor R_f and the capacitor C_f, as shown in Figure 17. The C_c is the interface cable capacitance. The R_i is the electrostatic discharge protection resistor, The Op-AMP gain is given by:

$$Gain = \frac{V_o}{qb} = -\frac{1}{C_f} \text{ (V/C)} \tag{10}$$

where V_o is the charge amplifier output, i.e., sensor output, qb is the total stacks charges due to the applied force. The typical normalized Bode plot of the amplifier is shown in Figure 18. The critical frequencies, f_L and f_H, are expressed by (Karki, 2000):

$$f_L = \frac{1}{2\pi R_f C_f} \tag{11}$$

$$f_H = \frac{1}{2\pi R_i (C_c + C_p)} \tag{12}$$

where C_p is the total stacks capacitance.

Figure 17. Charge amplifier circuit used in sensor design

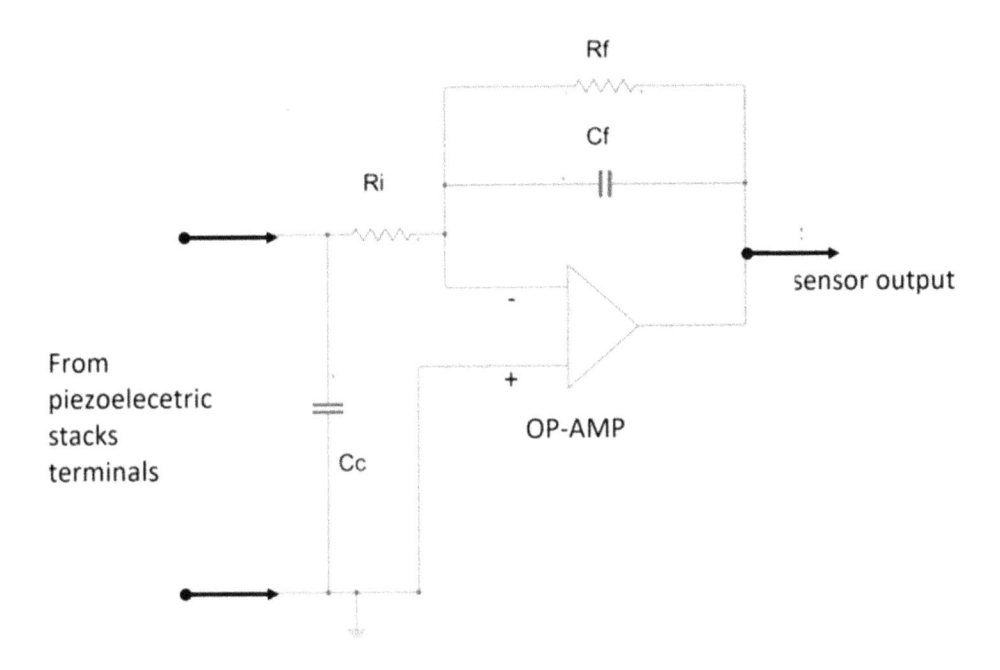

Figure 18. Normalized Bode plot of the charge amplifier

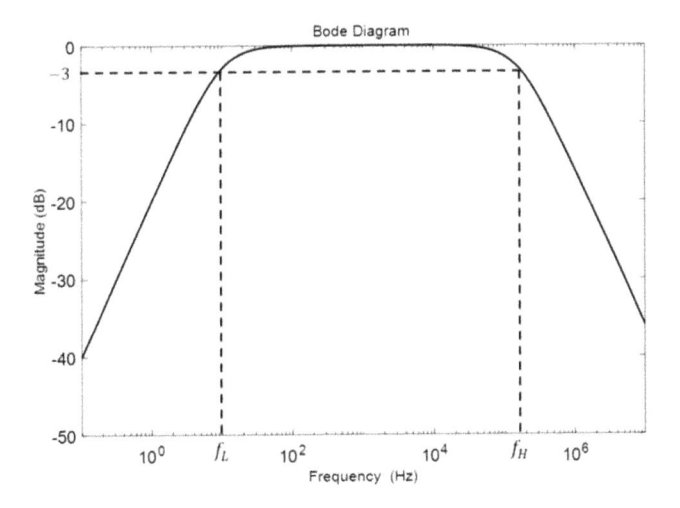

Piezoelectric Speed Sensor Simulation

The designed sensor shown in Figure 11 to Figure 17 is simulated using MATLAB/SIMULINK. Figure 19 illustrates the simulation circuit. The simulation has similar procedure as explained in *Example 4* and the speed is estimated by computing T, as illustrated in Figure 7. However, several modifications are carried out, as follows:

The pulse generator is used to simulate a vehicle pressing the designed sensor. Since the top plate has a diameter of 30 cm, then the generator pulse width can be adjusted as follows:

$$Pulse\ width = \frac{30 \times 10^{-2}}{Vehicle\ speed\ (m/s)} \qquad (13)$$

The pulse generator period is computed using (9), as explained in example 4. The electrical and mechanical connections are shown in Figure 19. The Asphalt layer is modelled using Burger's asphalt model (Lewandowski, 2013). Assuming that the Asphalt thickness is 5 cm, the model parameters ($K1$, $B1$, $K2$, $B2$) are illustrated in Table 5. The table also illustrates the charge amplifier parameters. Figure 20 and Figure 21 illustrate the sensor output pulses if the vehicle speeds are 120 km/h and 60 km/h, respectively. The generator pulse width varies from 9 ms to 18 ms, according to (13). However, the amplitude of the sensor output is kept constant at 5 V, regardless of the vehicle speed. The simulation is repeated for other values of vehicle speed (from 20 km/h to 140 km/h. The amplitude of the sensor output is also found constant as expected from Figure 20 and Figure 21. Figure 22 illustrates a plot of the amplitude of the sensor output versus the vehicle speed, assuming a constant vehicle weight of 40,000 kg. Figure 23 illustrates the effect of varying the vehicle weight on the amplitude of the sensor voltage, assuming a vehicle weight and 40,000 kg. Results indicates linear relationship. The sensitivity is equal to 5V/40,000*kg*, or 5V/392*kN*.

The circuit in Figure 19 is also simulated to plot the estimated speed vs. actual speed. This can be done by varying the vehicle speed, running simulation, estimating the time delay between the two output pulses, then computing the speed from (9). Figure 24 and Figure 25 illustrate the results for of estimating the speed, considering a noise free sensor output, and with a noise level equals to 10 dB. Results shows very acceptable estimation since the MSE is low in both cases.

Smart Highways-Based Piezoelectric Vehicle Speed Sensor

Figure 19. Piezoelectric speed sensor simulation circuit

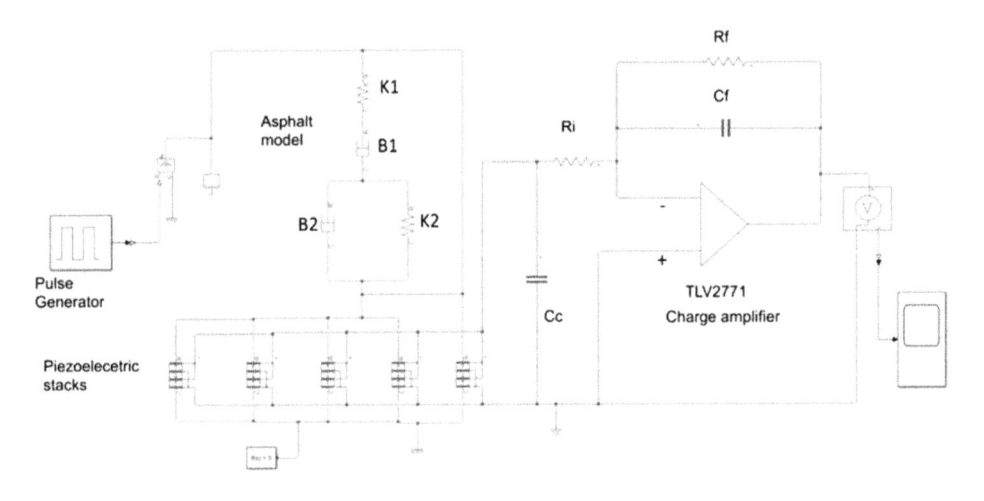

Table 5. Simulation parameters used to test the designed sensor

Asphalt Model Parameters				Charge Amplifier Parameters			
K1	B1	K2	B2	Cc	R_i	C_f	R_f
(N/m)	(N/m/s)	(N/m)	(N/m/s)	(pF)	(Ω)	(μF)	(MΩ)
12.35	39.25	308.9	9.3	10	10	6210	20

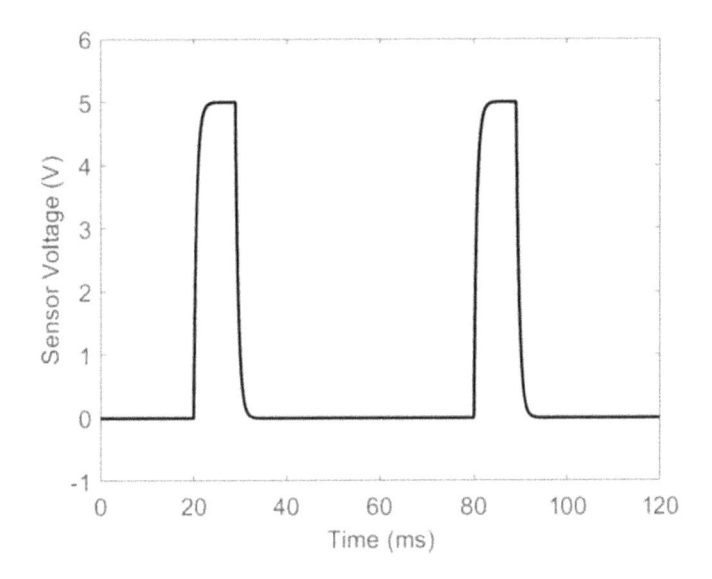

Figure 20. Sensor output pulse. The pulse width is 9ms, assuming vehicle speed of 120 km/h, and vehicle weight is 40,000 kg.

Figure 21. Sensor output pulse. The pulse width is 18ms, assuming vehicle speed of 60 km/h, and vehicle weight is 40,000 kg.

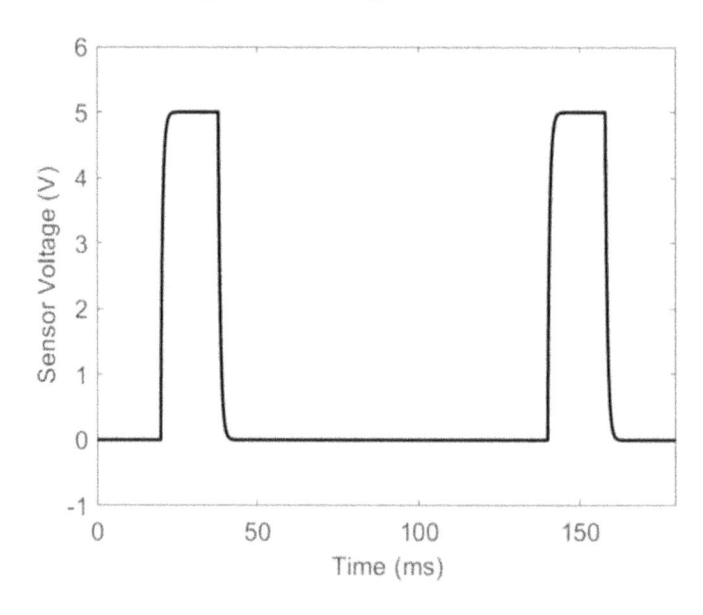

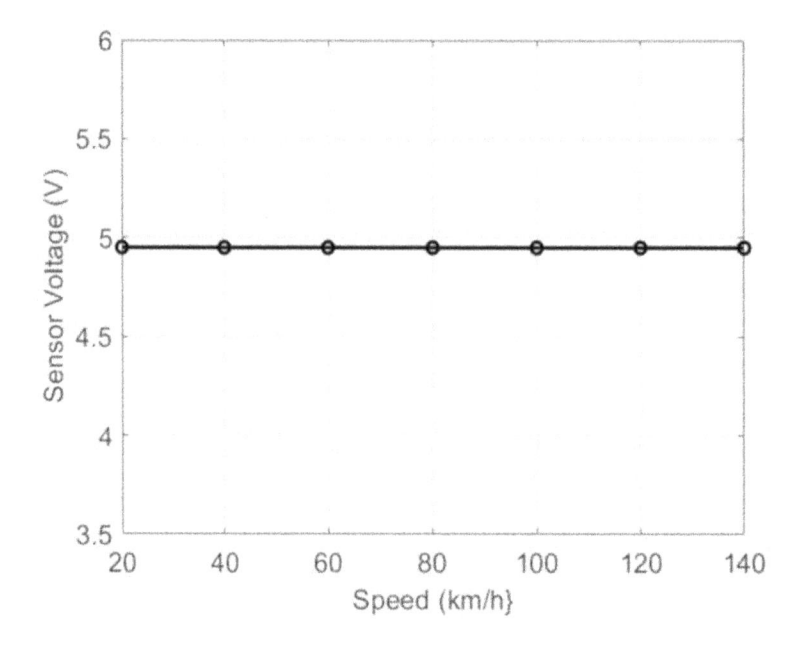

Figure 22. Effect of varying the vehicle speed on the amplitude of the sensor voltage, assuming a vehicle weight of 40,000 kg

Smart Highways-Based Piezoelectric Vehicle Speed Sensor

Figure 23. Effect of varying the vehicle weight on the amplitude of the sensor voltage V, assuming a vehicle speed of 120 km/h

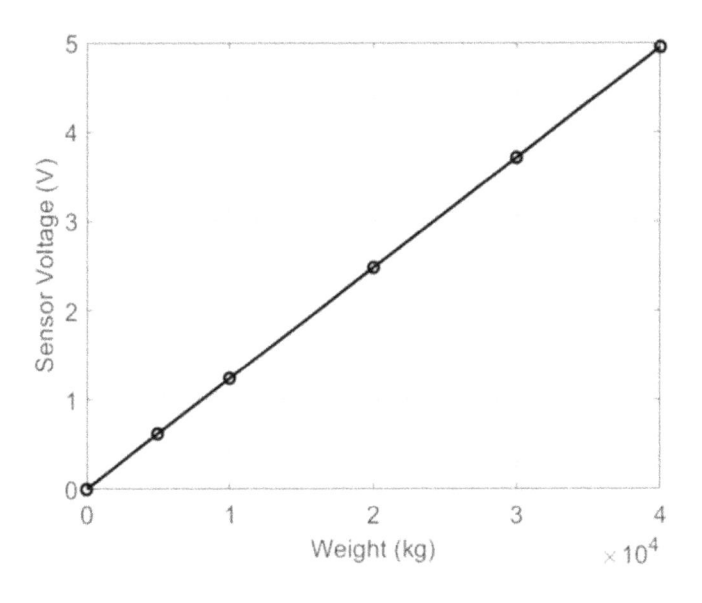

Figure 24. Estimated speed vs. actual speed using Simulink circuit shown in Figure 19, assuming noise free condition

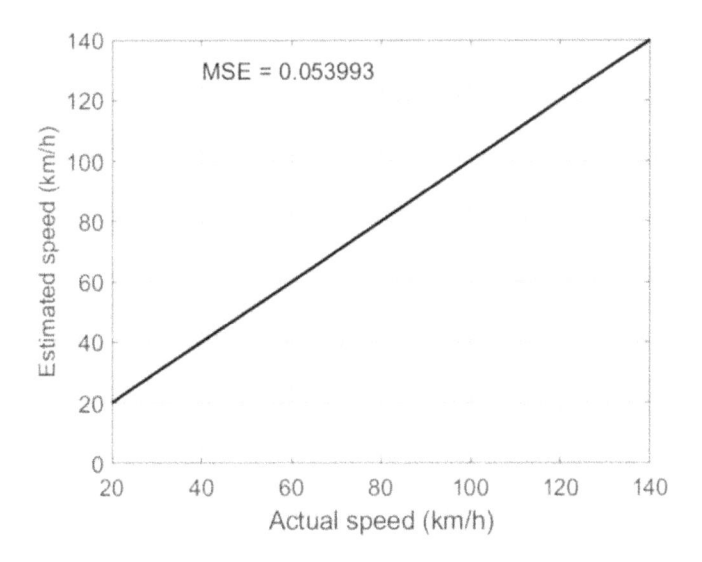

Figure 25. Estimated speed vs. actual speed using Simulink circuit shown in Figure 19, assuming 10 dB noise level

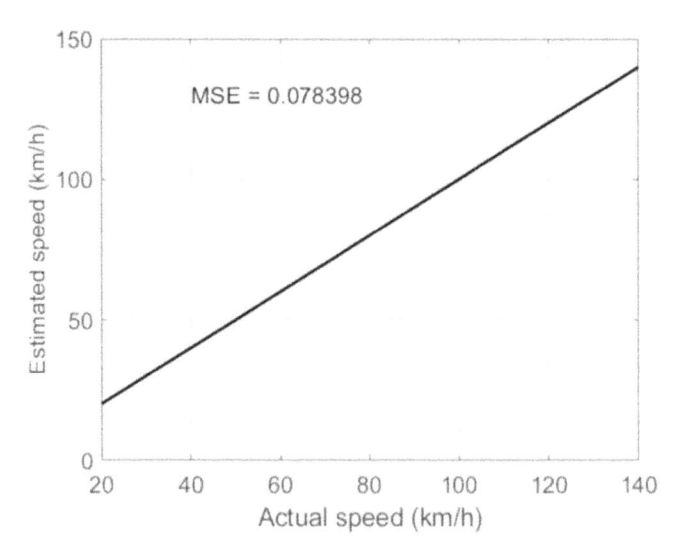

Piezoelectric Speed Sensor Experimental Testing

The designed sensor shown in Figure 11 to Figure 17 is also tested using experiment. Figure 26 illustrates the block diagram of the testing system. A variable impact force is generated (up to 1000 N) using free fall masses. The sensor output is connected to a digital storage oscilloscope SDS1000U-X having a high-speed data acquisition. The real-time sampling rate equals to up to 1 GSa/s. The captured sensor output pulses, after applying the force, are saved then transferred to the host computer to estimate the speed using MATLAB script. Both parallel resistor and charge amplifier are used in this experiment to investigate their effectiveness in producing good quality sensing pulses.

Figure 27 illustrates the study the effect of varying the load resistor R_L on the sensor output V_o, assuming an impact force of 980 N. The results show that the pulse amplitude is increased with the increase of R_L. However, the pulses are affected by noise and require further processing to remove the noise.

Figure 28 illustrates the effect of varying the impact force on the sensor output, using charge amplifier circuit in Figure 21. The results show that the pulse amplitude is increased with the increase in the impact force. Also, the sensor output pulses have good shape with negligible noise, as compared with load resistance method shown in Figure 27. Thus, in the following tests and analysis, only the charge amplifier method is used.

Smart Highways-Based Piezoelectric Vehicle Speed Sensor

The amplitude of the pulses in Figure 28 are utilized to estimate the sensitivity of the proposed sensor. Figure 29 show plots of the pulse amplitude V and the predicated pulse amplitude V^A versus the Force. Linear regression technique (Weisberg, 2005) is applied to extract the linear regression model of the V versus Force plot. Then, the model is applied to predict the values of V. The linear regression model is then applied to predict V for wide range of applied forces, such as from 0 N to 392 KN. This range covers most of the vehicle masses from 0 kg to 40,000 kg. Figure 30 shows a plot of V^A versus Force, using both simulation results, in Figure 23, and the results from linear regression model in Figure 29. The y axis in Figure 30 is a rescaled version of the y axis of Figure 29. This is required to have a common maximum voltage of 5V. The x-axis in Figure 23 is rescaled by multiplying by 9.8 then data is plotted in Figure 30. This is required to convert the x-axis unit from kg to N. Results indicate strong similarity between the two plots, with negligible MSE. The estimated sensitivity is equal to 5 V/40,000 kg, or 5 V/392 kN.

Figure 26. Block diagram of the testing system

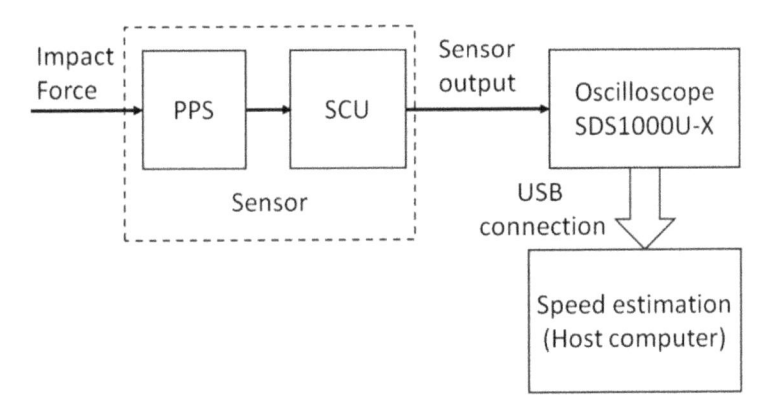

Speed Estimation

The experimental data obtained from Figure 28 are used to estimate the vehicle speed. The senor pulse corresponds to the impact force 588 N is selected in this experiment. Other pulses are also possible. A MATLAB script is used to generate a periodic sensor pulse, denoted by V_{op}, according to the selected speed, by rescaling the x-axis so that it complies with the actual sensor pulse width in equation (13). The sensor pulse amplitude is kept constant during the experiment. This is concluded from the previous experiment; results are illustrated in Figure 22. The resultant periodic signals V_{op} are illustrated in Figure 31, assuming the speed is changed from 10 km/h to 140 km/h. In each plot in Figure 31, two pulses are shown to represent its periodic nature. These pulses do not contain additive noise signal which exist normally in practice. Thus, the MATLAB script adds noise signal, having a known signal-to-noise ratio (SNR), prior to running the speed estimator algorithm. Figure 32 illustrates typical V_{op} pulses that contain noise. The speed u is computed according to (9) if a true time delay T between the two pulses in V_{op} is estimated.

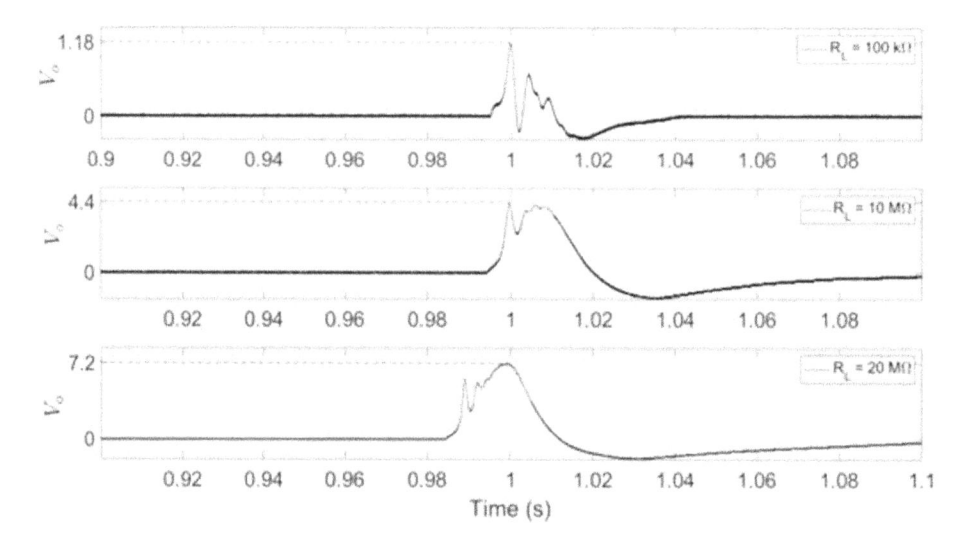

Figure 27. Effect of varying the load resistance on the sensor output, assuming an impact force of 980 N

The noisy V_{op} shown in Figure 32 contains the locations T2 and T6, needed to compute the true time delay T= T6–T1. The figure also shows false peaks and their locations at T1, T3, T4, T5, and T7. The estimation algorithm must differentiate between the two types of time locations, then addresses the true one used to estimate the speed and ignoring all false time delays.

The speed estimation algorithm is shown in ALGORITHM 1. The algorithm assumes that V_{op} is sampled by a sampling frequency equals to f_s Hz. The sampling time is equal to t_s seconds. Then, a frame of N_{op} samples is captured from V_{op}. The maximum time of the frame samples is denoted T_{frame} and computed as:

$$T_{frame} = N_{op} \times t_s \tag{14}$$

Smart Highways-Based Piezoelectric Vehicle Speed Sensor

Figure 28. Effect of varying the impact force on the sensor output, using charge amplifier circuit in Figure 19

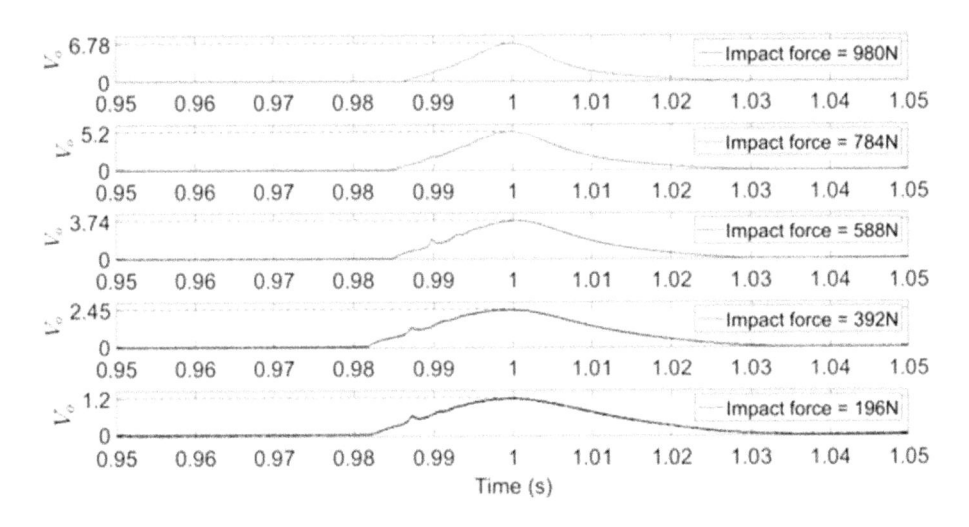

Figure 29. Plots of the pulse amplitude V and the predicated amplitude $\hat{V}$ versus the Force

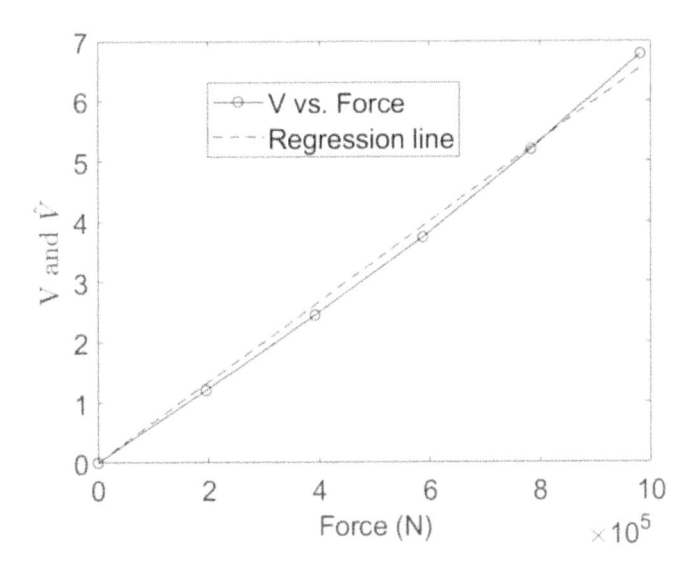

Figure 30. Plot of the predicated pulse amplitude $\hat{V}$ versus the Force using the linear regression model from Figure 29. Results are compared with the simulation results in Figure 23.

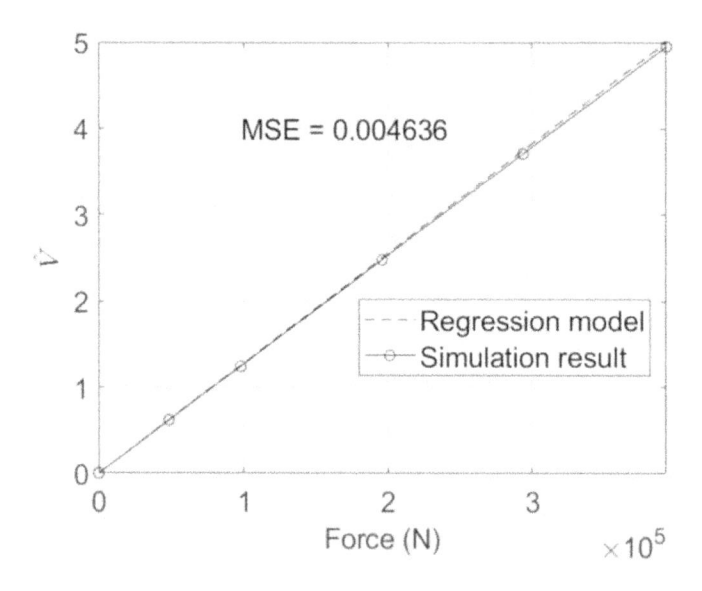

The frame is divided into segments of length equals to the pulse width, denoted by PW. The PW is computed according to (13). The PW has N_{PW} samples and calculated as follows:

$$N_{PW} = \frac{PW}{t_s} \quad (15)$$

Next, number of segments, denoted by N_{seg}, is computed as follows:

$$N_{seg} = \frac{N_{op}}{N_{PW}} \quad (16)$$

Smart Highways-Based Piezoelectric Vehicle Speed Sensor

Figure 31. Periodic sensor output signals generated from Figure 28 (impact force = 338 N), after rescaling the x-axis according to the vehicle speed

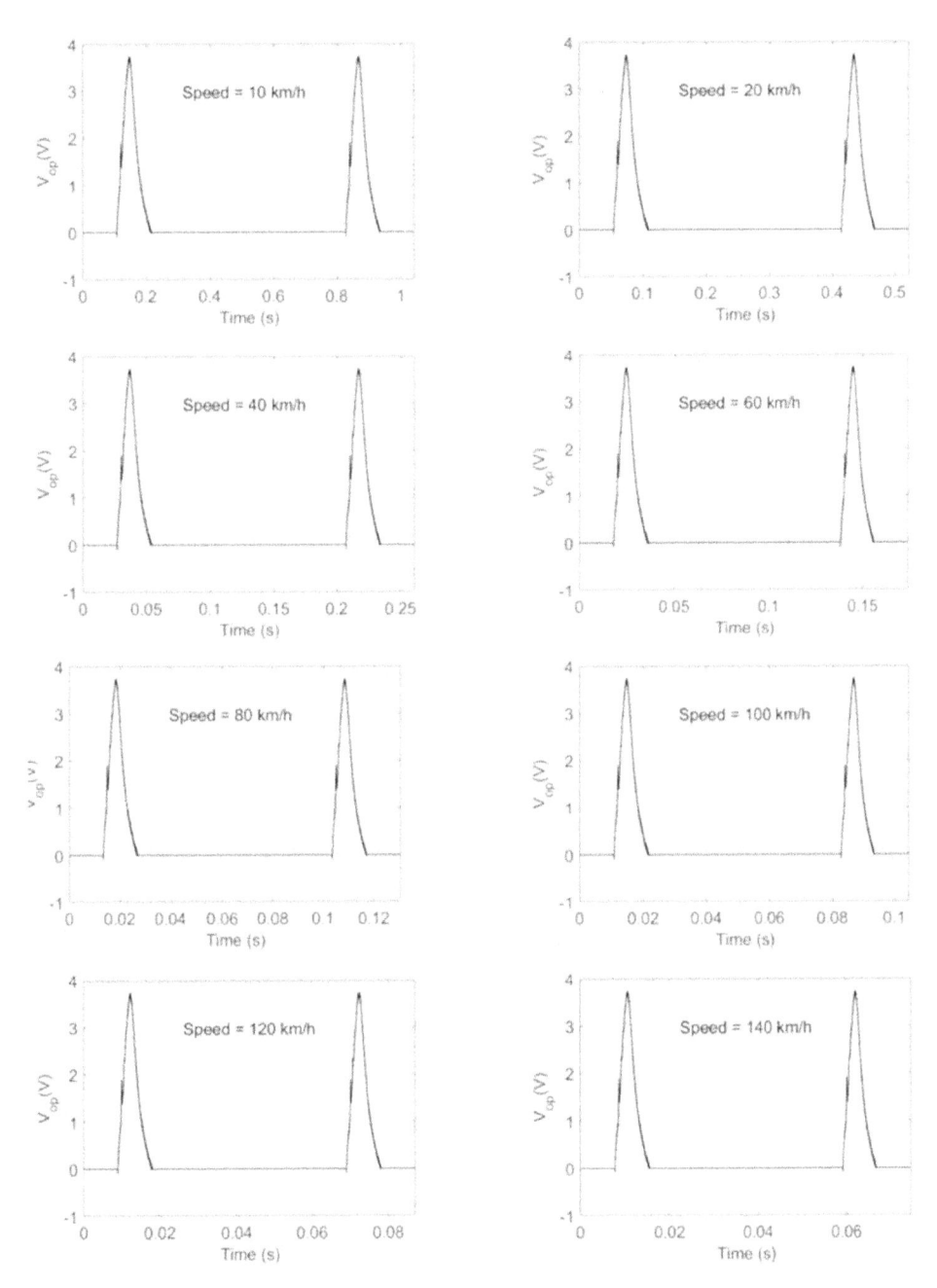

Figure 32. Typical noisy V_op. T1 through T6 are the locations at which V_op has peaks. T is the true time delay needed to estimate the vehicle speed u. V_TH is the threshold voltage used to detect V_op peaks beyond its value.

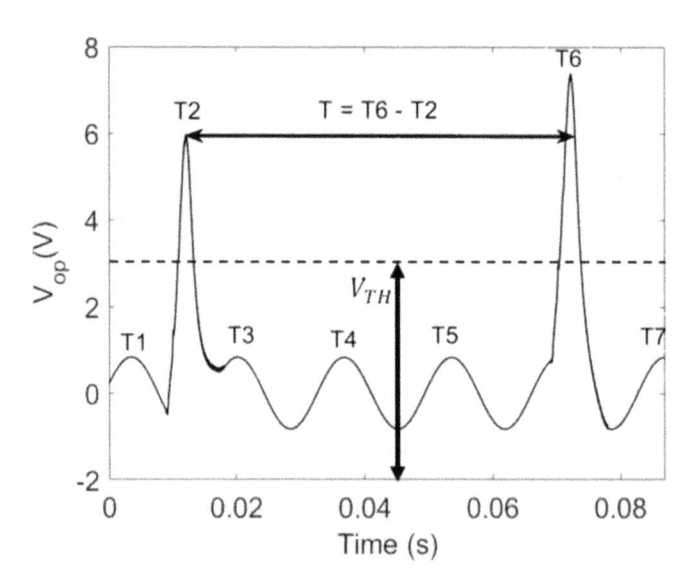

Then, for each segment, the algorithm is iterated to estimate all possible local maximums (also called peaks) and their time locations. The local peaks and their locations are denoted by L_{peaks} and L_{loc}, respectively. In each iteration, L_{peaks} and L_{loc} are saved to the global peaks vector G_{peaks} and the global locations vector G_{loc}, respectively. These global peaks are then sorted in descending order, then filtered according to selected threshold voltage V_{TH} shown in Figure 32. Any peak value less than V_{TH} will be removed from the G_{peaks} and G_{loc} vector. The remaining peaks are used to estimate the true time delay. Each peak is compared with all other peaks in the vector. If the time difference between two peaks locations is greater than *PW*, this indicates that the two peaks are candidates to utilize them in estimating the true peaks. The time delay can now be estimated from this difference and the value is saved in a vector denoted by T_{vec}. Note that there may be a chance of having other values of time delay from other peaks. Thus, all these values must be saved in the vector T_{vec}. Once all peaks are checked, the resultant true peaks vector, is averaged to estimate the true time delay T. Once the time delay is estimated, the speed is estimated according to (9).

Results of applying ALGORITHM 1 is shown in Figure 33 and Figure 34. Figure 33 shows the visualization of the estimated speed versus the actual speed, by applying ALGORITHM 1 on data from Figure 31. The plot shows linear relation. The error in estimating the speed is illustrated in Figure 34, by computing the MSE for different

Smart Highways-Based Piezoelectric Vehicle Speed Sensor

values of SNR. The results indicate that the MSE is decreased by increasing the SNR. This is an expected result since the sensor signal contains less noise thus the speed estimation will be improved. However, the MSE shows acceptable values for low SNR at 10 dB.

Algorithm 1. Vehicle speed estimation algorithm using time delay approach

1.	**Input**: PW, V_{op}, N_{op}, t_s, V_{TH}, L.		
2.	**Output**: True time delay T, estimated vehicle speed u_{est}.		
3.	Compute T_{frame} by (14).		
4.	Compute N_{PW} by (15).		
5.	Compute N_{seg} by (16).		
6.	$G_{peaks}=[]$. % Initialize the global peaks vector.		
7.	$G_{loc}=[]$. % Initialize the locations of the global peaks vector.		
8.	**for** i = 1 to N_{seg}		
9.		$L_{peaks} = \max\left(V_{op}((i-1) \times N_{PW} + 1 : N_{PW})\right)$. % Find local peaks	
10.		$L_{loc} = argmax\left(V_{op}((i-1) \times N_{PW} + 1 : N_{PW})\right)$. % Find locations of local peaks	
11.		$G_{peaks} = [G_{peaks} L_{peaks}]$. %Update the global peaks vector	
12.		$G_{loc} = [G_{loc} L_{loc}]$. %Update the locations of the global peaks vector	
13.	**end**		
14.	Sort G_{peaks} in descending order.		
15.	Reorder G_{loc} values according to the sorted G_{peaks}.		
16.	$N_G = dim(G_{peaks})$ % Find the length of G_{peaks} vector.		
17.	**for** j = 1 to N_G		
18.		**If** ($G_{peaks}(j) < V_{TH}$)	
19.			Remove $G_{peaks}(j)$ and $G_{loc}(j)$. % remove false peaks
20.		**End**	
21.	**end**		
22.	$N_G = dim(G_{peaks})$. % Find the new length of G_{peaks}		
23.	$T_{vex} = []$. % Initialize the true time delay vector.		
24.	**for** k = 1 to $N_G - 1$		
25.		**for** p = 2 to N_G	
26.			**If** $|G_{loc}(k) - G_{loc}(p)| > N_{PW}$ % check for true peak locations

Algorithm continued

27.			$T_{vec} = \left[T_{vec} \, \big| G_{loc}(k) - G_{loc}(p) \big| \right].$
28.		end	
29.		end	
30.	end		
31.	$T = \dfrac{1}{dim(T_{vec})} \sum\limits_{q=1}^{dim(T_{vec})} T_{vec}(q).$ % Estimate the true time delay between the peaks			
32.	Compute u_{est} by (9) % Compute the vehicle speed			

Figure 33. Estimated speed by applying Algorithm 1 on data from Figure 31

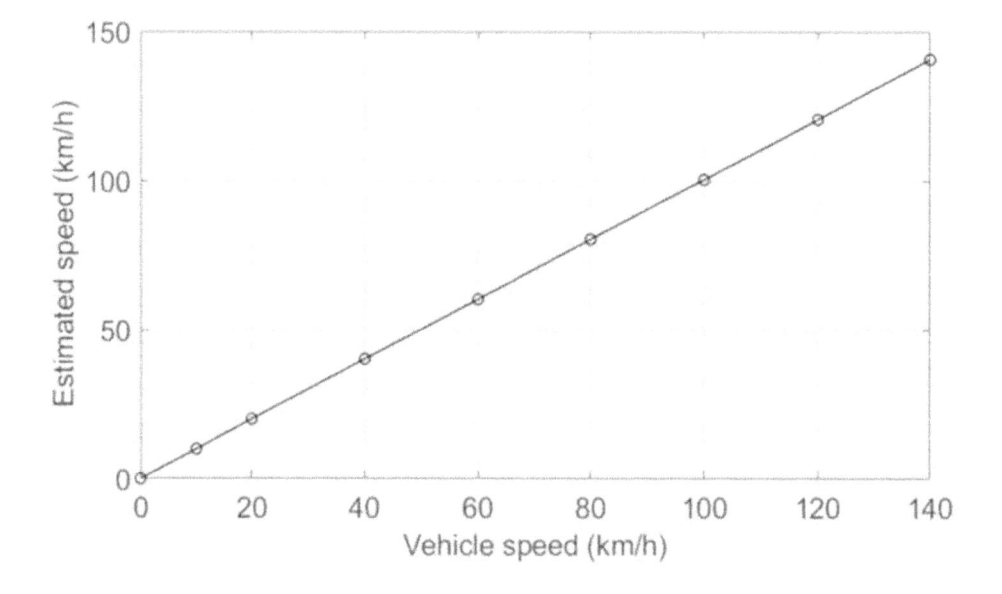

Piezoelectric Speed Array Sensor System

In this section, we shall explain the final piezoelectric speed sensor system using array sensors. So far, all discussions, simulation, and testing were carried out for one sensor. However, the system requires array sensors (front and back array sensors) as shown in Figure 1, to improve the sensing efficiency and to consider the vehicle location in the lane. Thus, one sensor may not be enough. The construction of the system with array sensors is very similar to that shown in Figure 11. However, all front and back array sensors must be considered. Figure 35 illustrates a block diagram

Smart Highways-Based Piezoelectric Vehicle Speed Sensor

of the piezoelectric speed array sensors system, The terminals of FAS and BAS are connected to front and back signal conditioning unit to postprocessing the sensor signals and end up with a clean pulse of 5V. The speed estimation algorithm 1 can be used for array sensors case. As the new system is like the previous system, we shall demonstrate one example to show the electro-mechanical connections then simulate the system to ensure that it is sensing properly.

Figure 34. MSE after applying Algorithm 1 on data in Figure 31. Noise signals are added to data so that SNR is equal to 10 dB, 20 dB, 30 dB, and 40 dB

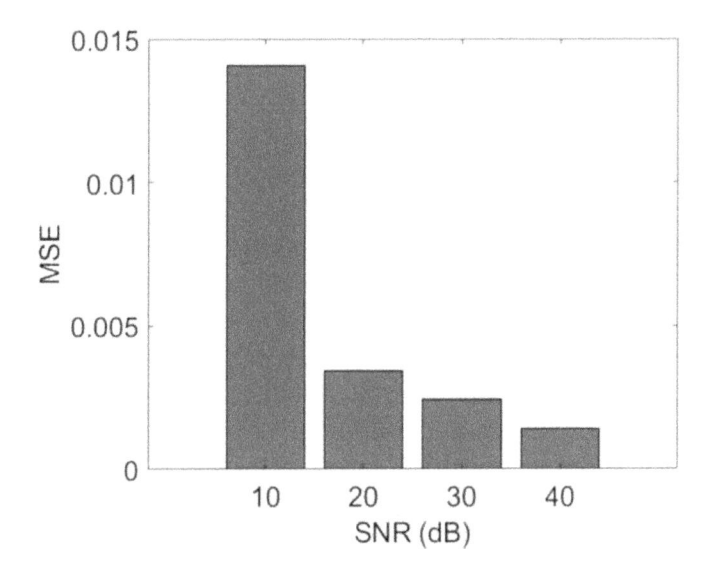

Figure 35. Block diagram of the piezoelectric speed array sensors system

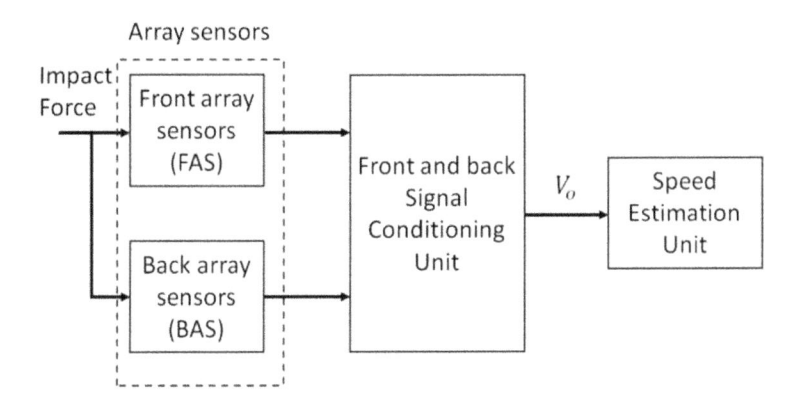

Example 5: The circuit shown in Figure 19 (for single sensor testing) is modified as follows:

1. Including front and back pulse generators. Both front and back pulse generators have a period that is doubled the period of the pulse generator used in the single sensor case. Also, the back pulse generator has a time delay determined from equation (9), i.e., it depends upon the vehicle speed. This is necessary since the vehicle first presses the front sensors, then after a time delay, it presses the back sensor.
2. Including front and back array sensors, and their corresponding charge amplifiers. We assumed five sensors embedded in each lane in the road, as shown in Figure 1. In this example, we only demonstrate the sensing of one lane in the road for simplicity. Adding the situation of other lanes requires more hardware and software and it's beyond the scope of this chapter.
3. Adjusting the charge amplifiers capacitances C_F so that the output voltage is within the range from 0-5V.
4. Including inverting amplifier as an output stage to add the two outputs from the front and back summing amplifiers, then inverts the sum.

The modified circuit is shown in Figure 36. The front and back summing amplifiers are identical. Figure 37 illustrates a typical unity gain summing amplifier used to add the charge amplifier outputs according to:

$$V_{o(sum)} = -\left(V_{o_1} + V_{o_2} + V_{o_3} + V_{o_4} + V_{o_5}\right) \tag{17}$$

where $V_{o(sum)}$ is the summing amplifier output, and $V_{o_1}, V_{o_2}, \ldots, V_{o_5}$ are the charge amplifier outputs. From (17), Figure 36, and Figure 37, the expressions of the front and back summing amplifiers are given by:

$$V_{of(sum)} = -\left(V_{of_1} + V_{of_2} + V_{of_3} + V_{of_4} + V_{of_5}\right) \tag{18}$$

$$V_{ob(sum)} = -\left(V_{ob_1} + V_{ob_2} + V_{ob_3} + V_{ob_4} + V_{ob_5}\right) \tag{19}$$

All resistors shown in Figure 37 are denoted by R_1, and its value is equal to $100k\Omega$.

The inverting amplifier is illustrated in Figure 38. The amplifier output V_o is given by:

Smart Highways-Based Piezoelectric Vehicle Speed Sensor

$$V_o = -\frac{R_2}{R_1}\left(V_{of(sum)} + V_{ob(sum)}\right) \tag{20}$$

where R_2 is a variable resistor. In this example, it is set to $38.68k\Omega$ to get a maximum output voltage of 5V. Equation (20) shows that the output V_o is the sum of all outputs from the front and back sensors. However, only the sensors (either front or back sensors but not both), that are stressed by a vehicle, will produce outputs. Others are not active. To understand this fact, the circuit in Figure 36 is simulated by assuming a vehicle of a mass 40,000 k/h stresses the front and the back array sensors in one lane at a speed of 120 km/h and 60 km/h. Results are shown in Figure 39 and Figure 40. Results indicate that the amplitude of V_o is constant at 5V, regardless of vehicle speed. However, the pulse width changes according to the speed. These results are the same as we got in the single sensor case, shown in Figure 20 and Figure 21. We conclude that array sensor system is identical in its operation as the single sensor system. The speed estimation algorithm can also be used for array sensors.

Figure 36. Array sensors simulation circuit for Example 5

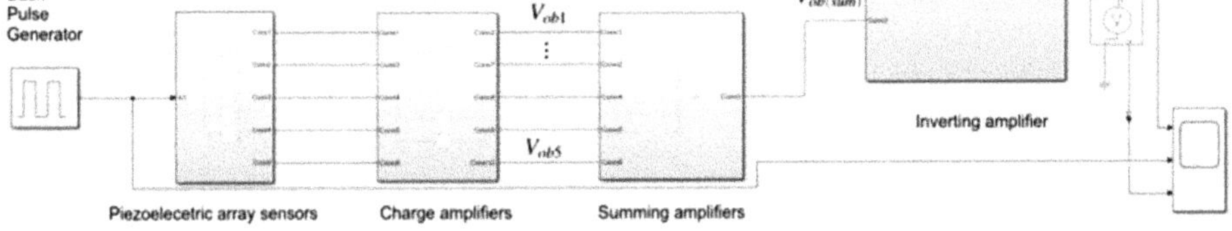

FUTURE RESEARCH DIRECTIONS

The work requires further development to test the prototype in an existing road and varying number of sensors arrays in all lanes, and their locations, then estimate the vehicle speed and type.

Furthermore, a more intensive study may be required to modify the speed estimation algorithm so that the effect of varying the PW and V_{TH} on the accuracy of estimated vehicle speed is investigated.

Figure 37. Summing amplifier used in Example 5

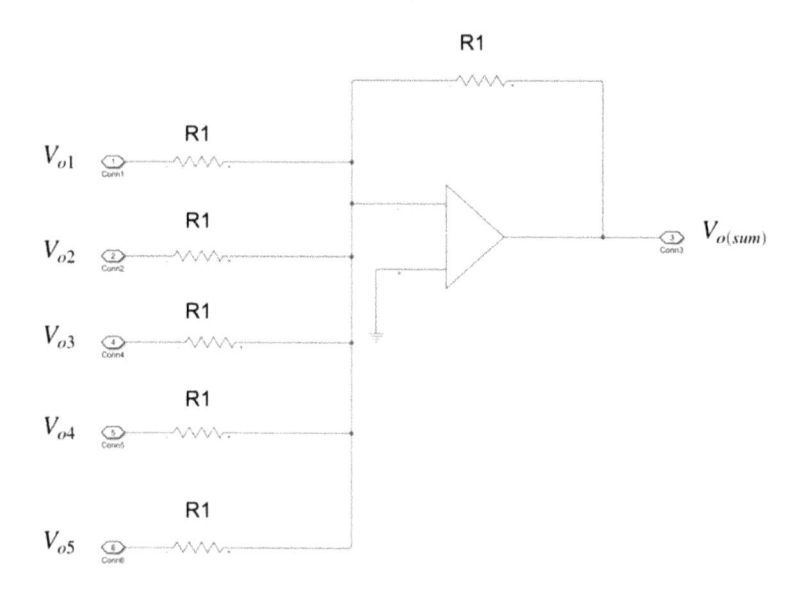

Figure 38. Inverting amplifier used at the output stage of the array sensor system in Example 5

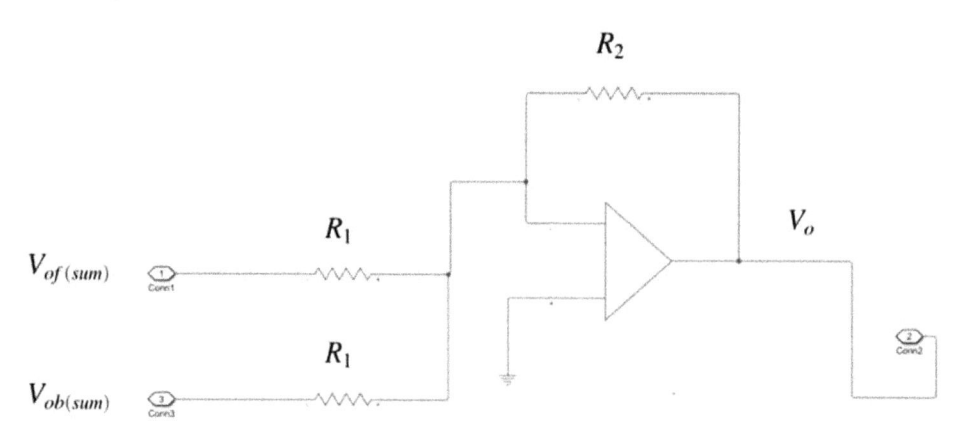

Furthermore, the effect of environmental conditions, such as temperature and moisture, require consideration. These factors may be required to address the sensor durability over time.

Smart Highways-Based Piezoelectric Vehicle Speed Sensor

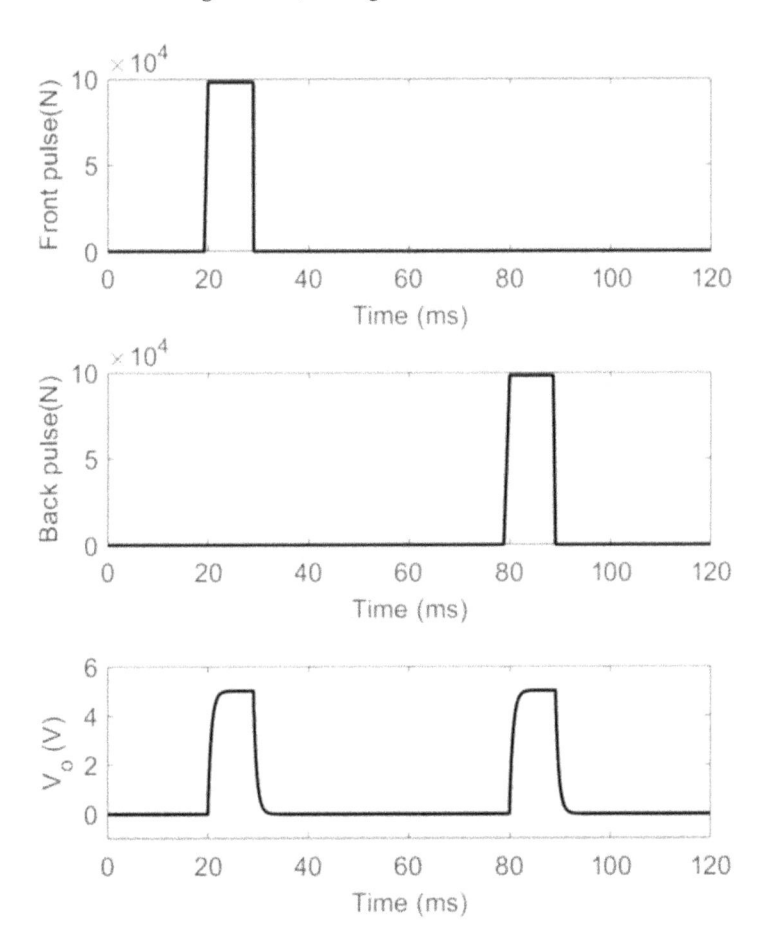

Figure 39. Visualization of the front /back force pulses and the array sensor output pulse V_o of amplitude 5V. The pulse width is 9ms, assuming vehicle speed of 120 km/h, and the vehicle weight is 40,000 kg.

Figure 40. Visualization of the front /back force pulses and the array sensor output pulse V_o of amplitude 5V. The pulse width is 18ms, assuming vehicle speed of 60 km/h, and the vehicle weight is 40,000 kg.

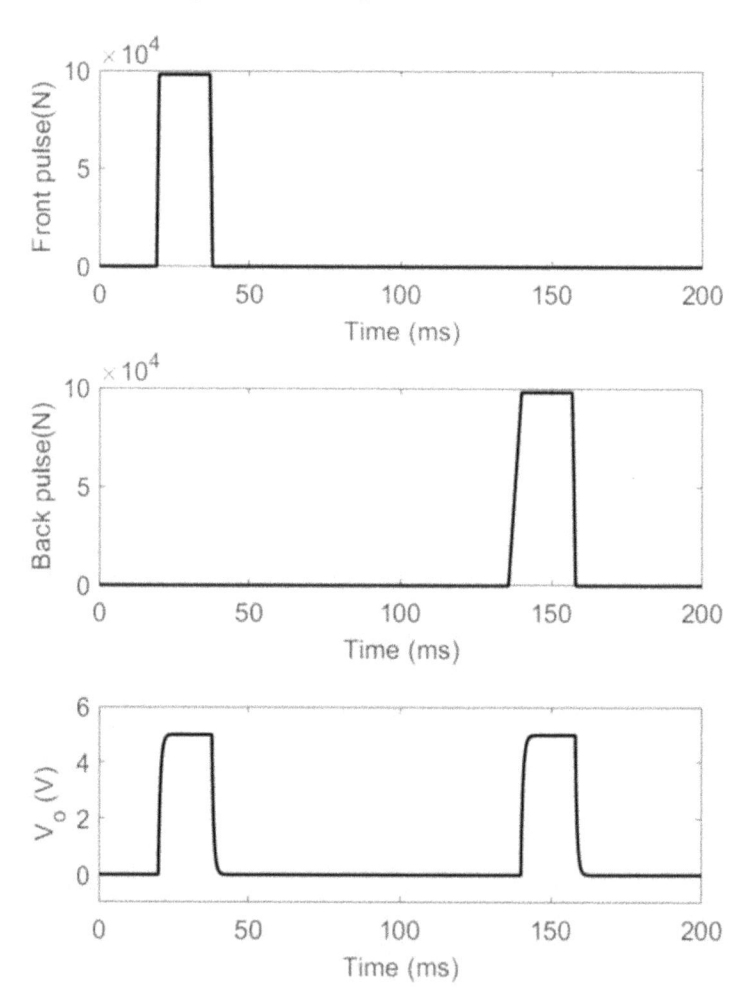

CONCLUSION

In this chapter, we have explained the piezoelectric direct effect and the materials used in piezoelectric sensors. A comparison between piezoelectric material properties has been made and decision was taken to use PZT material in the design of piezoelectric speed sensor, due to its highest strain constant, highest relative permittivity, and highest coupling factor. The mathematical modelling of the piezoelectric speed sensor has been explained and applied to compute the front and back sensed pulses according to the applied stressed force from the vehicle. The time delay (T) between the front and back pulses has been estimated then the sensed vehicle speed u is computed according to (9). The procedure of computing the speed is simulated using MATLAB/SIMULINK. The sensor system is constructed in SOLIDWORKS and 3D modelling. The constructed 3D prototype has been explained from mechanical and electrical point of views. Due to possible high voltages at output terminals, two approaches (load resistance approach and charge amplifier approach) have been proposed to control the level of the voltage. The first method has shown an increase in the pulse amplitude when the connected load resistance is increased. However, the pulses are affected by noise and require further processing to remove the noise. The testing of the second method, using charge amplifier, has shown excellent control of the output pulse amplitude using the feedback capacitance C_F. Results also have shown good shape of the output pulses with negligible noise. The sensor system is tested using simulation and experiment approaches. The sensitivity of the designed sensor has been investigated by simulation and experiment. Results shows comparable value of 5V/40,000 kg. The speed estimation algorithm has been demonstrated and applied to real data captured by a high-speed data acquisition system built in the digital storage oscilloscope SDS1000U-X. Results from experiment are compared with simulation and found very similar. Array sensor system has also been explained and demonstrated by example using five front sensors and five back sensors, placed in one lane in the road. Results have shown very strong matching with the results in single sensor system. As a results, the designed system is found capable of estimating the vehicle speed when the vehicle mass varies up to 40,000 Kg, and when the speed varies up to 140 km/h.

REFERENCES

Bartolome, E. (2010). Signal conditioning for piezoelectric sensors. *Texas Instruments Analog Applications Journal, 10*.

Cady, W. G. (2018). *Piezoelectricity: Volume Two: An Introduction to the Theory and Applications of Electromechanical Phenomena in Crystals*. Courier Dover Publications.

Ceramics, M. E. (2001). Introduction Piezoelectric Ceramics. *Technical Information*.

He-Shuai LTD. (2018). *Hard PZT & Soft PZT Data Sheet*. Author.

Jaffe, H. (1958). Piezoelectric ceramics. *Journal of the American Ceramic Society, 41*(11), 494–498. doi:10.1111/j.1151-2916.1958.tb12903.x

Juang, J. C., & Lin, C. F. (2015). A sensor fusion scheme for the estimation of vehicular speed and heading angle. *IEEE Transactions on Vehicular Technology*, *64*(7), 2773–2782.

Karki, J. (2000). *Signal conditioning piezoelectric sensors. App. rept. on mixed signal products (sloa033a)*. Texas Instruments Incorporated.

Keipour, A., Pereira, G. A., Bonatti, R., Garg, R., Rastogi, P., Dubey, G., & Scherer, S. (2022). Visual servoing approach to autonomous uav landing on a moving vehicle. *Sensors (Basel)*, *22*(17), 6549. doi:10.3390/s22176549 PMID:36081008

Lam, K. H. (2006). *Study of piezoelectric transducers in smart structure applications*. Hong Kong Polytechnic University.

Lewandowski, R., & Pawlak, Z. (2013). Optimal Placement of Viscoelastic Dampers Represented by the Classical and Fractional Rheological Models. In *Design Optimization of Active and Passive Structural Control Systems* (pp. 50–84). IGI Global. doi:10.4018/978-1-4666-2029-2.ch003

Liu, H., Ma, J., Xu, T., Yan, W., Ma, L., & Zhang, X. (2019). Vehicle detection and classification using distributed fiber optic acoustic sensing. *IEEE Transactions on Vehicular Technology*, *69*(2), 1363–1374. doi:10.1109/TVT.2019.2962334

Liu, H., Ma, J., Yan, W., Liu, W., Zhang, X., & Li, C. (2018). Traffic flow detection using distributed fiber optic acoustic sensing. *IEEE Access : Practical Innovations, Open Solutions*, *6*, 68968–68980. doi:10.1109/ACCESS.2018.2868418

Lu, S., Wang, Y., & Song, H. (2020). A high accurate vehicle speed estimation method. *Soft Computing*, *24*(2), 1283–1291. doi:10.1007/s00500-019-03965-w

Markevicius, V., Navikas, D., Idzkowski, A., Andriukaitis, D., Valinevicius, A., & Zilys, M. (2018). Practical methods for vehicle speed estimation using a microprocessor-embedded system with AMR Sensors. *Sensors (Basel)*, *18*(7), 2225. doi:10.3390/s18072225 PMID:29996564

Markevicius, V., Navikas, D., Idzkowski, A., Miklusis, D., Andriukaitis, D., Valinevicius, A., Zilys, Cepenas, & Walendziuk, W. (2019). Vehicle speed and length estimation errors using the intelligent transportation system with a set of anisotropic magneto-resistive (AMR) sensors. *Sensors (Basel)*, *19*(23), 5234. doi:10.3390/s19235234 PMID:31795212

Markevicius, V., Navikas, D., Miklusis, D., Andriukaitis, D., Valinevicius, A., Zilys, M., & Cepenas, M. (2020). Analysis of methods for long vehicles speed estimation using anisotropic magneto-resistive (AMR) sensors and reference piezoelectric sensor. *Sensors (Basel)*, *20*(12), 3541. doi:10.3390/s20123541 PMID:32580498

Piekarski, B. H. (2005). *Lead zirconate titanate thin films for piezoelectric actuation and sensing of MEMS resonators*. University of Maryland.

Platt, S. R., Farritor, S., & Haider, H. (2005). On low-frequency electric power generation with PZT ceramics. *IEEE/ASME Transactions on Mechatronics*, *10*(2), 240–252. doi:10.1109/TMECH.2005.844704

Rajab, S. A., Mayeli, A., & Refai, H. H. (2014, June). Vehicle classification and accurate speed calculation using multi-element piezoelectric sensor. In *2014 IEEE Intelligent Vehicles Symposium Proceedings* (pp. 894-899). IEEE. 10.1109/IVS.2014.6856432

Rall, K. C., Bailen, K. L., Bender, E. N., Taha, L. Y., Abdeltawab, H. M., & Anwar, S. (2023, April). Smart Highways Based Vehicle Speed Sensor With Piezoelectric Energy Harvesting: A Progress Report. In *2023 IEEE International Conference on Industrial Technology (ICIT)* (pp. 1-6). IEEE. 10.1109/ICIT58465.2023.10143163

Sherren, A., Fink, K., Eshelman, J., Taha, L. Y., Anwar, S., Brennecke, C., Abdeltawab, H. M., Shen, S., Ghofrani, F., & Zhang, C. (2022). Experimental and Simulation Validation of Piezoelectric Road Energy Harvesting. *Open Journal of Energy Efficiency*, *11*(3), 122–141. doi:10.4236/ojee.2022.113009

Shim, K. S., Park, N., Kim, J.-H., Jeon, O.-Y., & Lee, H. (2021). Vehicle Speed Measurement Methodology Robust to Playback Speed-Manipulated Video File. *IEEE Access : Practical Innovations, Open Solutions*, *9*, 132862–132874. doi:10.1109/ACCESS.2021.3115500

Taha, L. Y. (2009). *Design and Modelling of a MEMS Piezoelectric Microgenerator* (Doctoral dissertation, Ph. D. thesis, Universiti Kebangsaan Malaysia).

Underwood, I., Bailen, K. L., Dellapenna, F. J., Taha, L. Y., Abdeltawab, H. M., & Anwar, S. (2023, August). Experimental evaluation of piezoelectric Vehicle Speed Sensor for smart highways: A Progress Report. In *2023 12th International Conference on Renewable Energy Research and Applications (ICRERA)* (pp. 576-580). IEEE. 10.1109/ICRERA59003.2023.10269353

Weisberg, S. (2005). *Applied linear regression* (Vol. 528). John Wiley & Sons. doi:10.1002/0471704091

Zhang, C., Shen, S., Huang, H., & Wang, L. (2021). Estimation of the vehicle speed using cross-correlation algorithms and MEMS wireless sensors. *Sensors (Basel)*, *21*(5), 1721. doi:10.3390/s21051721 PMID:33801400

Zhang, Z., Zhao, T., Ao, X., & Yuan, H. (2017). A vehicle speed estimation algorithm based on dynamic time warping approach. *IEEE Sensors Journal*, *17*(8), 2456–2463. doi:10.1109/JSEN.2017.2672735

Chapter 8
The Principles and Applications of Electrostatic Transducers

Rita Tareq Aljadiri
Higher Colleges of Technology, UAE

ABSTRACT

This chapter provides an overview of electrostatic transducers, describing the fundamental principles of converting mechanical energy into electrical energy using variable capacitors. It explains the operation principle of electrostatic transducers, emphasizing the structure types and conversion mechanisms. The chapter outlines the variable capacitor factors, design considerations, and implementation requirements, followed by an analysis of electrostatic conversion mechanisms. A comparative analysis of capacitor structures and power processing circuits highlights the optimal design choices, considering efficiency, power output, and scalability. Furthermore, the chapter explores the applications of electrostatic harvesters, focusing on integrating them into smart road infrastructure. A case study on smart road development in the UAE showcases the prospects of using electrostatic transducers to enhance road connectivity, efficiency, and sustainability.

INTRODUCTION

This chapter concentrates on electrostatic transducers as energy harvesters, which are devices that convert electrical energy into mechanical energy. Electrostatic transducers are converters with a very simple structure that consists of two parallel plates separated by a dielectric material. The plates are charged with opposite polarities,

DOI: 10.4018/978-1-6684-9214-7.ch008

Copyright © 2024, IGI Global. Copying or distributing in print or electronic forms without written permission of IGI Global is prohibited.

creating an electric field that causes them to attract or repel each other, depending on the polarity of the charges. One of the main advantages of electrostatic harvesters is that the energy density of these devices can be easily modified by changing the structure of the transducer which is a variable capacitor, such as the size of the plates or the distance between them. This allows for a wide range of energy conversion applications, from high-power audio speakers to low-power micromechanical sensors. Another advantage of electrostatic transducers is that they have a high-frequency response and low distortion, making them ideal for applications that require accurate and fast responses. However, electrostatic harvesters are generally more expensive to manufacture than other types of transducers, such as electromagnetic harvesters. To address this issue, researchers have been investigating ways to develop low-cost energy conversion systems similar to electromagnetic transducers, using materials and processes that are compatible with large-scale manufacturing. These efforts have resulted in the development of new types of electrostatic transducers and harvesters.

In this chapter, the principles, design considerations, and applications of electrostatic harvesters will be explored, with a focus on emerging technologies and trends in the field. The challenges and opportunities in developing electrostatic harvesters for various energy conversion applications will also be discussed, highlighting the potential for these devices to play a key role in the transition to a more sustainable and energy-efficient future.

Electrostatic transducers have a wide range of applications, including audio speakers, sensors, actuators, and energy harvesters. For example, electrostatic transducers can be used to harvest energy from vibrations in machinery or the airflow in ventilation systems. They can also be used in micro-electromechanical systems MEMS to create microscale sensors and actuators. Additionally, electrostatic transducers have several potential applications in the context of smart roads, including energy harvesting, traffic monitoring, smart speed control, and road health monitoring. Overall, electrostatic harvesters have a good potential to play an important role in the development of smart roads, helping to improve safety, efficiency, and sustainability on the roads and highways.

In conclusion, designing an electrostatic harvester requires careful consideration of the capacitor structure and the conversion mechanism approach, as well as an understanding of the intended application and operating conditions. By optimizing these factors, electrostatic transducers can be developed for a wide range of applications and can play an important role in advancing the field of energy conversion and microscale systems. Accordingly, the main aim of the chapter is to educate readers about the principles of electrostatic transducers, including their operation, construction, and energy conversion mechanisms. Additionally, it provides readers with a thorough understanding of electrostatic energy harvesting and its role in

The Principles and Applications of Electrostatic Transducers

sustainable energy systems. This chapter is organized into 13 sections, each focusing on a different aspect as follows:

1. Basic Concepts of Electrostatic Transducers
2. Electrostatic Harvester's Principle of Operation
3. Variable Capacitor Structures for Electrostatic Harvesters
4. Electrostatic Energy Conversion Mechanisms
5. Comparative Analysis of Capacitor Structures, Power Processing, and Conversion Mechanisms
6. Benefits and Drawbacks of Electrostatic Harvesters
7. Applications of Electrostatic Harvesters
8. Electrostatic Transducers and Smart Roads
9. UAE Smart Road Case Study
10. Conclusion
11. References
12. Additional Reading
13. Key Terms and Definitions

BASIC CONCEPTS OF ELECTROSTATIC TRANSDUCERS

Electrostatic transducers or converters are devices that convert electrical energy into mechanical energy, or vice versa, through the principles of electrostatic attraction and repulsion. They consist of two or more parallel plates separated by an insulating material. The electrostatic force is generated due to attraction or repulsion between the electric charges. These can be very strong when the distance is short making them useful for many applications. This section provides an overview of electrostatic transducers, covering their structure, energy conversion process, basic applications, material selection criteria, and fabrication techniques.

STRUCTURE OF THE ELECTROSTATIC TRANSDUCERS

The structure of the electrostatic transducers is similar to capacitors as shown in Figure 1, made of conductive plates called electrodes that are separated by an insulating material called a dielectric. The dielectric stops direct electrical contact between the conductive plates, allowing the charges to impact each other through the insulating material.

Figure 1. Structure of electrostatic transducers (Zhou, Baker, & Hu, 2008)

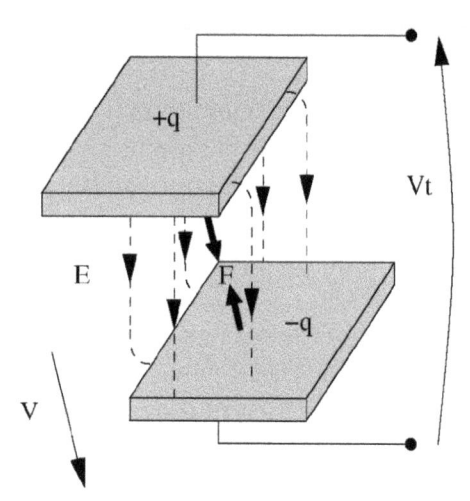

ENERGY CONVERSION PROCESS OF ELECTROSTATIC TRANSDUCERS

The energy conversion process in electrostatic transducers involves the conversion of electrical energy into mechanical energy and vice versa.

Electrical to Mechanical Conversion

Voltage needs to be applied across the electrodes, one electrode must be positively charged and the other must be negatively charged. The positively charged electrode generates a repulsion force towards the positively charged particles in the dielectric material, causing it to move away. At the same time, the negatively charged electrode produces a repulsive force toward the negatively charged particles in the dielectric, pushing them away as well. This produces a mechanical movement between the electrodes.

Mechanical to Electrical Conversion

When a mechanical force is applied to the electrodes of the transducers, the distance between them will change. This movement will affect the capacitance of the transducers which affects the ability of it to store electrical energy. When capacitance is modified, the distribution of charges on the electrodes changes accordingly. This change results in the generation of electrical energy as illustrated in Figure 2.

Figure 2. Electrostatic transducer as energy converter (New Atlas, 2014)

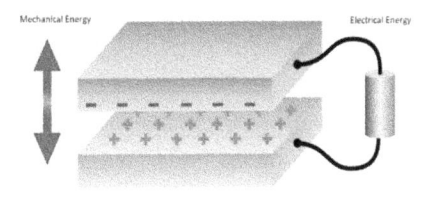

ELECTROSTATIC TRANSDUCERS APPLICATIONS

In this section, basic applications where electrostatic transducers play an essential role are explored. The applications are detailed below:

- Actuators: They can be used as actuators in robotic applications for accurate control of movement.
- Energy Harvesters: Electrostatic transducers can be used as energy harvesters from rotary or linear mechanical movements, converting them into usable electrical energy.
- Microphones and Loudspeakers: Electrostatic transducers can be used to convert sound vibration (mechanical energy) into electrical signals through a microphone and can convert electrical signals to back sound by generating mechanical vibrations through a speaker as shown in Figure 3.

Figure 3. Applications of electrostatic transducer (Electronics Tutorials, n.d.)

MATERIAL SELECTION AND FABRICATION TECHNIQUES OF ELECTROSTATIC TRANSDUCERS

The material selection plays an important role in determining the performance and efficiency of electrostatic transducers. The selection of materials is based on several key factors such as mechanical, electrical, and thermal features. These factors ensure that the materials used can handle the operational forces and environmental settings, while also providing suitable dielectric constant, low losses, thermal stability, and compatibility with other components. Many newly developed materials have been explored in recent years such as Nanocomposites. Nanocomposites are materials composed of a matrix reinforced with nanoparticles (particles with at least one dimension less than 100 nanometers). These materials offer enhanced performance characteristics, such as higher dielectric constants, lower dielectric losses, and improved mechanical properties. Various studies about Nanocomposites, specifically carbon nanomaterial-based composites (Jeong et al., 2020) and conducting polymers (Sonika et al., 2022) have received a lot of attention because of their unique advantages over popular materials, such as tunable electrical conductivity, high mechanical strength, low weight, low cost, and ease of processing. They have great promise for energy harvesting and sensing applications due to their ability to combine multiple desirable properties in a single material.

On the other hand, fabrication techniques play an essential role in determining the efficiency of electrostatic transducers such as additive manufacturing, lithography, etching, and deposition techniques (Ali et al., 2023). Additive manufacturing allows for the accurate deposition of materials layer by layer, enabling the creation of complex shapes and structures (Askari et al., 2020). While lithography (subtractive technology), uses light to transfer a pattern onto a substrate coated with a photosensitive material, this method is used for fabricating MEMs and nanoscale structures as shown in Figure 4.

Figure 4. Two methods of obtaining energy using a variable capacitor: (a) constant charge, (b) constant voltage (Mitcheson, 2008)

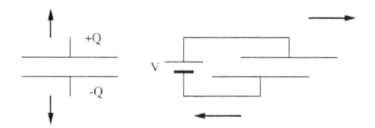

The etching techniques (Wolf & Tauber, 2018), such as wet etching and dry etching, selectively remove material from a substrate to create desired patterns or structures. Deposition techniques are also used to deposit thin films of material onto a substrate, creating thin-film transducers with exact thickness and composition. All the above-mentioned fabrication techniques could lead to increased customization options, cost-effectiveness, and scalability of electrostatic transducers. By understanding the type of material and the fabrication techniques, researchers and engineers can optimize the production process to achieve the required performance, cost reduction, and effectiveness of electrostatic transducers and harvesters.

ELECTROSTATIC HARVESTER'S PRINCIPLE OF OPERATION

Electrostatic harvesters are one of the applications of electrostatic transducers. They are capacitive devices in which energy conversion takes place as the plates of a capacitor separate or the area of the plates is modified in response to external mechanical energy (Boisseau, 2012). These devices can be divided into two categories:

- Electret-free electrostatic harvester that uses energy cycles to convert mechanical energy into electrical energy
- Electret-based electrostatic harvester that uses electrets to directly convert mechanical energy into electricity.

Electret-Free Electrostatic Harvester

Electret-free devices operate in either charge- or voltage-constrained systems, thus, there are two possible extremes of operation, as shown in Figure 5. In both cases, the mechanical movement is converted into electrical energy. Electret-free-based harvesters need an external power supply for pre-charging the capacitor to generate electricity.

Figure 5. Electret-based electrostatic harvesters: (a) dipole orientation and (b) charge injection (Boisseau, 2012)

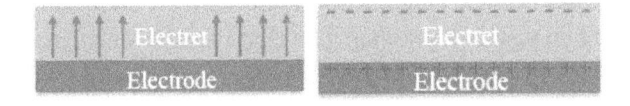

Electret-Based Electrostatic Harvester

Electret-based devices are similar to electret-free devices. The main difference is the additional electret layers that are applied to one or both plates of the capacitor for polarization purposes. However, due to the complexity of the manufacturing process, the electrostatic harvesters tend to use electret-based more. The manufacturing process requires injecting additional charge into the dielectric layer or heating the dielectric layer above its melting temperature. The layer is then left to cool down to maintain an electric field. This allows the dipoles of the dielectric material to be oriented in the same direction as the electric field (Boisseau, 2012), as shown in Figure 6.

In general, these electrostatic harvesters are more appropriate for small-scale harvesters. When compared to electromagnetic harvesters, electrostatic harvesters can be similar in power density to a conventional electromagnetic system, with fewer components that would reduce system cost and increase overall effectiveness. Despite advances in energy harvesting using electromagnetic approaches, the electrostatic approach has several advantages: it is compact, sensitive to low mechanical vibration, easier to integrate into small-scale systems, does not require smart materials, is simple to fabricate, its energy density can be modified geometrically and has a simple structure with fewer components.

However, enhancing the power density of electrostatic harvesters remains challenging (Boisseau,2012). Additionally, electrostatic harvesters are sensitive to environmental conditions, such as humidity and temperature, which can affect the overall performance. Another drawback is the complexity of the power processing and conditioning circuits required to process the harvested energy, which can add to the overall system cost and complexity (Al Rahis,2016).

VARIABLE CAPACITOR STRUCTURES FOR ELECTROSTATIC HARVESTERS

Variable capacitors play an important role in electrostatic harvesting. They are mainly used to store and release electrical charges, allowing for energy transfer and conversion. The basic principle behind variable capacitors in this setting is to control the capacitance and electric field to allow the flow of charges and convert them from one form of energy into another. The first variable capacitor for electrostatic energy harvesting was proposed in 1998 by Menninger and co-workers (Meninger 2001). The concept was later demonstrated by Roundy and co-workers utilizing large machined variable capacitors (Roundy, 2003). Since then, several variable capacitor structures for electrostatic harvesting have been reported in the literature. The structures of these capacitors depend mainly on changing capacitance parameters as shown in Figure 7. There are three types of variable capacitors: variable area, variable gap, and variable dielectric constant. These three types are considered next (Hayt & Kemmerly, 2001).

Variable Area Capacitor

This type is known as an overlap capacitor which operates on the principle of the capacitance that is proportional to the area of overlap between the conductive plates or electrodes of the variable capacitor. The higher the overlap, the higher the capacitance or ability to store energy.

Variable Gap Capacitor

The variable gap capacitors work by changing the distance between the capacitor plates. The capacitance is inversely proportional to the distance between the plates. The larger the gap between the plates, the lesser the capacitance, while the smaller gap increases the capacitance.

Variable Dielectric Constant Capacitor

The variable dielectric constant capacitor uses different materials to fill the gap between the capacitor plates. The dielectric constant of a material affects the ability of the capacitor to store electric charge at a certain voltage. By changing the dielectric material between the plates, the capacitance can be increased or decreased.

Figure 6. Factors affecting variable capacitor structure (Hayt & Kemmerly, 2001)

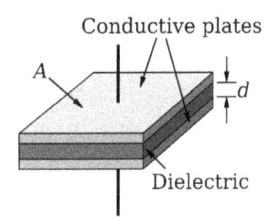

For further details on previous research on variable capacitor structures, refer to additional readings.

ELECTROSTATIC ENERGY CONVERSION MECHANISMS

The use of variable capacitors in the electrostatic energy harvester plays an important role in deciding the energy conversion process. The capacitance can be adjusted based on changes in operating conditions, environmental factors, or mechanical inputs. The electrostatic harvesters can change the amount of energy harvested and their efficiencies based on the controlled capacitance. The variable capacitance mechanism is particularly useful in situations where the energy source is not constant or has varying intensity over time. By adjusting the capacitance, the electrostatic energy harvester can match the varying input conditions and ensure efficient energy extraction. Here are the methods related to the utilization of variable capacitance in electrostatic energy harvesters:

Mechanical Capacitance Adjustment Method

One of the most common methods is the mechanical method, such as changing the overlap area or changing the gap between capacitor plates. In this method, mechanical actuators are required. The mechanical actuators can be designed to respond to external sources of energy, such as wind energy, linear vibrations, or rotational mechanical motion, allowing the energy harvester to change to external conditions.

Electrical Capacitance Adjustment Method

The capacitance can be adjusted through electrical methods. One of the examples is the use of voltage-controlled varactors. The varactor is a semiconductor device that changes its capacitance because of the variation in the width of the depletion layer in the PN junction. By changing the amount of reverse voltage applied across the varactors, the gap width changes, and as a result capacitance value changes.

Structural Capacitance Adjustment Method

In this method, the entire capacitor structure can be adjusted to modify capacitance, such as the geometry of the capacitor plates, by applying mechanical stretching or compression, leading to capacitance changes.

Overall, the concept of variable capacitance in electrostatic energy harvesters improves their versatility, adaptability, and efficiency. By adjusting the capacitance using mechanical, electrical, or structural methods, the electrostatic harvester can optimize its performance under varying input conditions and maximize the energy conversion process. As research and technological advancements continue, the integration of variable capacitance mechanisms in energy harvesters holds significant potential for enabling sustainable and autonomous power solutions across a wide range of applications.

Apart from the three main conversion mechanisms mentioned, there are three main techniques used for the electrostatic microgenerator conversion mechanism as reported in the literature. The switched constant charge systems, the switched constant voltage systems, and the continuous electret-based system. These three approaches are reviewed next (Allen, 2000).

Switched Constant Charge System

Under this method, the capacitor during operation maintains a constant change. To achieve constant change, the change is applied to the capacitor and then switched to another capacitor allowing the charge to stay constant during the process. This results in a mechanical displacement being consistent through the switching process. This method is used in MEMS, for example, MEMS actuators in which precise and repeatable mechanical movement is desired (Senturia, 2006).

Switched Constant Voltage System

The switched constant voltage systems keep a constant voltage across a capacitor during the conversion process. When the voltage is kept constant, the electric field strength remains constant, resulting in even mechanical displacements. This method is mainly used when the focus is on achieving an expectable and controlled displacement. This method is used in micro-positioning systems and microactuators such as in optical devices and medical instruments (Senturia, 2006).

Continuous Electret-Based Systems

Continuous electret-based systems use materials with permanent electric charge, known as electrets. These materials hold a fixed charge distribution that continues over time. The electret layers provide a continuous electric field, allowing a constant force for conversion between electrical and mechanical energy. This method is used in energy harvesting and sensors. Energy harvesters harvest mechanical vibrations and convert them into usable electrical power also, electret-based sensors use the change distribution for accurate sensing such as pressure sensors (White & Sessler,1999).

For further information regarding research conducted on the method described above, please refer to the additional readings.

COMPARATIVE ANALYSIS OF CAPACITOR STRUCTURES, POWER PROCESSING, AND CONVERSION MECHANISMS

Comparative Study of Capacitor Structures

This section provides a summary of different capacitor structures used in electrostatic energy harvesters reported in literature such as in-plane overlap converters, in-plane gap-closing converters, and out-of-plane gap-closing converters. It evaluates these designs based on structure, power density, and special features.

In-Plane Overlap Converter

This converter was developed in 2009 as an in-plane overlap harvester (Figure 8) (Le & Zuo, 2009). In this type of converter, the capacitance variation is due to the change in the overlap area between the electrodes. It operates in a low capacitance range of 203.7 - 134.7 pF and has a power capability of 0.6 μW only. This harvester is resonance-based, meaning it operates best at a specific frequency. However, one drawback is that the fringing field affects its power output, which is a common challenge with this type of design. One of the main characteristics of this harvester is that it requires power conversion and conditioning before it can be used to power electronic devices or systems. This type represents an important advancement in the field of energy harvesting and offers a positive opportunity for further improvement and optimization (Le & Zuo, 2009), (Li, 2021).

Figure 7. In-plane overlap converter (Le & Zuo, 2009)

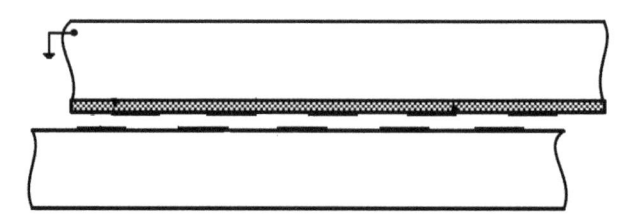

In-Plane Gap Closing Converter

The in-plane gap-closing converter shown in Figure 9 was developed in 2008 and has a power capability of 16 μW only. This converter operates based on the principle of the variable gap between the fingers of the capacitor structure as shown in Figure 8.

One of the drawbacks of this converter is that it requires an initial energy source to be polarized. This means that an external energy source or mechanism is needed to set the converter in motion before it can start harvesting energy (Li, 2021).

Figure 8. In-plane gap closing converter (Despesse et al., 2008)

Rolling Rod Harvester

The rolling rod harvester was developed in 2009 as illustrated in Figure 10 by Kiziroglou and operates in a capacitance range of 10 - 2 pF and has a voltage gain of 2.4, with a high output power capability of 250 µW. This high-power output is achieved when the harvester is operated at a voltage of 50 V and a load resistance of 10 MΩ.

The rolling rod harvester shows very impressive power output capabilities but requires careful consideration of its power processing and fabrication costs in practical implementations. This type of harvester also requires power processing and a step-down converter, to convert the harvested energy into a usable form. Additionally, the fabrication processes for this harvester are considered to be high-cost, which is a limiting factor in its practical application and scalability.

Figure 9. Rolling rod harvester (Kiziroglou et al., 2009)

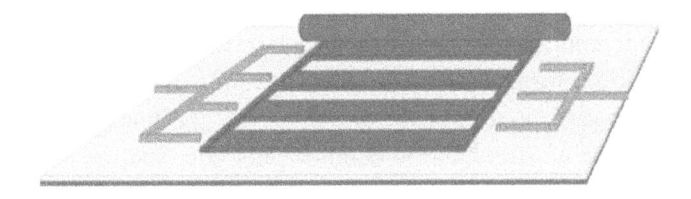

Varying Capacitance Machine

The varying capacitance machine shown in Figure 11 operates in a capacitance range of 40 - 9 pF and has a power capability of 100 µW. This machine is designed with capacitance poles, the capacitance can increase with the increase in the number of poles, which allows for greater power generation capability.

This machine is used for high-voltage direct current applications. Additionally, the researchers who developed this machine used 3D finite-element analysis software to model the capacitor, which brings a high level of detail and accuracy to the design and analysis of this energy harvester.

Figure 10. Varying capacitance machine (O'Donnell et al., 2009)

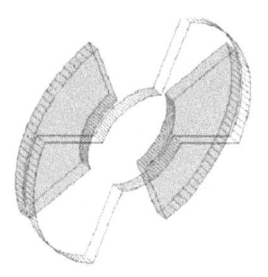

CYTOP Electret Generator

The CYTOP electret generator shown in Figure 12 was developed in 2006 and operates in a capacitance range of 25.4 - 39.8 pF and has a power capability of 280 µW. The CYTOP electret generator shows significant power generation capabilities but requires careful optimization and specific operating conditions to achieve optimal performance.

This generator requires power processing and rectification to convert the harvested energy into a usable form. One of the main features of this generator is that its power output depends on several factors, including the thickness of the electret, the gap between the electret and the counter electrode, and the overlapping area. These parameters need to be carefully adjusted to increase power generation (Lu & Suzuki, 2021).

Figure 11. CYTOP electret generator (Tsutsumino et al., 2006)

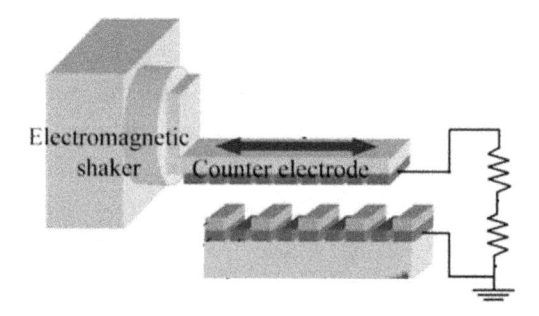

Single Wafer Floating Electrode Generator

The single-wafer floating electrode generator (Figure 13) was developed in 2007 and operates with a capacitance of 16.82 pF and has a power capability of 60 nW. This generator demonstrates a low power output but showcases innovative design features and fabrication techniques for energy-harvesting applications. This generator requires a power processing circuit, sensing transistor, and diode bridge for converting the harvested energy into a usable form. The main feature of this generator is its use of a CMOS floating gate and post-CMOS photo-resist process for its operation. This is considered a specialized fabrication process custom-made for this specific generator design.

Figure 12. Single wafer floating electrode generator (Ma et al., 2007)

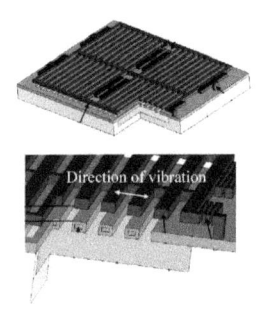

Tunable Capacitor

The tunable capacitor in Figure 14 was developed in 2002 and operates with a capacitance range of 1.34 - 0.32 pF. The tunable capacitor demonstrates a specific design focus on reducing parasitic capacitances and employing advanced actuation mechanisms for tuning, but specific power capabilities are not provided. This capacitor design utilizes angular vertical comb-drive actuators, which are commonly used in MEMS devices for actuation. Additionally, the capacitor uses a glass substrate to reduce parasitic capacitances, which can improve its performance in high-frequency applications (Arussy Ruth & Bao, 2020).

Figure 13. Tunable capacitor (Nguyen et al., 2002)

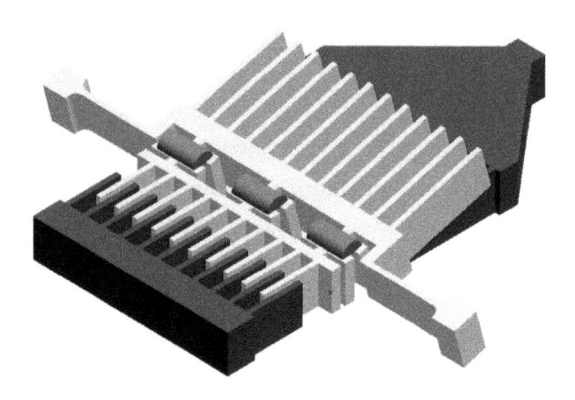

Electrostatic Swing Harvester

The electrostatic swing harvester developed by Reznikov et al. (2010) operates with a capacitance range of 0.2 - 4.5 pF and has a power capability of 0.75 µW (Figure 15). the electrostatic swing harvester shows a moderate power output and utilizes innovative design features for efficient energy harvesting.

This harvester is used with an oscillating, polarized charge pump for converting the harvested energy into usable electrical power. One of the main features of this harvester is the use of a homo-polar electret, which contains real charges of only one sign and induces a charge of the opposite sign in available electrodes. This design helps in generating a DC voltage greater than 3 V, making it suitable for portable electronic devices.

Figure 14. Electrostatic swing harvester (Reznikov et al., 2010)

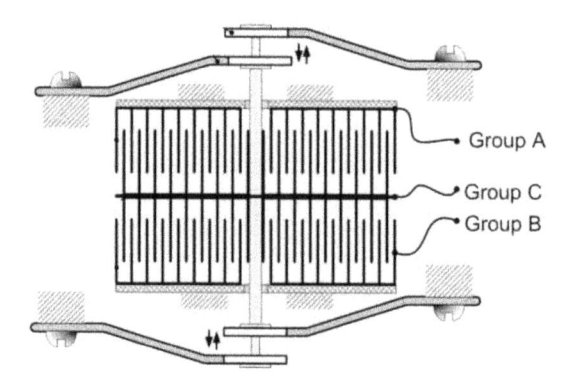

Multi-Pole Variable Capacitor Harvester

The multi-pole capacitor harvester (Figure 16) was developed in 2014 and operates with a capacitance range from 2.5 to 0.5 n F. It is a variable capacitor made of two parallel plates: rotor and stator. The stator is the stationary part of the capacitor while the rotor is the rotating part that rotates according to the applied wind energy. The number of poles of the rotor and the stator determines the amount of capacitance variation within a single rotation. This capacitor was chosen for the proposed wind harvesting system for several reasons. It is compact, can be easily coupled with the microturbine system, has a simple structure that is simple to fabricate and has a capacitance that can be modified geometrically, which can allow the harvester to collect more energy per cycle. It also requires a power processing circuit compared to other types.

Figure 15. Multi-pole variable capacitor plates (Aljadiri, 2014)

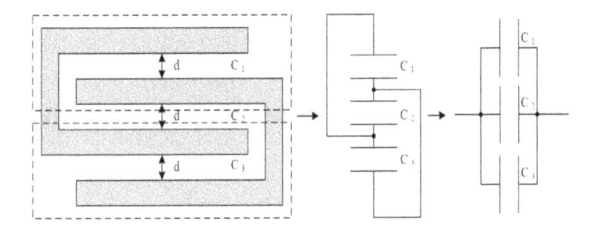

In summary, among these harvesters, the choice of the most suitable one depends on the specific application requirements, such as available mechanical movement, space restrictions, power requirements, and power processing needs. Each harvester has its advantages and disadvantages, and the selection should be based on a detailed analysis of these factors.

COMPARATIVE ANALYSIS OF POWER PROCESSING CIRCUITS

In some capacitor structures presented in the previous section, power processing, and conditioning circuits are required to improve the quality of the power delivered to a load or a storage device (Mitcheson 2006).

In a constant charge system, the main difficulty is processing the output power that works with a small charge at a high voltage. The voltage generated can be in the range of hundred volts, which must be stepped down to a lower voltage to be suitable for powering loads.

Many power processing circuits for harvesters have been reported in the literature, such as buck converters, buck-boost converters, linear voltage regulators, classical switching techniques, and switched-capacitor regulators. The selection of a suitable power processing circuit for the electrostatic harvester depends on many factors such as the output voltage, current, frequency, resonance behavior of the harvester, size of the harvester, and the mechanical movement that moves the capacitor (D'hulst n.d.). Three main types of regulators used for energy harvesters are explained next.

Linear Voltage Regulator

A linear voltage regulator is a type of voltage regulator that uses transistors in its circuit to regulate the output voltage to a fixed value. It is relatively simple and inexpensive compared to other types of voltage regulators. However, it is not very efficient because it dissipates excess energy as heat. This inefficiency limits its use in harvesting energy applications where power efficiency is important. Additionally, linear voltage regulators cannot achieve optimal load conditions, they are less effective when the load changes over time. Furthermore, the output voltage of a linear regulator is always lower than the input voltage, which is a limitation in some applications.

Classical Switching Regulators

Classical switching regulators, such as buck converters and flyback converters, are more efficient than linear regulators because they switch the input voltage on and off to regulate the output voltage. This switching process reduces energy losses and heat dissipation compared to linear regulators. However, classical switching regulators have other limitations. They typically operate in the low power range of micro-watts to milli-watts, which is not suitable for high-power applications. They also require a large inductor or transformer in their circuitry, which can increase the size and cost of the regulator. Additionally, they need a high switching frequency, which may lead to energy losses and reduced efficiency.

Switched-Capacitor Regulators

Switched-capacitor regulators use capacitors to store and transfer energy, providing step-up or step-down voltage conversion. They offer advantages such as better energy efficiency and a simpler control circuit compared to other types of regulators. However, switched-capacitor regulators have limitations. For example, if two capacitors with different initial energy levels are connected, energy losses can occur. They may also struggle to handle high deviations in vibration or frequency, which can affect their performance. Additionally, switched-capacitor regulators require a more complex control circuit compared to linear regulators.

When comparing the three types of power processing regulators, switched-capacitor regulators may offer a better solution for energy harvesting applications where size, efficiency, and cost are important factors.

COMPARATIVE ANALYSIS OF CONVERSION MECHANISMS

The comparative analysis of the conversion mechanism provides insights into the operation and efficiency of different electrostatic harvester systems focusing on constant charge and constant voltage systems. This information is valuable for researchers and engineers working on energy harvesting systems, helping them design more efficient and effective systems for various applications.

Constant Charge vs. Constant Voltage

Menninger and the MIT researchers compared constant charge and constant voltage approaches for electrostatic microgenerators. They achieved constant voltage operation by adding a large fixed capacitor in parallel with the variable capacitor. However, they noted that this arrangement leads to more initial energy loss. They also found that constant voltage operation is more efficient for maximizing power generation (Menninger, 1999).

Power Electronics for Constant Charge vs. Constant Voltage

Miranda's Ph.D. thesis concluded that power electronics for constant charge systems are easier to implement than for constant voltage systems. Constant voltage systems tend to generate low-level voltages within the breakdown limits of standard IC process technologies. The results were based on simulating variable capacitors and power electronics using MATLAB (Miranda, 2004).

Comparison of Electrostatic Microgenerator Conversion Mechanisms

Mitcheson et al. compared three main types of electrostatic microgenerator conversion mechanisms: constant voltage, constant charge switched, and electret-based continuous systems. For large electrostatic forces, constant charge designs are preferable while constant voltage is suitable for small electrostatic forces.

They identified that switched types can be controlled easily for maximal power point conversion but that will increase the complexity of the processing circuits (Mitcheson et al., 2004).

Power Processing Requirements for Constant Charge Systems

The study highlighted that the main issue with constant charge systems is that they work with small amounts of charge at high voltage. When the generator experiences acceleration, the capacitance of the variable capacitor decreases, and the voltage increases. As a result, the harvested voltage needs to be stepped down to lower voltages for powering low-power loads using a step-down converter (Mitcheson et al., 2004).

Review of Kinetic Energy Harvesting

The study highlighted three types of electrostatic harvesters: in-plane overlap converters, in-plane gap-closing converters, and out-of-plane gap-closing converters. They found that all three types can operate in both constant charge and constant voltage systems, with the constant voltage approach generating more energy. However, the use of a parallel capacitor in constant charge systems can increase energy levels to match constant voltage systems (Beeby et al., 2008).

BENEFITS AND DRAWBACKS OF ELECTROSTATIC HARVESTERS

Electrostatic energy harvesters present several advantages and disadvantages compared to other conventional energy harvesting methods. These features are essential in assessing the suitability of electrostatic energy harvesting for specific applications.

One of the key advantages of electrostatic energy harvesters is the compact design, making it suitable for situations where space is limited. Additionally, they are sensitive to low and high levels of mechanical vibration or rotation, enabling them to be used in both environments. Their ease of integration into small-scale systems further enhances their suitability for microelectronic devices (Daneshvar, Yuce, & Redouté, 2022).

Another benefit is the lack of a requirement for smart materials, which are often costly and challenging to manufacture. Furthermore, their energy density can be modified by changing the geometry of the variable capacitor structure, by modifying the size, shape, or arrangement of the capacitor's plates or the spacing between them. This alteration can affect the capacitance, sensitivity, and other performance characteristics that offer flexibility in design and optimization for specific applications. This, combined with their simple structure, contributes to improved reliability and reduced maintenance requirements (Aljadiri, 2014), (Daneshvar, Yuce, & Redouté, 2022).

Even with these benefits, electrostatic energy harvesting has some drawbacks. One significant limitation is its low power density compared to other energy harvesting approaches, which can be challenging to overcome, especially for applications requiring higher power output. Additionally, these harvesters may be sensitive to environmental conditions such as temperature and humidity, impacting their performance and reliability. The complexity of the electronics needed to condition and process the harvested energy also adds to the overall system efficiency, cost, and complexity (Zhu et al., 2021).

Another challenge is the restricted scalability of electrostatic energy harvesting systems. Scaling up these systems to harvest larger amounts of energy is possible but can be an issue due to limitations in the design and fabrication of larger-scale transducers. Additionally, the manufacturing processes for electrostatic transducers require precise control over factors such as electrode geometry, dielectric material properties, and gap distances. Reliability is also another significant challenge for electrostatic energy harvesting systems. The performance of these systems can be affected by factors such as mechanical deterioration, environmental conditions, and aging effects, which can lead to a decrease in efficiency and effectiveness over time. Addressing these challenges requires collaboration and integration of different disciplines that combine expertise in materials science and manufacturing technologies to improve the performance and reliability of electrostatic energy harvesting systems (Daneshvar, Yuce, & Redouté, 2022).

ELECTROSTATIC ENERGY HARVESTER APPLICATIONS

The applications of electrostatic energy harvesters are varied, ranging from powering small portable devices to powering monitoring and control systems in various industries((Roundy & Wright, 2004). Their ability to harvest energy from the environment makes them a valuable technology for creating sustainable and energy-efficient systems.

- **Wearable Electronics:** In wearable electronic devices, electrostatic energy harvesters can convert mechanical vibrations from body movements into electrical energy. The harvested energy can power sensors, displays, and

wireless communication modules in wearables, extending their battery life or eliminating the need for batteries.
- **Internet of Things:** Electrostatic energy harvesters can be used for powering low-power IoT devices, especially in inaccessible and remote areas. They can be built into smart home devices, trackers, and IoT sensors to provide sustainable and maintenance-free power sources.
- **Wireless Sensor Networks:** WSNs can be powered by Electrostatic energy harvesters, especially the WSNs used for environmental monitoring, smart infrastructure, and agriculture. These harvesters can harvest energy from ambient sources like vibrations, enabling long-term self-sufficient operation of sensor networks without the need for battery replacements.
- **Structural Health Monitoring:** Electrostatic energy harvesters can be used in structural systems to power sensors that monitor the condition of structures such as bridges, buildings, and aircraft. They allow continuous monitoring of structural integrity without the need for external power sources or frequent maintenance.
- **Emergency Backup Power:** In case of emergencies where access to power sources is interrupted, electrostatic energy harvesters can provide a reliable backup power source for critical systems. These harvesters can be integrated into emergency lighting, communication devices, and other essential equipment to ensure continuous operation during emergencies.
- **Smart roads:** Smart roads can utilize electrostatic transducers to convert mechanical energy from passing vehicles into electrical energy. This harvested energy can be used to power roadside sensors, lighting systems, and other smart infrastructure components, reducing the dependence on conventional power sources.

ELECTROSTATIC TRANSDUCERS AND SMART ROADS

Smart roads or smart highways are roadways that incorporate the use of various digital and electronic systems to improve efficiency, safety, user experience, and sustainability. The smart roads represent a key component of smart cities and future transportation systems, they offer benefits to both road users and the environment. Some of the technologies used can include communication infrastructure, energy harvesting, sensor systems, smart lighting, traffic management systems, environmental monitoring, and vehicle detection and identification. Electrostatic transducers can play an important role in the development of smart roadways. They can offer advanced solutions for sensing and energy harvesting applications. The list of applications can be summarized to:

- **Traffic Monitoring and Management:** Smart roads can use electrostatic transducers to detect and monitor traffic flow parameters such as vehicle speed, weight, and occupancy. This data can help in optimizing traffic flow, reducing congestion, and improving safety. Electrostatic transducers can sense the presence of vehicles and adjust traffic signals accordingly to minimize the waiting time and improve the efficiency of smart roads.
- **Energy Harvesting:** Smart roads can use electrostatic transducers to convert mechanical energy from traffic into electrical energy. Electrostatic harvesters can be embedded in the road and can generate electricity when subjected to vibrations caused by passing vehicles. The harvested energy can be used to power lighting systems roadside sensors, and other smart infrastructure components, reducing the dependence on conventional power sources.
- **Environmental Monitoring:** Smart roads can use electrostatic sensors to monitor environmental conditions such as temperature, air quality, and humidity. For example, Electrostatic sensors can detect temperature through changes in electrical properties that occur with temperature variations. When the temperature changes, it affects the dielectric constant of the material between the parallel plates of the capacitor which in turn changes the capacitance of the electrostatic sensor. This data can be used to implement adaptive environmental control systems that help improve the overall environmental sustainability of the road infrastructure.
- **Communication and Connectivity:** Smart roads can use electrostatic transducers as communication hubs, allowing for vehicle-to-infrastructure and vehicle-to-vehicle communication. This system can improve road safety, allow autonomous driving structures, and improve overall transportation efficiency.
- **Structural Health Monitoring:** smart roads can use Electrostatic transducers for structural health monitoring of bridges, tunnels, and other infrastructure components. By detecting mechanical stress which results in changes in capacitance of the transducers. They can provide early warnings of potential structural issues, enabling active maintenance and preventing disastrous failures.

Overall, the use of electrostatic transducers in smart roads can transform roads into smart and sustainable infrastructure systems. By harvesting energy from the environment and enabling modern sensing and communication systems, these transducers contribute to the development of smarter, safer, and more efficient transportation networks. A case study related to UAE smart roads is presented in the next section.

UAE SMART ROAD CASE STUDY

In UAE, especially Dubai, smart roads are a key component of the smart city initiatives, aiming to enhance road safety, traffic efficiency, and sustainability. Smart roads use various technologies such as cameras, sensors, and data analytics to monitor traffic conditions and manage congestion effectively (Figure 17).

Dubai roads use smart signage and digital message boards to provide drivers with real-time information about traffic conditions, road closures, and alternative route options (AESYS, 2019). Additionally, Dubai has implemented the use of smart traffic management systems that use sensors, and cameras to monitor traffic conditions in real-time. These systems can detect congestion, accidents, and other traffic violations, allowing authorities to take practical measures to manage traffic flow and improve road safety (Khaleej Times, 2020). Another initiative is the Smart lighting systems which are arranged to adjust street light intensity based on the presence of vehicles and pedestrians to save energy and reduce light pollution (Smart Lighting, 2014).

Connected vehicles and parking management is another initiative implemented in Dubai roads using connected vehicle technology, which allows vehicles to communicate with each other and with infrastructure. This improves safety by providing drivers with real-time information about potential hazards and road conditions and the Smart parking systems use sensors to detect free parking spaces and provide this information to drivers via mobile applications. This helps to reduce traffic congestion and improve the overall parking experience.

Furthermore, as part of the city's goal to develop into a smart and futuristic city, Dubai has been working on testing autonomous vehicles. To make the placement of autonomous vehicles easier, the Dubai Roads and Transport Authority has been actively involved in drafting laws, infrastructure, and policies for autonomous transportation. These regulations ensure the safety of passengers, pedestrians, and other road users while allowing the use of smart innovation and technological advancements in transportation as shown in Figure 18 (HiDubai, 2023).

Drawing from prior research on the utilization of electrostatic transducers, they can potentially be used to improve road sustainability and safety in the UAE. This can be through various applications such as:

- **Solar energy harvesting** is particularly successful in the UAE and Dubai due to their ideal location for abundant sunlight. This makes the conversion of solar energy into electrical energy highly efficient. The energy harvested from sunlight can power various smart road features like lighting, signage, and sensors. Research studies have demonstrated the feasibility of using electrostatic transducers to separate and collect charges within solar cells.

The Principles and Applications of Electrostatic Transducers

These transducers play a crucial role in creating and sustaining the electric field needed for efficient charge separation and collection, thereby enhancing the overall efficiency of solar cells (Li & Wu, 2019).

Figure 16. UAE smart road initiatives (smart traffic management, smart signage, smart lighting system, connected vehicles) (Khaleej Times, 2020)

Figure 17. Redefining transportation in Dubai using autonomous vehicles (HiDubai, 2023)

- **Air quality monitoring:** Electrostatic sensors can be used to monitor air quality by detecting pollutants. These sensors work by using an electric field to attract and collect particles from the air. When pollutants such as nitrogen dioxide pass through the sensor, they become electrically charged, allowing the sensor to detect their presence. This data can be used to develop strategies to improve air quality, such as implementing emission controls, adjusting traffic flow, or identifying sources of pollution (Al-Jaroodi & Mohamed, 2016).
- **Real-time data collection:** Dubai smart roads require real-time data on traffic flow, vehicle speed, and road conditions to optimize traffic signals and improve traffic flow. This can be done through the use of electrostatic sensors embedded in the roads that can collect real-time data by detecting the changes in the electric field caused by the pressure applied by the passing vehicles. Electrostatic sensors can measure the capacitance between electrodes embedded in the road surface and the ground. When a vehicle passes over the sensor, it changes the electric field, which causes a change in capacitance that can be measured. These changes can be collected and sent to a central control system, where they can be processed and used to optimize traffic signals, monitor road conditions, and provide valuable information for traffic management (Tientrakool & Rakphongphairoj, 2017).
- **Structural health monitoring:** With the continuous development of modern infrastructure in Dubai, the structural health monitoring of roads and bridges is a necessity. Electrostatic transducers can be used for the health monitoring. By monitoring vibrations, these sensors can detect irregularities that may indicate structural damage or deterioration. Similarly, by measuring strain and deformation, they can provide a valuable understanding of the overall health of the bridge and help identify areas that may require maintenance or repair (Zhang et al., 2018).
- **Vehicle-to-infrastructure communication:** since Dubai roads are pioneering autonomous vehicle technology, communication between vehicles and infrastructure is required to improve road safety and traffic efficiency. The Data Transmission can be done through the use of Electrostatic sensor communication systems by modulating the electric field around the sensor, they can transmit information wirelessly to nearby infrastructure or vehicles, enabling real-time communication. The other advantage of electrostatic transducers is their low power consumption. This is crucial in this communication, where devices may need to operate continuously without access to external power sources. Electrostatic transducers can operate efficiently on harvested energy, ensuring reliable communication (Li, 2019).

- **Adaptive lighting systems:** Electrostatic energy harvesting can be used for adaptive lighting systems. These lights can adjust their brightness based on ambient light conditions and traffic density, improving energy efficiency. Electrostatic energy harvester can capture energy from the surrounding environment, such as vibrations or changes in pressure from vehicles passing by or wind. This method provides a cost-effective solution for powering adaptive lighting systems. Once the harvesting system is in place, it can generate energy continuously without the need for external power sources or regular maintenance (Kazanci & Olesen, 2017).

Based on the potential applications mentioned above, further future research in electrostatic transducer applications and smart roads is required to explore several promising directions for UAE roads. One key area is improving energy harvesting efficiency by optimizing the conversion of mechanical vibrations into electrical energy. Researchers need to find the best possible ways of integrating energy-harvesting transducers into road surfaces, creating sustainable smart roads that contribute to renewable energy sources. Additionally, there should be further research focused on real data traffic management, using data from transducers to optimize traffic flow and safety. Advanced structural health monitoring is another area of interest, exploring how changes in the electrostatic field can help in the early detection of road defects. Another area of research is integrating multiple functions into a single transducer using nanotechnology for better overall system performance. Ensuring privacy and data security in smart road systems, integrating with 5G and IoT networks, and developing algorithms for real-time data processing are crucial for the future of these technologies. These research directions reflect the ongoing development of technology and its potential to transform environmental and infrastructure connections. This future research journey is about shaping a future where roads are more than just paths; they are part of developing cities. It shows how human intelligence can change the world. With each step forward, we are painting a picture of tomorrow, where roads are safe, brighter, and more connected than ever before.

CONCLUSION

The chapter starts with a detailed introduction to electrostatic transducers, providing an overview of their development and explaining their fundamental principles. It discusses how electrostatic transducers convert mechanical energy into electrical energy through the use of variable capacitors. The principle of operation for electrostatic harvesters is then explained in detail, outlining the key components and conversion mechanisms required for energy harvesting.

Variable capacitor structures used in electrostatic harvesters are explained with a focus on their design considerations and practical implementation requirements. An in-depth analysis of electrostatic conversion mechanisms follows, comparing the different approaches to energy conversion based on previous research studies. A comparative study of capacitor structures and processing circuits provides valuable insights into the optimal design choices for electrostatic harvesters. Factors such as efficiency, power output, and scalability are considered, helping readers understand the adjustments required in designing electrostatic energy harvesting systems.

The chapter then shifts its focus to the applications of electrostatic harvesters, with a specific emphasis on their integration into smart road infrastructure. It discusses how electrostatic harvesters can be implanted in roads to harvest energy from passing vehicles, providing a sustainable power source for smart road systems. The chapter includes a case study on smart road development in the UAE, highlighting the possible uses of electrostatic transducers in the country's smart road infrastructure. The case study examines how the UAE can benefit from energy harvesting technology to enhance its road networks' efficiency and sustainability. The UAE's smart road initiatives, such as the Dubai Autonomous Transportation Strategy, are discussed. These initiatives aim to transform the country's transportation system by integrating advanced technologies, including electrostatic transducers, to improve traffic management, reduce congestion, and enhance road safety. The case study also explores specific areas in the UAE where electrostatic transducers can be implemented. For example, in Dubai, electrostatic harvesters can be used for powering adaptive lighting systems for roads, solar energy harvesting, road health monitoring, real-time data collection, and many more. These areas can lead to significant energy savings and improved sustainability.

Overall, the UAE smart road case study serves as a real-world example of how energy harvesting systems and smart sensing can be effectively utilized to create smarter, more sustainable road infrastructure. It demonstrates the potential of this technology to transform transportation systems and improve the quality of life for residents and visitors. In conclusion, the chapter offers an overview of electrostatic transducers, from their basic principles to their practical applications. It serves as a valuable resource for researchers and engineers interested in the field of energy harvesting and smart infrastructure.

REFERENCES

AESYS. (2019). *Variable Message Sign - VMS*. Retrieved from https://www.aesys.com/products-solutions/traffic/variable-message-sign-vms.html

Al-Jaroodi, J., & Mohamed, N. (2016). Smart city architecture and its applications based on IoT. *Procedia Computer Science*, *83*, 36–43.

Al Rahis, L. (2016). *Electrostatic Energy Harvesting Interface Circuits for Microsystems* (Master's thesis). Khalifa University.

Ali, A., Shaukat, H., Bibi, S., Altabey, W. A., Noori, M., & Kouritem, S. A. (2023). Recent progress in energy harvesting systems for wearable technology. *Energy Strategy Reviews*, *49*, 101124. doi:10.1016/j.esr.2023.101124

Allen, M. G. (2000). *Introduction to Micromachining*. CRC Press.

Arussy Ruth, S. R., & Bao, Z. (2020). Designing tunable capacitive pressure sensors based on material properties and microstructure geometry. *ACS Applied Materials & Interfaces*, *12*(52), 58301–58316. doi:10.1021/acsami.0c19196 PMID:33345539

Askari, M., Hutchins, D. A., Thomas, P. J., Astolfi, L., Watson, R. L., Abdi, M., Ricci, M., Laureti, S., Nie, L., Freear, S., Wildman, R., Tuck, C., Clarke, M., Woods, E., & Clare, A. T. (2020). Additive manufacturing of metamaterials: A review. *Additive Manufacturing*, *36*, 101562. doi:10.1016/j.addma.2020.101562

Atlas, N. (2014, July 15). *Mechanical vibrations used to generate electricity*. Retrieved from https://newatlas.com/mechanical-vibration-generate-electricity/34701/

Beeby, S. P., Torah, R. N., & Tudor, M. J. (2008). Kinetic energy harvesting. In *Act Workshop on Innovative Concepts*. Esa-Estec.

Boisseau, S., & Despesse, G. (2012). Energy harvesting, wireless sensor networks & opportunities for industrial applications. *EE Times*. Retrieved February 15, 2024, from http://www.eetimes.com/design/smart-energy-design/4237022/Energy-harvesting-wireless-sensor-networks-opportunities-for-industrial-applications

D'hulst, R., Sterken, T., Puers, R., & Driesen, J. (n.d.). *Requirements for power electronics used for energy harvesting devices*. Retrieved January 14, 2024, from https://www.researchgate.net/publication/242096692_Requirements_for_Power_Electronics_used_for_Energy_Harvesting_Devices

Daneshvar, S. H., Yuce, M. R., & Redouté, J. M. (2022). Electrostatic Harvesters Overview and Applications. In *Design of Miniaturized Variable-Capacitance Electrostatic Energy Harvesters*. Springer. doi:10.1007/978-3-030-90252-0_2

Despesse, G. (2008). In-plane gap closing capacitor for energy harvesting. *Journal of Energy Harvesting and Systems*, *4*(3), 16.

Despesse, G., Jager, T., Condemine, C., & Berger, P. D. (2008). Mechanical vibrations energy harvesting and power management. In *IEEE Sensors Conference* (pp. 29-32). 10.1109/ICSENS.2008.4716375

Espera, A. H. Jr, Dizon, J. R. C., Chen, Q., & Advincula, R. C. (2019). 3D-printing and advanced manufacturing for electronics. *Progress in Additive Manufacturing*, *4*(3), 245–267. doi:10.1007/s40964-019-00077-7

Hayt, W. H., & Kemmerly, J. E. (2001). *Engineering Circuit Analysis*. McGraw-Hill Education.

HiDubai. (2023). *Dubai's Autonomous Vehicles: Redefining Transportation in the City*. Retrieved from https://focus.hidubai.com/driverless-cars-dubai/

Jeong, C., Joung, C., Lee, S., Feng, M. Q., & Park, Y.-B. (2020). Carbon nanocomposite based mechanical sensing and energy harvesting. *International Journal of Precision Engineering and Manufacturing-Green Technology*, *7*(1), 247–267. doi:10.1007/s40684-019-00154-w

Kazanci, O. B., & Olesen, B. W. (2017). Review of occupant sensing for smart lighting systems. *Energy and Building*, *139*, 607–620.

Khaleej Times. (2020). *Exclusive: Inside centre that monitors Dubai traffic 24/7*. Retrieved from https://www.khaleejtimes.com/transport/exclusive-inside-centre-that-monitors-dubai-traffic-24-7

Kiziroglou, M. E., & .... (2009). Rolling rod harvester for energy harvesting. *Journal of Energy Harvesting and Systems*, *2*(1), 2–10.

Kiziroglou, M. E., He, C., & Yeatman, E. M. (2009). Rolling rod electrostatic microgenerator. *IEEE Transactions on Industrial Electronics*, *56*(4), 1101–1108. doi:10.1109/TIE.2008.2004381

Le, H. V., Le, T. P., & Le, T. T. (2009). In-plane overlap capacitor for energy harvesting. *Journal of Energy Harvesting and Systems*, *1*(2), 134.7-203.7.

Le, T. T., Halvorsen, E., & Zuo, L. (2009). In-plane overlap variable capacitor for electrostatic energy harvesting. In *2009 IEEE 22nd International Conference on Micro Electro Mechanical Systems* (pp. 753-756). IEEE.

Li, G., & Wu, J. (2019). Solar energy harvesting by using electrostatic transducers. *Energy Conversion and Management*, *195*, 96–104. doi:10.1016/j.enconman.2019.05.051

Li, J. (2021). *Electrostatic MEMS Energy Harvesting for Sensor Powering* (Publication No. 28414789) [Doctoral dissertation, Rensselaer Polytechnic Institute]. ProQuest Dissertations Publishing

Li, L. (2019). Vehicle-to-infrastructure communication for connected and autonomous vehicles: A survey. *IEEE Transactions on Intelligent Transportation Systems*, 20(1), 378–403.

Lighting, S. (2014, July 15). *Dubai to expand smart LED lighting project*. Retrieved from https://smart-lighting.es/dubai-smart-led-lighting/

Lu, J., & Suzuki, Y. (2021). Push-button energy harvester with ultra-soft all-polymer piezoelectret. In *2021 20th International Conference on Solid-State Sensors, Actuators, and Microsystems (TRANSDUCERS)*. IEEE. 10.1109/PowerMEMS54003.2021.9658397

Ma, S. K., Zhang, X., Xu, Y., & Miao, J. (2007). Single wafer floating electrode for energy harvesting. In *Proceedings of the PowerMEMS 2007 Workshop* (pp. 104-107). Academic Press.

Ma, Y. (2007). Single wafer floating electrode for energy harvesting. *Journal of Energy Harvesting and Systems,* 6(2), 16.82.

Meninger, S., Miranda, J. M., Chandrakasan, J. L. A., Slocum, A., Schmidt, M., & Amirtharajah, R. (2001). Vibration to electric energy conversion. *IEEE Transactions Very Large-Scale Integration (VLSI) Systems*, 9, 64–76.

Menninger, S., Miranda, J. O. M., Amirtharajah, R., Chandrakasan, A., & Lang, J. (1999). Vibration-to-electric energy conversion. In *Low Power Electronics and Design Proceedings. 1999 International Symposium* (pp. 48-53). 10.1145/313817.313840

Miranda, J. O. M. (2004). *Electrostatic vibration-to-electric energy conversion* (PhD Thesis). Massachusetts Institute of Technology, Massachusetts, USA

Mitcheson, P. D., Green, T. C., & Yeatman, E. M. (2004). Power processing circuits for electromagnetic, electrostatic, and piezoelectric inertial energy scavengers. *Microsystem Technologies*, 10(3-4), 1629–1635.

Mitcheson, P. D., Sterken, T., He, C., Kiziroglou, M., Yeatman, E. M., & Puers, R. (2008). *Electrostatic micro-generators*. Retrieved February 3, 2014, from https://journals.sagepub.com/doi/10.1177/002029400804100404

Nguyen, H., Hah, D., Patterson, P. R., Piywattanametha, W., & Wu, M. C. (2002). A novel MEMS tunable capacitor based on angular vertical comb drive actuators. In *Solid-State Sensor, Actuator and Microsystems Workshop* (pp. 277-280). 10.31438/trf.hh2002.69

Nguyen, N. V. (2002). Tunable capacitor for energy harvesting. *Journal of Energy Harvesting and Systems*, 7(3), 32-34.

O'Donnell, R., Schofield, N., Smith, A. C., & Cullen, J. (2009). Design concepts for high-voltage variable-capacitance DC generators. *IEEE Transactions on Industry Applications*, 45(5), 1778–1784. doi:10.1109/TIA.2009.2027545

O'Donnell, T. (2009). Varying capacitance machine for energy harvesting. *Journal of Energy Harvesting and Systems*, 3(4), 9–40.

Reznikov, M. (2010). Electrostatic swing energy harvester. *Proceedings of ESA Annual Meeting on Electrostatics*. Retrieved from https://www.electrostatics.org/images/ESA2010_G3_Reznikov.pdf

Reznikov, M. (2010). Electrostatic swing harvester for energy harvesting. *Journal of Energy Harvesting and Systems*, 8(4), 2-5.

Roundy, S. (2003). *Energy scavenging for wireless sensor nodes with a focus on vibration to electricity conversion* (PhD Thesis). University of California, Berkeley.

Roundy, S., Wright, P. K., & Rabaey, J. M. (2004). *Energy scavenging for wireless sensor networks with a special focus on vibrations*. Kluwer Academic Publishers. doi:10.1007/978-1-4615-0485-6

Senturia, S. D. (2006). *Microsystem Design*. Springer Science & Business Media.

Sonika, S. K., Samanta, S., Srivastava, A. K., Biswas, S., Alsharabi, R. M., & Rajput, S. (2022). *Conducting polymer nanocomposite for energy storage and energy harvesting systems*. Review Article, 2266899.

Tientrakool, P., & Rakphongphairoj, V. (2017). Real-time traffic data collection using electrostatic sensors. *Transportation Research Procedia*, 25, 2561–2574. doi:10.1016/j.trpro.2017.05.495

Tsutsumino, T. (2006). CYTOP electret generator for energy harvesting. *Journal of Energy Harvesting and Systems*, 5(1), 25.4-39.8.

Tsutsumino, T. S., Suzuki, Y., Kasagi, N., Kashiwagi, K., & Morizawa, Y. (2006). Efficiency evaluation of microseismic electret power generator. In *Proceedings of the 23rd Sensor Symposium* (pp. 521-524). Academic Press.

Tutorials, E. (n.d.). *Introduction to Digital Logic Circuits*. Retrieved from https://www.electronics-tutorials.ws/io/io_1.html

White, R. M., & Sessler, G. M. (1999). Handbook of Sensors and Actuators: Electrostatic Actuators and Sensors. Elsevier.

Wolf, S., & Tauber, R. N. (2018). *Silicon Processing for the VLSI Era: Process Integration* (Vol. 1). Lattice Press.

Zhang, J. (2018). Structural health monitoring of bridges using electrostatic sensors. *Structural Health Monitoring*, *17*(1), 88–100.

Zhou, J., Baker, M., & Hu, Y. (2008). Energy harvesting for pervasive computing. *IEEE Pervasive Computing*, *7*(4), 14–21. doi:10.1109/MPRV.2008.78

Zhu, D., Bai, Y., Li, Y., & Qin, Y. (2021). Recent progress and perspectives on electrostatic energy harvesting: From fundamentals to applications. *Nano Energy*, *86*, 106075.

ADDITIONAL READING

Aljadiri, R. T. (2014). *Modeling and design of electrostatic-based wind energy harvester* [Doctoral dissertation, Coventry University].

Hayt, W. H., & Kemmerly, J. E. (2001). *Engineering Circuit Analysis*. McGraw-Hill Education.

KEY TERMS AND DEFINITIONS

Adaptive Lighting Systems: Lighting systems that adjust their brightness based on ambient light conditions and traffic density.

Electret-Based Systems: Systems that use continuous electrets, which are materials with a quasi-permanent electric charge, in sensors and energy harvesting.

Electrostatic Transducers: Devices that convert mechanical energy into electrical energy through the use of variable capacitors.

Energy Harvesting: The process of capturing and converting ambient energy into usable electrical energy.

IoT (Internet of Things): The network of interconnected devices that can communicate and exchange data.

MEMS Devices: Microelectromechanical systems that integrate mechanical and electrical components on a microscale to create sensors, actuators, and other devices.

Smart Roads: Road infrastructure embedded with sensors and smart technology to improve traffic management, reduce congestion, and enhance safety.

Structural Health Monitoring: The process of monitoring the structural condition of infrastructure, such as bridges, roads, and tunnels using sensors to detect damage and prevent failure.

Variable Capacitor Structures: Capacitor designs that can change their capacitance are often used in energy harvesting applications to optimize energy conversion.

Vehicle-to-Infrastructure Communication: Communication between vehicles and infrastructure to improve road safety and traffic efficiency.

Compilation of References

Abid, M. S., Ahshan, R., Al Abri, R., Al-Badi, A., & Albadi, M. (2024). Techno-economic and environmental assessment of renewable energy sources, virtual synchronous generators, and electric vehicle charging stations in microgrids. *Applied Energy*, *353*, 122028. doi:10.1016/j.apenergy.2023.122028

AbuElrub, A., Hamed, F., & Saadeh, O. (2020). Microgrid integrated electric vehicle charging algorithm with photovoltaic generation. *Journal of Energy Storage*, *32*, 101858. doi:10.1016/j.est.2020.101858

AESYS. (2019). *Variable Message Sign - VMS*. Retrieved from https://www.aesys.com/products-solutions/traffic/variable-message-sign-vms.html

Agrawal, A. V., Magulur, L. P., Priya, S. G., Kaur, A., Singh, G., & Boopathi, S. (2023). Smart Precision Agriculture Using IoT and WSN. In Advances in Information Security, Privacy, and Ethics (pp. 524–541). IGI Global. doi:10.4018/978-1-6684-8145-5.ch026

Agrawal, A. V., Shashibhushan, G., Pradeep, S., Padhi, S. N., Sugumar, D., & Boopathi, S. (2024). Synergizing Artificial Intelligence, 5G, and Cloud Computing for Efficient Energy Conversion Using Agricultural Waste. In B. K. Mishra (Ed.), Practice, Progress, and Proficiency in Sustainability. IGI Global. doi:10.4018/979-8-3693-1186-8.ch026

Ahmad, S., Abdul Mujeebu, M., & Farooqi, M. A. (2019). Energy harvesting from pavements and roadways: A comprehensive review of technologies, materials, and challenges. *International Journal of Energy Research*, *43*(6), 1974–2015. doi:10.1002/er.4350

Al Rahis, L. (2016). *Electrostatic Energy Harvesting Interface Circuits for Microsystems* (Master's thesis). Khalifa University.

Al Wahedi, A., & Bicer, Y. (2022). Techno-economic optimization of novel stand-alone renewables-based electric vehicle charging stations in Qatar. *Energy*, *243*, 123008. doi:10.1016/j.energy.2021.123008

Ali, A., Shaukat, H., Bibi, S., Altabey, W. A., Noori, M., & Kouritem, S. A. (2023). Recent progress in energy harvesting systems for wearable technology. *Energy Strategy Reviews*, *49*, 101124. doi:10.1016/j.esr.2023.101124

Compilation of References

Al-Jaroodi, J., & Mohamed, N. (2016). Smart city architecture and its applications based on IoT. *Procedia Computer Science*, *83*, 36–43.

Allen, M. G. (2000). *Introduction to Micromachining*. CRC Press.

Alsharoa, A., Ghazzai, H., Kamal, A. E., & Kadri, A. (2017). Optimization of a power splitting protocol for two-way multiple energy harvesting relay system. *IEEE Transactions on Green Communications and Networking*, *1*(4), 444–457. doi:10.1109/TGCN.2017.2724438

AltEnergyMag. (2018). *Piezoelectric Power Generation Automotive Tires*. https://www.altenergymag.com/article/2017/12/1-article-for-2018-piezoelectric-power-generation-in-automotive-tires/27642/

Andriopoulou, S. (2012). *A review on energy harvesting from roads*. Academic Press.

Arussy Ruth, S. R., & Bao, Z. (2020). Designing tunable capacitive pressure sensors based on material properties and microstructure geometry. *ACS Applied Materials & Interfaces*, *12*(52), 58301–58316. doi:10.1021/acsami.0c19196 PMID:33345539

Askari, M., Hutchins, D. A., Thomas, P. J., Astolfi, L., Watson, R. L., Abdi, M., Ricci, M., Laureti, S., Nie, L., Freear, S., Wildman, R., Tuck, C., Clarke, M., Woods, E., & Clare, A. T. (2020). Additive manufacturing of metamaterials: A review. *Additive Manufacturing*, *36*, 101562. doi:10.1016/j.addma.2020.101562

Atlas, N. (2014, July 15). *Mechanical vibrations used to generate electricity*. Retrieved from https://newatlas.com/mechanical-vibration-generate-electricity/34701/

Baballe, M. A., & Bello, M. I. (2022). Gas leakage detection system with alarming system. *Review of Computer Engineering Research*, *9*(1), 30–43. doi:10.18488/76.v9i1.2984

Bai, S., & Liu, C. (2021). Overview of energy harvesting and emission reduction technologies in hybrid electric vehicles. *Renewable & Sustainable Energy Reviews*, *147*, 111188. doi:10.1016/j.rser.2021.111188

Bartolome, E. (2010). Signal conditioning for piezoelectric sensors. *Texas Instruments Analog Applications Journal, 10*.

Beeby, S. P., Torah, R. N., & Tudor, M. J. (2008). Kinetic energy harvesting. In *Act Workshop on Innovative Concepts*. Esa-Estec.

Bikash Chandra Saha, M. S., Deepa, R., Akila, A., & Sai Thrinath, B. V. (2022). IOT based smart energy meter for smart grid. Academic Press.

Bilal, M., Ahmad, F., & Rizwan, M. (2023). Techno-economic assessment of grid and renewable powered electric vehicle charging stations in India using a modified metaheuristic technique. *Energy Conversion and Management*, *284*, 116995. doi:10.1016/j.enconman.2023.116995

Boisseau, S., & Despesse, G. (2012). Energy harvesting, wireless sensor networks & opportunities for industrial applications. *EE Times*. Retrieved February 15, 2024, from http://www.eetimes.com/design/smart-energy-design/4237022/Energy-harvesting-wireless-sensor-networks-opportunities-for-industrial-applications

Boopathi, S., Kumar, P. K. S., Meena, R. S., Sudhakar, M., & Associates. (2023). Sustainable Developments of Modern Soil-Less Agro-Cultivation Systems: Aquaponic Culture. In Human Agro-Energy Optimization for Business and Industry (pp. 69–87). IGI Global.

Boopathi, S. (2023). Internet of Things-Integrated Remote Patient Monitoring System: Healthcare Application. In A. Suresh Kumar, U. Kose, S. Sharma, & S. Jerald Nirmal Kumar (Eds.), Advances in Healthcare Information Systems and Administration. IGI Global., doi:10.4018/978-1-6684-6894-4.ch008

Boopathi, S. (2023). Securing Healthcare Systems Integrated With IoT: Fundamentals, Applications, and Future Trends. In A. Suresh Kumar, U. Kose, S. Sharma, & S. Jerald Nirmal Kumar (Eds.), Advances in Healthcare Information Systems and Administration. IGI Global. doi:10.4018/978-1-6684-6894-4.ch010

Boopathi, S., & Davim, J. P. (2023). Applications of Nanoparticles in Various Manufacturing Processes. In S. Boopathi & J. P. Davim (Eds.), Advances in Chemical and Materials Engineering. IGI Global. doi:10.4018/978-1-6684-9135-5.ch001

Boopathi, S., & Kanike, U. K. (2023). Applications of Artificial Intelligent and Machine Learning Techniques in Image Processing. In B. K. Pandey, D. Pandey, R. Anand, D. S. Mane, & V. K. Nassa (Eds.), Advances in Computational Intelligence and Robotics. IGI Global. doi:10.4018/978-1-6684-8618-4.ch010

Boopathi, S., Khare, R., Jaya Christiyan, K. G., Muni, T. V., & Khare, S. (2023). Additive Manufacturing Developments in the Medical Engineering Field. In R. Keshavamurthy, V. Tambrallimath, & J. P. Davim (Eds.), Advances in Chemical and Materials Engineering. IGI Global. doi:10.4018/978-1-6684-6009-2.ch006

Boopathi, S., Pandey, B. K., & Pandey, D. (2023). Advances in Artificial Intelligence for Image Processing: Techniques, Applications, and Optimization. In B. K. Pandey, D. Pandey, R. Anand, D. S. Mane, & V. K. Nassa (Eds.), Advances in Computational Intelligence and Robotics. IGI Global. doi:10.4018/978-1-6684-8618-4.ch006

Boopathi, S., & Sivakumar, K. (2016). Optimal parameter prediction of oxygen-mist near-dry wire-cut EDM. *Inderscience: International Journal of Manufacturing Technology and Management*, 30(3–4), 164–178. doi:10.1504/IJMTM.2016.077812

Boopathi, S., Sureshkumar, M., & Sathiskumar, S. (2022). Parametric Optimization of LPG Refrigeration System Using Artificial Bee Colony Algorithm. *International Conference on Recent Advances in Mechanical Engineering Research and Development*, 97–105.

Boopathi, S., Sureshkumar, M., & Sathiskumar, S. (2023). Parametric Optimization of LPG Refrigeration System Using Artificial Bee Colony Algorithm. In S. Tripathy, S. Samantaray, J. Ramkumar, & S. S. Mahapatra (Eds.), *Recent Advances in Mechanical Engineering* (pp. 97–105). Springer Nature Singapore. doi:10.1007/978-981-19-9493-7_10

Boopathi, S., Umareddy, M., & Elangovan, M. (2023). Applications of Nano-Cutting Fluids in Advanced Machining Processes. In S. Boopathi & J. P. Davim (Eds.), Advances in Chemical and Materials Engineering. IGI Global. doi:10.4018/978-1-6684-9135-5.ch009

Bosso, N., Magelli, M., & Zampieri, N. (2021). Application of low-power energy harvesting solutions in the railway field: A review. *Vehicle System Dynamics*, *59*(6), 841–871. doi:10.1080/00423114.2020.1726973

Budhiraja, I., Kumar, N., Tyagi, S., Tanwar, S., & Guizani, M. (2021). SWIPT-enabled D2D communication underlaying NOMA-based cellular networks in imperfect CSI. *IEEE Trans. on Vech. Tech*, *70*(1), 692–699.

Cady, W. G. (2018). *Piezoelectricity: Volume Two: An Introduction to the Theory and Applications of Electromechanical Phenomena in Crystals*. Courier Dover Publications.

Ceramics, M. E. (2001). Introduction Piezoelectric Ceramics. *Technical Information*.

Chandrika, V. S., Sivakumar, A., Krishnan, T. S., Pradeep, J., Manikandan, S., & Boopathi, S. (2023). Theoretical Study on Power Distribution Systems for Electric Vehicles. In B. K. Mishra (Ed.), Advances in Civil and Industrial Engineering. IGI Global. doi:10.4018/979-8-3693-0044-2.ch001

Chataut, R., & Akl, R. (2020). Massive MIMO Systems for 5G and beyond Networks—Overview, Recent Trends, Challenges, and Future Research Direction. *Sensors (Basel)*, *20*(10), 2753. doi:10.3390/s20102753 PMID:32408531

Chen, C., Xu, T. B., Yazdani, A., & Sun, J. Q. (2021). A high density piezoelectric energy harvesting device from highway traffic - System design and road test. *Applied Energy*, *299*, 117331. doi:10.1016/j.apenergy.2021.117331

Chen, G., Xiao, P., Kelly, J. R., Li, B., & Tafazolli, R. M. (2017). Full-duplex wireless-powered relay in two way cooperative networks. *IEEE Access : Practical Innovations, Open Solutions*, *5*, 1548–1558. doi:10.1109/ACCESS.2017.2661378

Chen, H., Li, Y., Rebelatto, J. L., Filho, B. F. U., & Vucetic, B. (2015). Harvest-then-cooperate: Wireless-powered cooperative communications. *IEEE Transactions on Signal Processing*, *63*(7), 1700–1711. doi:10.1109/TSP.2015.2396009

Cunha, Á., Brito, F. P., Martins, J., Rodrigues, N., Monteiro, V., Afonso, J. L., & Ferreira, P. (2016). Assessment of the use of vanadium redox flow batteries for energy storage and fast charging of electric vehicles in gas stations. *Energy*, *115*, 1478–1494. doi:10.1016/j.energy.2016.02.118

D'hulst, R., Sterken, T., Puers, R., & Driesen, J. (n.d.). *Requirements for power electronics used for energy harvesting devices*. Retrieved January 14, 2024, from https://www.researchgate.net/publication/242096692_Requirements_for_Power_Electronics_used_for_Energy_Harvesting_Devices

Daneshvar, S. H., Yuce, M. R., & Redouté, J. M. (2022). Electrostatic Harvesters Overview and Applications. In *Design of Miniaturized Variable-Capacitance Electrostatic Energy Harvesters*. Springer. doi:10.1007/978-3-030-90252-0_2

Das, S., Lekhya, G., Shreya, K., Lydia Shekinah, K., Babu, K. K., & Boopathi, S. (2024). Fostering Sustainability Education Through Cross-Disciplinary Collaborations and Research Partnerships: Interdisciplinary Synergy. In P. Yu, J. Mulli, Z. A. S. Syed, & L. Umme (Eds.), Advances in Higher Education and Professional Development. IGI Global. doi:10.4018/979-8-3693-0487-7.ch003

De Fazio, R., De Giorgi, M., Cafagna, D., Del-Valle-Soto, C., & Visconti, P. (2023). Energy Harvesting Technologies and Devices from Vehicular Transit and Natural Sources on Roads for a Sustainable Transport: State-of-the-Art Analysis and Commercial Solutions. *Energies*, *16*(7), 3016. doi:10.3390/en16073016

Despesse, G. (2008). In-plane gap closing capacitor for energy harvesting. *Journal of Energy Harvesting and Systems*, *4*(3), 16.

Despesse, G., Jager, T., Condemine, C., & Berger, P. D. (2008). Mechanical vibrations energy harvesting and power management. In *IEEE Sensors Conference* (pp. 29-32). 10.1109/ICSENS.2008.4716375

Dhanya, D., Kumar, S. S., Thilagavathy, A., Prasad, D. V. S. S. S. V., & Boopathi, S. (2023). Data Analytics and Artificial Intelligence in the Circular Economy: Case Studies. In B. K. Mishra (Ed.), Advances in Civil and Industrial Engineering. IGI Global. doi:10.4018/979-8-3693-0044-2.ch003

Ding, H., Wang, X., Costa, D. B. D., Chen, Y., & Gong, F. (2017). Adaptive time-switching based energy harvesting relaying protocols. *IEEE Transactions on Communications*, *65*(7), 2821–2837. doi:10.1109/TCOMM.2017.2693358

Ding, X., & Liu, X. (2023). Renewable energy development and transportation infrastructure matters for green economic growth? Empirical evidence from China. *Economic Analysis and Policy*, *79*, 634–646. doi:10.1016/j.eap.2023.06.042

Domakonda, V. K., Farooq, S., Chinthamreddy, S., Puviarasi, R., Sudhakar, M., & Boopathi, S. (2022). Sustainable Developments of Hybrid Floating Solar Power Plants: Photovoltaic System. In Human Agro-Energy Optimization for Business and Industry (pp. 148–167). IGI Global.

Domakonda, V. K., Farooq, S., Chinthamreddy, S., Puviarasi, R., Sudhakar, M., & Boopathi, S. (2023). Sustainable Developments of Hybrid Floating Solar Power Plants: Photovoltaic System. In P. Vasant, R. Rodríguez-Aguilar, I. Litvinchev, & J. A. Marmolejo-Saucedo (Eds.), Advances in Environmental Engineering and Green Technologies. IGI Global. doi:10.4018/978-1-6684-4118-3.ch008

Du, R., Xiao, J., Chang, S., Zhao, L.-C., Wei, K.-X., Zhang, W.-M., & Zou, H.-X. (2023). Mechanical energy harvesting in traffic environment and its application in smart transportation. *Journal of Physics. D, Applied Physics*, *56*(37), 373002. doi:10.1088/1361-6463/acdadb

Elahi, H., Munir, K., Eugeni, M., Atek, S., & Gaudenzi, P. (2020). Energy harvesting towards self-powered IoT devices. *Energies*, *13*(21), 5528. doi:10.3390/en13215528

Electrek, (2022). *BMW designs EV suspension system that turns bumpy roads into usable power*. https://electrek.co/2022/12/01/bmw-designs-ev-suspension-system-that-generates-usable-energy/

Engineering New-Record. (2022). *Mixed Results as Smart Road Testing Begins*. https://www.enr.com/articles/54723-q3-tech-focus-mixed-results-as-smart-roading-testing-begins

Espera, A. H. Jr, Dizon, J. R. C., Chen, Q., & Advincula, R. C. (2019). 3D-printing and advanced manufacturing for electronics. *Progress in Additive Manufacturing*, *4*(3), 245–267. doi:10.1007/s40964-019-00077-7

Fantin Irudaya Raj, E., & Appadurai, M. (2022). Internet of things-based smart transportation system for smart cities. In Intelligent Systems for Social Good: Theory and Practice (pp. 39–50). Springer. doi:10.1007/978-981-19-0770-8_4

Fayyazi, M., Sardar, P., Thomas, S. I., Daghigh, R., Jamali, A., Esch, T., Kemper, H., Langari, R., & Khayyam, H. (2023). Artificial Intelligence/Machine Learning in Energy Management Systems, Control, and Optimization of Hydrogen Fuel Cell Vehicles. *Sustainability (Basel)*, *15*(6), 5249. doi:10.3390/su15065249

Ganesh, A. H., & Xu, B. (2022). A review of reinforcement learning based energy management systems for electrified powertrains: Progress, challenge, and potential solution. *Renewable & Sustainable Energy Reviews*, *154*, 111833. doi:10.1016/j.rser.2021.111833

Gao, X., Li, X., Han, C., Zeng, M., Liu, H., Mumtaz, S., & Nallanathan, A. (2024). Rate-splitting multiple access-based cognitive radio network with ipSIC and CEEs. *IEEE Transactions on Vehicular Technology*, *73*(1), 1430–1434. doi:10.1109/TVT.2023.3305960

Ghosh, A., & Aggarwal, V. (2018). Menu-based pricing for charging of electric vehicles with vehicle-to-grid service. *IEEE Transactions on Vehicular Technology*, *67*(11), 10268–10280. doi:10.1109/TVT.2018.2865706

Gnanaprakasam, C., Vankara, J., Sastry, A. S., Prajval, V., Gireesh, N., & Boopathi, S. (2023). Long-Range and Low-Power Automated Soil Irrigation System Using Internet of Things: An Experimental Study. In G. S. Karthick (Ed.), Advances in Environmental Engineering and Green Technologies. IGI Global. doi:10.4018/978-1-6684-7879-0.ch005

Government Technology. (2016). *California to Test Road Vibrations as an Energy Source*. https://www.govtech.com/fs/california-to-test-road-vibrations-as-an-energy-source.html

Guo, J., Zhang, S., Zhao, N., & Wang, X. (2020). Performance of SWIPT for full-duplex relay system with co-channel interference. *IEEE Transactions on Vehicular Technology*, *67*(2), 2311–2315. doi:10.1109/TVT.2019.2958626

Compilation of References

Ha, N., Xu, K., Ren, G., Mitchell, A., & Ou, J. Z. (2020). Machine learning-enabled smart sensor systems. *Advanced Intelligent Systems*, 2(9), 2000063. doi:10.1002/aisy.202000063

Hanumanthakari, S., Gift, M. D. M., Kanimozhi, K. V., Bhavani, M. D., Bamane, K. D., & Boopathi, S. (2023). Biomining Method to Extract Metal Components Using Computer-Printed Circuit Board E-Waste. In P. Srivastava, D. Ramteke, A. K. Bedyal, M. Gupta, & J. K. Sandhu (Eds.), Practice, Progress, and Proficiency in Sustainability. IGI Global. doi:10.4018/978-1-6684-8117-2.ch010

Haribalaji, V., Boopathi, S., & Asif, M. M. (2021). Optimization of friction stir welding process to join dissimilar AA2014 and AA7075 aluminum alloys. *Materials Today: Proceedings*, 50, 2227–2234. doi:10.1016/j.matpr.2021.09.499

Hasan, S., Zeyad, M., Ahmed, S. M., Mahmud, D. M., Anubhove, M. S. T., & Hossain, E. (2023). Techno-economic feasibility analysis of an electric vehicle charging station for an International Airport in Chattogram, Bangladesh. *Energy Conversion and Management*, 293.

Hayt, W. H., & Kemmerly, J. E. (2001). *Engineering Circuit Analysis*. McGraw-Hill Education.

Hema, N., Krishnamoorthy, N., Chavan, S. M., Kumar, N. M. G., Sabarimuthu, M., & Boopathi, S. (2023). A Study on an Internet of Things (IoT)-Enabled Smart Solar Grid System. In P. Swarnalatha & S. Prabu (Eds.), Advances in Computational Intelligence and Robotics. IGI Global. doi:10.4018/978-1-6684-8098-4.ch017

He-Shuai LTD. (2018). *Hard PZT & Soft PZT Data Sheet*. Author.

HiDubai. (2023). *Dubai's Autonomous Vehicles: Redefining Transportation in the City*. Retrieved from https://focus.hidubai.com/driverless-cars-dubai/

Huang, G., Zhang, Q., & Qin, J. (2015). Joint time switching and power allocation for multicarrier decode-and-forward relay networks with SWIPT. *IEEE Signal Processing Letters*, 22(12), 2284–2288. doi:10.1109/LSP.2015.2477424

Huang, X., & Ansari, N. (2016). Optimal cooperative power allocation for energy-harvesting-enabled relay networks. *IEEE Transactions on Vehicular Technology*, 65(4), 2424–2434. doi:10.1109/TVT.2015.2424218

Hussain, Z., Babe, M., Saravanan, S., Srimathy, G., Roopa, H., & Boopathi, S. (2023). Optimizing Biomass-to-Biofuel Conversion: IoT and AI Integration for Enhanced Efficiency and Sustainability. In N. Cobîrzan, R. Muntean, & R.-A. Felseghi (Eds.), Advances in Finance, Accounting, and Economics. IGI Global. doi:10.4018/978-1-6684-8238-4.ch009

Ingle, R. B., Swathi, S., Mahendran, G., Senthil, T. S., Muralidharan, N., & Boopathi, S. (2023). Sustainability and Optimization of Green and Lean Manufacturing Processes Using Machine Learning Techniques. In N. Cobîrzan, R. Muntean, & R.-A. Felseghi (Eds.), Advances in Finance, Accounting, and Economics. IGI Global. doi:10.4018/978-1-6684-8238-4.ch012

Jaffe, H. (1958). Piezoelectric ceramics. *Journal of the American Ceramic Society*, 41(11), 494–498. doi:10.1111/j.1151-2916.1958.tb12903.x

Jayakody, D. N. K., Perera, T. D. P., Ghrayeb, A., & Hasna, O. M. (2020). Self-energized UAV-assisted scheme for cooperative wireless relay networks. *IEEE Trans. on Vech. Tech*, *69*(1), 578–592.

Jeon, D. H., Cho, J. Y., Jhun, J. P., Ahn, J. H., Jeong, S., Jeong, S. Y., Kumar, A., Ryu, C. H., Hwang, W., Park, H., Chang, C., Lee, H., & Sung, T. H. (2021). A lever-type piezoelectric energy harvester with deformation-guiding mechanism for electric vehicle charging station on smart road. *Energy*, *218*, 119540. doi:10.1016/j.energy.2020.119540

Jeong, C., Joung, C., Lee, S., Feng, M. Q., & Park, Y.-B. (2020). Carbon nanocomposite based mechanical sensing and energy harvesting. *International Journal of Precision Engineering and Manufacturing-Green Technology*, *7*(1), 247–267. doi:10.1007/s40684-019-00154-w

Jordaan, G. J., & Steyn, W. J. (2022). Practical Application of Nanotechnology Solutions in Pavement Engineering: Construction Practices Successfully Implemented on Roads (Highways to Local Access Roads) Using Marginal Granular Materials Stabilised with New-Age (Nano) Modified Emulsions (NME). *Applied Sciences (Basel, Switzerland)*, *12*(3), 1332. doi:10.3390/app12031332

Joseph, P. K., & Devaraj, E. (2019). Design of hybrid forward boost converter for renewable energy powered electric vehicle charging applications. *IET Power Electronics*, *12*(8), 2015–2021. doi:10.1049/iet-pel.2019.0151

Juang, J. C., & Lin, C. F. (2015). A sensor fusion scheme for the estimation of vehicular speed and heading angle. *IEEE Transactions on Vehicular Technology*, *64*(7), 2773–2782.

Kang, T., Li, H., Zheng, L., Li, J., Xia, D., Ji, L., Shi, Y., Wang, H., & Chen, M. (2023). Distributed plug-in electric vehicles charging strategy considering driver behaviours and load constraints. *Electric Power Systems Research*, *220*, 109367. doi:10.1016/j.epsr.2023.109367

Karki, J. (2000). *Signal conditioning piezoelectric sensors. App. rept. on mixed signal products (sloa033a)*. Texas Instruments Incorporated.

Karthik, S. A., Hemalatha, R., Aruna, R., Deivakani, M., Reddy, R. V. K., & Boopathi, S. (2023). Study on Healthcare Security System-Integrated Internet of Things (IoT). In M. K. Habib (Ed.), Advances in Systems Analysis, Software Engineering, and High Performance Computing. IGI Global. doi:10.4018/978-1-6684-7684-0.ch013

Kashef, M., & Ephremides, A. (2016). Optimal partial relaying for energy-harvesting wireless networks. *IEEE/ACM Transactions on Networking*, *24*(1), 113–122. doi:10.1109/TNET.2014.2361683

Kavitha, C. R., Varalatchoumy, M., Mithuna, H. R., Bharathi, K., Geethalakshmi, N. M., & Boopathi, S. (2023). Energy Monitoring and Control in the Smart Grid: Integrated Intelligent IoT and ANFIS. In M. Arshad (Ed.), Advances in Bioinformatics and Biomedical Engineering. IGI Global. doi:10.4018/978-1-6684-6577-6.ch014

Kazanci, O. B., & Olesen, B. W. (2017). Review of occupant sensing for smart lighting systems. *Energy and Building*, *139*, 607–620.

Keipour, A., Pereira, G. A., Bonatti, R., Garg, R., Rastogi, P., Dubey, G., & Scherer, S. (2022). Visual servoing approach to autonomous uav landing on a moving vehicle. *Sensors (Basel)*, *22*(17), 6549. doi:10.3390/s22176549 PMID:36081008

Khaleej Times. (2020). *Exclusive: Inside centre that monitors Dubai traffic 24/7*. Retrieved from https://www.khaleejtimes.com/transport/exclusive-inside-centre-that-monitors-dubai-traffic-24-7

Khandaker, M. R. A., & Kai-Kit Wong. (2014). SWIPT in MISO multicasting systems. *IEEE Wireless Communications Letters*, *3*(3), 277–280. doi:10.1109/WCL.2014.030514.140057

Khan, K. A., Quamar, M. M., Al-Qahtani, F. H., Asif, M., Alqahtani, M., & Khalid, M. (2023). Smart grid infrastructure and renewable energy deployment: A conceptual review of Saudi Arabia. *Energy Strategy Reviews.*, *50*, 101247. doi:10.1016/j.esr.2023.101247

Khan, M. M. (2020). Sensor-based gas leakage detector system. *Engineering Proceedings*, *2*(1), 28.

Khan, S., Sudhakar, K., Yusof, M. H. B., Azmi, W. H., & Ali, H. M. (2023). Roof integrated photovoltaic for electric vehicle charging towards net zero residential buildings in Australia. *Energy for Sustainable Development*, *73*, 340–354. doi:10.1016/j.esd.2023.02.005

Kima, S., Sternb, I., Shenc, J., Ahadd, M., & Baie, Y. (2018). Energy harvesting assessment using PZT sensors and roadway materials. *Int. J. of Thermal & Environmental Engineering*, *16*(1), 19–25. doi:10.5383/ijtee.16.01.003

Kim, H., Tai, W. C., Parker, J., & Zuo, L. (2019). Self-tuning stochastic resonance energy harvesting for rotating systems under modulated noise and its application to smart tires. *Mechanical Systems and Signal Processing*, *122*, 769–785. doi:10.1016/j.ymssp.2018.12.040

Kiziroglou, M. E., & .... (2009). Rolling rod harvester for energy harvesting. *Journal of Energy Harvesting and Systems*, *2*(1), 2–10.

Kiziroglou, M. E., He, C., & Yeatman, E. M. (2009). Rolling rod electrostatic microgenerator. *IEEE Transactions on Industrial Electronics*, *56*(4), 1101–1108. doi:10.1109/TIE.2008.2004381

Kong, C., & Lu, H. (2023). Cooperative rate-splitting multiple access in heterogeneous networks. *IEEE Communications Letters*, *27*(10), 2807–2811. doi:10.1109/LCOMM.2023.3309818

Koshariya, A. K., Kalaiyarasi, D., Jovith, A. A., Sivakami, T., Hasan, D. S., & Boopathi, S. (2023). AI-Enabled IoT and WSN-Integrated Smart Agriculture System. In R. K. Gupta, A. Jain, J. Wang, S. K. Bharti, & S. Patel (Eds.), Practice, Progress, and Proficiency in Sustainability. IGI Global. doi:10.4018/978-1-6684-8516-3.ch011

Kumar Reddy, R. V., Rahamathunnisa, U., Subhashini, P., Aancy, H. M., Meenakshi, S., & Boopathi, S. (2023). Solutions for Software Requirement Risks Using Artificial Intelligence Techniques. In Advances in Information Security, Privacy, and Ethics (pp. 45–64). IGI Global. doi:10.4018/978-1-6684-8145-5.ch003

Kumara, V., Mohanaprakash, T., Fairooz, S., Jamal, K., Babu, T., & Sampath, B. (2023). Experimental Study on a Reliable Smart Hydroponics System. In *Human Agro-Energy Optimization for Business and Industry* (pp. 27–45). IGI Global. doi:10.4018/978-1-6684-4118-3.ch002

Kumar, B. M., Kumar, K. K., Sasikala, P., Sampath, B., Gopi, B., & Sundaram, S. (2024). Sustainable Green Energy Generation From Waste Water: IoT and ML Integration. In B. K. Mishra (Ed.), Practice, Progress, and Proficiency in Sustainability. IGI Global. doi:10.4018/979-8-3693-1186-8.ch024

Kumar, D., Singya, P. K., Krejcar, O., & Bhatia, V. (2023). On performance of a SWIPT enabled FD CRN with HIs and imperfect SIC over α–μ fading channel. *IEEE Transactions on Cognitive Communications and Networking*, *9*(1), 99–113. doi:10.1109/TCCN.2022.3220791

Kumar, D., Singya, P. K., Nebhen, J., & Bhatia, V. (2023). Performance of SWIPT-enabled FD TWR network with hardware impairments and imperfect CSI. *IEEE Systems Journal*, *17*(1), 1224–1234. doi:10.1109/JSYST.2022.3183501

Kumar, P. R., Meenakshi, S., Shalini, S., Devi, S. R., & Boopathi, S. (2023). Soil Quality Prediction in Context Learning Approaches Using Deep Learning and Blockchain for Smart Agriculture. In R. Kumar, A. B. Abdul Hamid, & N. I. Binti Ya'akub (Eds.), Advances in Computational Intelligence and Robotics. IGI Global. doi:10.4018/978-1-6684-9151-5.ch001

Kumar, R., & Hossain, A. (2018). Experimental performance and study of low power strain gauge based wireless sensor node for structure health monitoring. *Wireless Personal Communications*, *101*(3), 1657–1669. doi:10.1007/s11277-018-5782-6

Kumar, R., & Hossain, A. (2020). Full-duplex wireless information and power transfer in two-way relaying networks with self-energy recycling. *Wireless Networks*, *26*(8), 6139–6154. doi:10.1007/s11276-020-02432-x

Kurup, R. R., & Babu, A. V. (2020). Power adaptation for improving the performance of time switching SWIPT-based full-duplex cooperative NOMA network. *IEEE Communications Letters*, *24*(12), 2956–2960. doi:10.1109/LCOMM.2020.3017624

Lam, K. H. (2006). *Study of piezoelectric transducers in smart structure applications*. Hong Kong Polytechnic University.

Le, H. V., Le, T. P., & Le, T. T. (2009). In-plane overlap capacitor for energy harvesting. *Journal of Energy Harvesting and Systems*, *1*(2), 134.7-203.7.

Le, T. T., Halvorsen, E., & Zuo, L. (2009). In-plane overlap variable capacitor for electrostatic energy harvesting. In *2009 IEEE 22nd International Conference on Micro Electro Mechanical Systems* (pp. 753-756). IEEE.

Le, Q. N., Bao, V. N. Q., & An, B. (2018). Full-duplex distributed switch-and-stay energy harvesting selection relaying networks with imperfect CSI: Design and outage analysis. *Journal of Communications and Networks (Seoul)*, *20*(1), 29–46. doi:10.1109/JCN.2018.000004

Lewandowski, R., & Pawlak, Z. (2013). Optimal Placement of Viscoelastic Dampers Represented by the Classical and Fractional Rheological Models. In *Design Optimization of Active and Passive Structural Control Systems* (pp. 50–84). IGI Global. doi:10.4018/978-1-4666-2029-2.ch003

Li, J. (2021). *Electrostatic MEMS Energy Harvesting for Sensor Powering* (Publication No. 28414789) [Doctoral dissertation, Rensselaer Polytechnic Institute]. ProQuest Dissertations Publishing

Li, G., & Wu, J. (2019). Solar energy harvesting by using electrostatic transducers. *Energy Conversion and Management*, *195*, 96–104. doi:10.1016/j.enconman.2019.05.051

Lighting, S. (2014, July 15). *Dubai to expand smart LED lighting project*. Retrieved from https://smart-lighting.es/dubai-smart-led-lighting/

Li, J., Herdem, M. S., Nathwani, J., & Wen, J. Z. (2023). Methods and applications for Artificial Intelligence, Big Data, Internet of Things, and Blockchain in smart energy management. *Energy and AI*, *11*, 100208. doi:10.1016/j.egyai.2022.100208

Li, K., Chen, J., Sun, X., Lei, G., Cai, Y., & Chen, L. (2023). Application of wireless energy transmission technology in electric vehicles. *Renewable & Sustainable Energy Reviews*, *184*, 113569. doi:10.1016/j.rser.2023.113569

Li, K., Chen, J., Sun, X., Lei, G., Cai, Y., & Chen, L. (2023). Transportation systems management considering dynamic wireless charging electric vehicles: Review and prospects. *Transportation Research Part E, Logistics and Transportation Review*, *163*, 102761.

Li, L. (2019). Vehicle-to-infrastructure communication for connected and autonomous vehicles: A survey. *IEEE Transactions on Intelligent Transportation Systems*, *20*(1), 378–403.

Lin, H., Zhou, Y., Li, Y., & Zheng, H. (2024). Aggregator pricing and electric vehicles charging strategy based on a two-layer deep learning model. *Electric Power Systems Research*, *227*, 109971. doi:10.1016/j.epsr.2023.109971

Li, S., & Murch, R. D. (2014). An investigation into baseband techniques for single-channel full-duplex wireless communication systems. *IEEE Transactions on Wireless Communications*, *13*(9), 4794–4806. doi:10.1109/TWC.2014.2341569

Li, T., Zhang, H., Zhou, X., & Yuan, D. (2022). Full-duplex cooperative rate-splitting for multigroup multicast with SWIPT. *IEEE Transactions on Wireless Communications*, *21*(6), 4379–4393. doi:10.1109/TWC.2021.3129881

Liu, H., Kim, K. J., Kwak, K. S., & Vincent Poor, H. (2016). Power splitting-based SWIPT with decode-and-forward full-duplex relaying. *IEEE Transactions on Wireless Communications*, *5*(11), 7561–7577. doi:10.1109/TWC.2016.2604801

Liu, H., Ma, J., Xu, T., Yan, W., Ma, L., & Zhang, X. (2019). Vehicle detection and classification using distributed fiber optic acoustic sensing. *IEEE Transactions on Vehicular Technology*, *69*(2), 1363–1374. doi:10.1109/TVT.2019.2962334

Liu, H., Ma, J., Yan, W., Liu, W., Zhang, X., & Li, C. (2018). Traffic flow detection using distributed fiber optic acoustic sensing. *IEEE Access : Practical Innovations, Open Solutions*, *6*, 68968–68980. doi:10.1109/ACCESS.2018.2868418

Liu, J., Xiong, K., Lu, Y., Fan, P., Zhong, Z., & Letaief, B. K. (2020). SWIPT-enabled full-duplex NOMA networks with full and partial CSI. *IEEE Transactions on Green Communications and Networking*, *4*(3), 804–818. doi:10.1109/TGCN.2020.2977611

Liu, P., Gazor, S., Kim, I. M., & Kim, D. I. (2015). Noncoherent relaying in energy harvesting communication systems. *IEEE Transactions on Wireless Communications*, *14*(12), 6940–6954. doi:10.1109/TWC.2015.2462838

Liu, T. (2022). *Reinforcement learning-enabled intelligent energy management for hybrid electric vehicles*. Springer Nature.

Liu, Y., Ding, Z., Elkashlan, M., & Poor, H. V. (2016). Cooperative nonorthogonal multiple access with simultaneous wireless information and power transfer. *IEEE Journal on Selected Areas in Communications*, *34*(4), 938–953. doi:10.1109/JSAC.2016.2549378

Liu, Y., Wang, L., Elkashlan, M., Duong, T. Q., & Nallanathan, A. (2016). Two-way relay networks with wireless power transfer: Design and performance analysis. *IET Communications*, *10*(14), 1810–1819. doi:10.1049/iet-com.2015.0728

Liu, Z., Gao, Y., & Liu, B. (2022). An artificial intelligence-based electric multiple units using a smart power grid system. *Energy Reports*, *8*, 13376–13388. doi:10.1016/j.egyr.2022.09.138

Li, Z., Xu, Z., Chen, Z., Xie, C., Chen, G., & Zhong, M. (2023). An empirical analysis of electric vehicles' charging patterns. *Transportation Research Part D, Transport and Environment*, *117*, 103651. doi:10.1016/j.trd.2023.103651

Lu, J., & Suzuki, Y. (2021). Push-button energy harvester with ultra-soft all-polymer piezoelectret. In *2021 20th International Conference on Solid-State Sensors, Actuators, and Microsystems (TRANSDUCERS)*. IEEE. 10.1109/PowerMEMS54003.2021.9658397

Lu, S., Wang, Y., & Song, H. (2020). A high accurate vehicle speed estimation method. *Soft Computing*, *24*(2), 1283–1291. doi:10.1007/s00500-019-03965-w

Lyu, F., Cai, T., & Huang, F. (2023). A universal wireless charging platform with novel bulged-structure transmitter design for multiple heterogeneous autonomous underwater vehicles (AUVs). *IET Power Electronics*, *16*(13), 2162–2177. doi:10.1049/pel2.12536

Ma, S. K., Zhang, X., Xu, Y., & Miao, J. (2007). Single wafer floating electrode for energy harvesting. In *Proceedings of the PowerMEMS 2007 Workshop* (pp. 104-107). Academic Press.

Ma, Y. (2007). Single wafer floating electrode for energy harvesting. *Journal of Energy Harvesting and Systems*, *6*(2), 16.82.

Maguluri, L. P., Ananth, J., Hariram, S., Geetha, C., Bhaskar, A., & Boopathi, S. (2023). Smart Vehicle-Emissions Monitoring System Using Internet of Things (IoT). In Handbook of Research on Safe Disposal Methods of Municipal Solid Wastes for a Sustainable Environment (pp. 191–211). IGI Global.

Maguluri, L. P., Ananth, J., Hariram, S., Geetha, C., Bhaskar, A., & Boopathi, S. (2023). Smart Vehicle-Emissions Monitoring System Using Internet of Things (IoT). In P. Srivastava, D. Ramteke, A. K. Bedyal, M. Gupta, & J. K. Sandhu (Eds.), Practice, Progress, and Proficiency in Sustainability. IGI Global. doi:10.4018/978-1-6684-8117-2.ch014

Maguluri, L. P., Arularasan, A. N., & Boopathi, S. (2023). Assessing Security Concerns for AI-Based Drones in Smart Cities. In R. Kumar, A. B. Abdul Hamid, & N. I. Binti Ya'akub (Eds.), Advances in Computational Intelligence and Robotics. IGI Global. doi:10.4018/978-1-6684-9151-5.ch002

Maheswari, B. U., Imambi, S. S., Hasan, D., Meenakshi, S., Pratheep, V. G., & Boopathi, S. (2023). Internet of Things and Machine Learning-Integrated Smart Robotics. In M. K. Habib (Ed.), Advances in Computational Intelligence and Robotics. IGI Global. doi:10.4018/978-1-6684-7791-5.ch010

Mahmood, Z. (2021). Connected vehicles: A vital component of smart transportation in an intelligent city. In *Developing and Monitoring Smart Environments for Intelligent Cities* (pp. 198–215). IGI Global. doi:10.4018/978-1-7998-5062-5.ch008

Mao, Y., Dizdar, O., Clerckx, B., Schober, R., Popovski, P., & Poor, H. V. (2022). Rate-splitting multiple access: Fundamentals, survey, and future research trends. *IEEE Communications Surveys and Tutorials*, *24*(4), 2073–2126. doi:10.1109/COMST.2022.3191937

Ma, R., Wu, H., Ou, J., Yang, S., & Gao, Y. (2020). Power splitting-based SWIPT systems with full-duplex jamming. *IEEE Trans. on Vech. Tech*, *69*(9), 9822–9836.

Markevicius, V., Navikas, D., Idzkowski, A., Andriukaitis, D., Valinevicius, A., & Zilys, M. (2018). Practical methods for vehicle speed estimation using a microprocessor-embedded system with AMR Sensors. *Sensors (Basel)*, *18*(7), 2225. doi:10.3390/s18072225 PMID:29996564

Markevicius, V., Navikas, D., Idzkowski, A., Miklusis, D., Andriukaitis, D., Valinevicius, A., Zilys, Cepenas, & Walendziuk, W. (2019). Vehicle speed and length estimation errors using the intelligent transportation system with a set of anisotropic magneto-resistive (AMR) sensors. *Sensors (Basel)*, *19*(23), 5234. doi:10.3390/s19235234 PMID:31795212

Markevicius, V., Navikas, D., Miklusis, D., Andriukaitis, D., Valinevicius, A., Zilys, M., & Cepenas, M. (2020). Analysis of methods for long vehicles speed estimation using anisotropic magneto-resistive (AMR) sensors and reference piezoelectric sensor. *Sensors (Basel)*, *20*(12), 3541. doi:10.3390/s20123541 PMID:32580498

Mehrjerdi, H. (2020). Dynamic and multi-stage capacity expansion planning in microgrid integrated with electric vehicle charging station. *Journal of Energy Storage*, *29*, 101351. doi:10.1016/j.est.2020.101351

Meninger, S., Miranda, J. M., Chandrakasan, J. L. A., Slocum, A., Schmidt, M., & Amirtharajah, R. (2001). Vibration to electric energy conversion. *IEEE Transactions Very Large-Scale Integration (VLSI) Systems, 9*, 64–76.

Menninger, S., Miranda, J. O. M., Amirtharajah, R., Chandrakasan, A., & Lang, J. (1999). Vibration-to-electric energy conversion. In *Low Power Electronics and Design Proceedings. 1999 International Symposium* (pp. 48-53). 10.1145/313817.313840

Miller, K., Reichert, C. L., & Schmid, M. (2023). Biogenic amine detection systems for intelligent packaging concepts: Meat and Meat Products. *Food Reviews International, 39*(5), 2543–2567. doi:10.1080/87559129.2021.1961270

Miranda, J. O. M. (2004). *Electrostatic vibration-to-electric energy conversion* (PhD Thesis). Massachusetts Institute of Technology, Massachusetts, USA

Mitcheson, P. D., Sterken, T., He, C., Kiziroglou, M., Yeatman, E. M., & Puers, R. (2008). *Electrostatic micro-generators.* Retrieved February 3, 2014, from https://journals.sagepub.com/doi/10.1177/002029400804100404

Mitcheson, P. D., Green, T. C., & Yeatman, E. M. (2004). Power processing circuits for electromagnetic, electrostatic, and piezoelectric inertial energy scavengers. *Microsystem Technologies, 10*(3-4), 1629–1635.

Mohammed, B. K., Mortatha, M. B., Abdalrada, A. S., & ALRikabi, H. T. H. S. (2021). A comprehensive system for detection of flammable and toxic gases using IoT. *Periodicals of Engineering and Natural Sciences, 9*(2), 702–711. doi:10.21533/pen.v9i2.1894

Mohanty, A., Venkateswaran, N., Ranjit, P. S., Tripathi, M. A., & Boopathi, S. (2023). Innovative Strategy for Profitable Automobile Industries: Working Capital Management. In Y. Ramakrishna & S. N. Wahab (Eds.), Advances in Finance, Accounting, and Economics. IGI Global. doi:10.4018/978-1-6684-7664-2.ch020

Murphey, Y. L., Park, J., Kiliaris, L., Kuang, M. L., Masrur, M. A., Phillips, A. M., & Wang, Q. (2012). Intelligent hybrid vehicle power control—Part II: Online intelligent energy management. *IEEE Transactions on Vehicular Technology, 62*(1), 69–79. doi:10.1109/TVT.2012.2217362

Nasir, A. A., Tuan, H. D., Ngo, D. T., Duong, T. Q., & Poor, H. V. (2017). Beamforming design for wireless information and power transfer systems: Receive power-splitting versus transmit time-switching. *IEEE Transactions on Communications, 65*(2), 876–889. doi:10.1109/TCOMM.2016.2631465

Nasir, A. A., Zhou, X., Durrani, S., & Kennedy, R. A. (2013). Relaying protocols for wireless energy harvesting and information processing. *IEEE Transactions on Wireless Communications, 12*(7), 3622–3636. doi:10.1109/TWC.2013.062413.122042

Nasir, A. A., Zhou, X., Durrani, S., & Kennedy, R. A. (2015). Wireless-powered relays in cooperative communications: Time-switching relaying protocols and throughput analysis. *IEEE Transactions on Communications, 63*(5), 1607–1622. doi:10.1109/TCOMM.2015.2415480

Compilation of References

Nasr, T., Torabi, S., Bou-Harb, E., Fachkha, C., & Assi, C. (2022). Power jacking your station: In-depth security analysis of electric vehicle charging station management systems. *Computers & Security*, *112*, 102511. doi:10.1016/j.cose.2021.102511

Next Move Strategy Consulting. (2023). https://www.nextmsc.com/report/india-electric-vehicle-ev-charging-market

Nguyen, H., Hah, D., Patterson, P. R., Piywattanametha, W., & Wu, M. C. (2002). A novel MEMS tunable capacitor based on angular vertical comb drive actuators. In *Solid-State Sensor, Actuator and Microsystems Workshop* (pp. 277-280). 10.31438/trf.hh2002.69

Nguyen, N. V. (2002). Tunable capacitor for energy harvesting. *Journal of Energy Harvesting and Systems*, *7*(3), 32-34.

Nishanth, J. R., Deshmukh, M. A., Kushwah, R., Kushwaha, K. K., Balaji, S., & Sampath, B. (2023). Particle Swarm Optimization of Hybrid Renewable Energy Systems. In B. K. Mishra (Ed.), Advances in Civil and Industrial Engineering. IGI Global. doi:10.4018/979-8-3693-0044-2.ch016

O'Donnell, R., Schofield, N., Smith, A. C., & Cullen, J. (2009). Design concepts for high-voltage variable-capacitance DC generators. *IEEE Transactions on Industry Applications*, *45*(5), 1778–1784. doi:10.1109/TIA.2009.2027545

O'Donnell, T. (2009). Varying capacitance machine for energy harvesting. *Journal of Energy Harvesting and Systems*, *3*(4), 9–40.

Ozkan, H. A., & Erol-Kantarci, M. (2022). A novel Electric Vehicle Charging/Discharging Scheme with incentivization and complementary energy sources. *Journal of Energy Storage*, *51*, 104493. doi:10.1016/j.est.2022.104493

Pachiappan, K., Anitha, K., Pitchai, R., Sangeetha, S., Satyanarayana, T. V. V., & Boopathi, S. (2023). Intelligent Machines, IoT, and AI in Revolutionizing Agriculture for Water Processing. In B. B. Gupta & F. Colace (Eds.), Advances in Computational Intelligence and Robotics. IGI Global. doi:10.4018/978-1-6684-9999-3.ch015

Palaniappan, M., Tirlangi, S., Mohamed, M. J. S., Moorthy, R. M. S., Valeti, S. V., & Boopathi, S. (2023). Fused Deposition Modelling of Polylactic Acid (PLA)-Based Polymer Composites: A Case Study. In R. Keshavamurthy, V. Tambrallimath, & J. P. Davim (Eds.), Advances in Chemical and Materials Engineering. IGI Global. doi:10.4018/978-1-6684-6009-2.ch005

Pan, H., Qi, L., Zhang, Z., & Yan, J. (2021). Kinetic energy harvesting technologies for applications in land transportation: A comprehensive review. *Applied Energy*, *286*, 116518. doi:10.1016/j.apenergy.2021.116518

Pei, J., Guo, F., Zhang, J., Zhou, B., Bi, Y., & Li, R. (2021). Review and analysis of energy harvesting technologies in roadway transportation. *Journal of Cleaner Production*, *288*, 125338. doi:10.1016/j.jclepro.2020.125338

Pérez, D. E., López, Q. L. A., Alves, H., & Latva-aho, M. (2022). Self-energy recycling for low-power reliable networks: Half-duplex or full-duplex? *IEEE Systems Journal, 16*(3), 4780–4791. doi:10.1109/JSYST.2021.3127266

Petru, J., & Krivda, V. (2021). The transport of oversized cargoes from the perspective of sustainable transport infrastructure in cities. *Sustainability (Basel), 13*(10), 5524. doi:10.3390/su13105524

Piekarski, B. H. (2005). *Lead zirconate titanate thin films for piezoelectric actuation and sensing of MEMS resonators*. University of Maryland.

Platt, S. R., Farritor, S., & Haider, H. (2005). On low-frequency electric power generation with PZT ceramics. *IEEE/ASME Transactions on Mechatronics, 10*(2), 240–252. doi:10.1109/TMECH.2005.844704

Pompigna, A., & Mauro, R. (2022). Smart roads: A state of the art of highways innovations in the Smart Age. *Engineering Science and Technology, an International Journal, 25*, 100986.

Pramila, P. V., Amudha, S., Saravanan, T. R., Sankar, S. R., Poongothai, E., & Boopathi, S. (2023). Design and Development of Robots for Medical Assistance: An Architectural Approach. In G. S. Karthick & S. Karupusamy (Eds.), Advances in Healthcare Information Systems and Administration. IGI Global. doi:10.4018/978-1-6684-8913-0.ch011

Prus, P., & Sikora, M. (2021). The impact of transport infrastructure on the sustainable development of the region—Case study. *Agriculture, 11*(4), 279. doi:10.3390/agriculture11040279

Puri, V., Mondal, S., Das, S., & Vrana, V. G. (2023). Blockchain propels tourism industry—An attempt to explore topics and information in smart tourism management through text mining and machine learning. *Informatics (MDPI), 10*(1), 9. doi:10.3390/informatics10010009

Qin, Y., Kishk, M. A., & Alouini, M. S. (2022). Performance Analysis of Charging Infrastructure Sharing in UAV and EV-involved Networks. *IEEE Transactions on Vehicular Technology, 72*(3), 3973–3988. doi:10.1109/TVT.2022.3219764

Rabie, K., Adebisi, B., Nauryzbayev, G., Badarneh, Q. S., Li, X., & Alouini, M. S. (2019). Full-duplex energy harvesting enabled relay networks in generalized fading channels. *IEEE Wireless Communications Letters, 8*(2), 384–387. doi:10.1109/LWC.2018.2873360

Rahamathunnisa, U., Sudhakar, K., Murugan, T. K., Thivaharan, S., Rajkumar, M., & Boopathi, S. (2023). Cloud Computing Principles for Optimizing Robot Task Offloading Processes. In S. Kautish, N. K. Chaubey, S. B. Goyal, & P. Whig (Eds.), Advances in Computational Intelligence and Robotics. IGI Global. doi:10.4018/978-1-6684-8171-4.ch007

Rahamathunnisa, U., Sudhakar, K., Padhi, S. N., Bhattacharya, S., Shashibhushan, G., & Boopathi, S. (2023). Sustainable Energy Generation From Waste Water: IoT Integrated Technologies. In A. S. Etim (Ed.), Advances in Human and Social Aspects of Technology. IGI Global. doi:10.4018/978-1-6684-5347-6.ch010

Compilation of References

Rajab, S. A., Mayeli, A., & Refai, H. H. (2014, June). Vehicle classification and accurate speed calculation using multi-element piezoelectric sensor. In *2014 IEEE Intelligent Vehicles Symposium Proceedings* (pp. 894-899). IEEE. 10.1109/IVS.2014.6856432

Rall, K. C., Bailen, K. L., Bender, E. N., Taha, L. Y., Abdeltawab, H. M., & Anwar, S. (2023, April). Smart Highways Based Vehicle Speed Sensor With Piezoelectric Energy Harvesting: A Progress Report. In *2023 IEEE International Conference on Industrial Technology (ICIT)* (pp. 1-6). IEEE. 10.1109/ICIT58465.2023.10143163

Ramudu, K., Mohan, V. M., Jyothirmai, D., Prasad, D., Agrawal, R., & Boopathi, S. (2023a). Machine Learning and Artificial Intelligence in Disease Prediction: Applications, Challenges, Limitations, Case Studies, and Future Directions. In Contemporary Applications of Data Fusion for Advanced Healthcare Informatics (pp. 297–318). IGI Global.

Ramudu, K., Mohan, V. M., Jyothirmai, D., Prasad, D. V. S. S. S. V., Agrawal, R., & Boopathi, S. (2023). Machine Learning and Artificial Intelligence in Disease Prediction: Applications, Challenges, Limitations, Case Studies, and Future Directions. In G. S. Karthick & S. Karupusamy (Eds.), Advances in Healthcare Information Systems and Administration. IGI Global. doi:10.4018/978-1-6684-8913-0.ch013

Rashid, R. A., Chin, L., Sarijari, M. A., Sudirman, R., & Ide, T. (2019). Machine learning for smart energy monitoring of home appliances using IoT. *2019 Eleventh International Conference on Ubiquitous and Future Networks (ICUFN)*, 66–71. 10.1109/ICUFN.2019.8806026

Rathor, K., Vidya, S., Jeeva, M., Karthivel, M., Ghate, S. N., & Malathy, V. (2023). Intelligent System for ATM Fraud Detection System using C-LSTM Approach. *2023 4th International Conference on Electronics and Sustainable Communication Systems (ICESC)*, 1439–1444.

Ravisankar, A., Sampath, B., & Asif, M. M. (2023). Economic Studies on Automobile Management: Working Capital and Investment Analysis. In C. S. V. Negrão, I. G. P. Maia, & J. A. F. Brito (Eds.), Advances in Logistics, Operations, and Management Science. IGI Global. doi:10.4018/978-1-7998-9213-7.ch009

Rebecca, B., Kumar, K. P. M., Padmini, S., Srivastava, B. K., Halder, S., & Boopathi, S. (2023). Convergence of Data Science-AI-Green Chemistry-Affordable Medicine: Transforming Drug Discovery. In B. B. Gupta & F. Colace (Eds.), Advances in Computational Intelligence and Robotics. IGI Global. doi:10.4018/978-1-6684-9999-3.ch014

Reddy, M. A., Reddy, B. M., Mukund, C. S., Venneti, K., Preethi, D. M. D., & Boopathi, S. (2023). Social Health Protection During the COVID-Pandemic Using IoT. In F. P. C. Endong (Ed.), Advances in Electronic Government, Digital Divide, and Regional Development. IGI Global. doi:10.4018/978-1-7998-8394-4.ch009

Reznikov, M. (2010). Electrostatic swing energy harvester. *Proceedings of ESA Annual Meeting on Electrostatics*. Retrieved from https://www.electrostatics.org/images/ESA2010_G3_Reznikov.pdf

Reznikov, M. (2010). Electrostatic swing harvester for energy harvesting. *Journal of Energy Harvesting and Systems*, *8*(4), 2-5.

Rincón-Morantes, J. F., Reyes-Ortiz, O. J., & Ruge-Cardenas, J. C. (2020). Review of the use of nanomaterials in soils for construction of roads. *Respuestas*, *25*(2), 213–223. doi:10.22463/0122820X.2959

Roshani, M., Phan, G., Faraj, R. H., Phan, N.-H., Roshani, G. H., Nazemi, B., Corniani, E., & Nazemi, E. (2021). Proposing a gamma radiation based intelligent system for simultaneous analyzing and detecting type and amount of petroleum by-products. *Nuclear Engineering and Technology*, *53*(4), 1277–1283. doi:10.1016/j.net.2020.09.015

Roundy, S. (2003). *Energy scavenging for wireless sensor nodes with a focus on vibration to electricity conversion* (PhD Thesis). University of California, Berkeley.

Roundy, S., Wright, P. K., & Rabaey, J. M. (2004). *Energy scavenging for wireless sensor networks with a special focus on vibrations*. Kluwer Academic Publishers. doi:10.1007/978-1-4615-0485-6

Saad, W., Bennis, M., & Chen, M. (2020). A Vision of 6G Wireless Systems: Applications, Trends, Technologies, and Open Research Problems. *IEEE Network*, *34*(3), 134–142. doi:10.1109/MNET.001.1900287

Sadeghian, O., Oshnoei, A., Mohammadi-Ivatloo, B., Vahidinasab, V., & Anvari-Moghaddam, A. (2022). A comprehensive review on electric vehicles smart charging: Solutions, strategies, technologies, and challenges. *Journal of Energy Storage*, *54*, 105241. doi:10.1016/j.est.2022.105241

Sagaria, S., Duarte, G., Neves, D., & Baptista, P. (2022). Photovoltaic integrated electric vehicles: Assessment of synergies between solar energy, vehicle types and usage patterns. *Journal of Cleaner Production*, *348*, 131402. doi:10.1016/j.jclepro.2022.131402

Salem, A., & Hamdi, K. A. (2016). Wireless power transfer in multi-pair two-way AF relaying networks. *IEEE Transactions on Communications*, *64*(11), 4578–4591. doi:10.1109/TCOMM.2016.2607751

Samikannu, R., Koshariya, A. K., Poornima, E., Ramesh, S., Kumar, A., & Boopathi, S. (2022). Sustainable Development in Modern Aquaponics Cultivation Systems Using IoT Technologies. In *Human Agro-Energy Optimization for Business and Industry* (pp. 105–127). IGI Global.

Samikannu, R., Koshariya, A. K., Poornima, E., Ramesh, S., Kumar, A., & Boopathi, S. (2023). Sustainable Development in Modern Aquaponics Cultivation Systems Using IoT Technologies. In P. Vasant, R. Rodríguez-Aguilar, I. Litvinchev, & J. A. Marmolejo-Saucedo (Eds.), Advances in Environmental Engineering and Green Technologies. IGI Global. doi:10.4018/978-1-6684-4118-3.ch006

Sampath, B. C. S., & Myilsamy, S. (2022). Application of TOPSIS Optimization Technique in the Micro-Machining Process. In M. A. Mellal (Ed.), Advances in Mechatronics and Mechanical Engineering. IGI Global. doi:10.4018/978-1-6684-5887-7.ch009

Sampath, B., Pandian, M., Deepa, D., & Subbiah, R. (2022). Operating parameters prediction of liquefied petroleum gas refrigerator using simulated annealing algorithm. *AIP Conference Proceedings*, *2460*(1), 070003. doi:10.1063/5.0095601

Sanislav, T., Mois, G. D., Zeadally, S., & Folea, S. C. (2021). Energy harvesting techniques for internet of things (IoT). *IEEE Access : Practical Innovations, Open Solutions*, *9*, 39530–39549. doi:10.1109/ACCESS.2021.3064066

Saravanan, M., Vasanth, M., Boopathi, S., Sureshkumar, M., & Haribalaji, V. (2022). Optimization of Quench Polish Quench (QPQ) Coating Process Using Taguchi Method. *Key Engineering Materials*, *935*, 83–91. doi:10.4028/p-z569vy

Satav, S. D., Lamani, D. K. G., H., Kumar, N. M. G., Manikandan, S., & Sampath, B. (2024). Energy and Battery Management in the Era of Cloud Computing: Sustainable Wireless Systems and Networks. In B. K. Mishra (Ed.), Practice, Progress, and Proficiency in Sustainability (pp. 141–166). IGI Global. doi:10.4018/979-8-3693-1186-8.ch009

Satav, S. D., Hasan, D. S., Pitchai, R., Mohanaprakash, T. A., Sultanuddin, S. J., & Boopathi, S. (2024). Next Generation of Internet of Things (NGIoT) in Healthcare Systems. In B. K. Mishra (Ed.), Practice, Progress, and Proficiency in Sustainability. IGI Global. doi:10.4018/979-8-3693-1186-8.ch017

Senthil, T. S., Ohmsakthi Vel, R., Puviyarasan, M., Babu, S. R., Surakasi, R., & Sampath, B. (2023). Industrial Robot-Integrated Fused Deposition Modelling for the 3D Printing Process. In R. Keshavamurthy, V. Tambrallimath, & J. P. Davim (Eds.), Advances in Chemical and Materials Engineering. IGI Global. doi:10.4018/978-1-6684-6009-2.ch011

Senturia, S. D. (2006). *Microsystem Design*. Springer Science & Business Media.

Sharma, D. M., Venkata Ramana, K., Jothilakshmi, R., Verma, R., Uma Maheswari, B., & Boopathi, S. (2024). Integrating Generative AI Into K-12 Curriculums and Pedagogies in India: Opportunities and Challenges. In P. Yu, J. Mulli, Z. A. S. Syed, & L. Umme (Eds.), Advances in Higher Education and Professional Development. IGI Global. doi:10.4018/979-8-3693-0487-7.ch006

Sharma, M., Sharma, M., Sharma, N., & Boopathi, S. (2023). Building Sustainable Smart Cities Through Cloud and Intelligent Parking System. In B. B. Gupta & F. Colace (Eds.), Advances in Computational Intelligence and Robotics. IGI Global. doi:10.4018/978-1-6684-9999-3.ch009

Sherren, A., Fink, K., Eshelman, J., Taha, L. Y., Anwar, S., Brennecke, C., Abdeltawab, H. M., Shen, S., Ghofrani, F., & Zhang, C. (2022). Experimental and Simulation Validation of Piezoelectric Road Energy Harvesting. *Open Journal of Energy Efficiency*, *11*(3), 122–141. doi:10.4236/ojee.2022.113009

Shim, K. S., Park, N., Kim, J.-H., Jeon, O.-Y., & Lee, H. (2021). Vehicle Speed Measurement Methodology Robust to Playback Speed-Manipulated Video File. *IEEE Access : Practical Innovations, Open Solutions*, *9*, 132862–132874. doi:10.1109/ACCESS.2021.3115500

Singh, A. K., Kulshreshtha, A., Banerjee, A., & Singh, B. R. (2020). Nanotechnology in road construction. *AIP Conference Proceedings*, *2224*(1).

Singh, R., Sharma, R., Akram, S. V., Gehlot, A., Buddhi, D., Malik, P. K., & Arya, R. (2021). Highway 4.0: Digitalization of highways for vulnerable road safety development with intelligent IoT sensors and machine learning. *Safety Science*, *143*, 105407. doi:10.1016/j.ssci.2021.105407

Smida, P., Sabharwal, A., Fodor, G., Alexandropoulos, G. C., Suraweera, H. A., & Chae, C. B. (2023). Full-duplex wireless for 6G: Progress brings new opportunities and challenges. *IEEE Journal on Selected Areas in Communications*, *41*(9), 2729–2750. doi:10.1109/JSAC.2023.3287612

Solar Power World. (2019). *4 ways solar is contributing to smart road technologies*. https://www.solarpowerworldonline.com/2019/03/4-ways-solar-is-contributing-to-smart-road-technologies/

Song, Z., Ye, W., Chen, Z., Chen, Z., Li, M., Tang, W., Wang, C., Wan, Z., Poddar, S., Wen, X., Pan, X., Lin, Y., Zhou, Q., & Fan, Z. (2021). Wireless self-powered high-performance integrated nanostructured-gas-sensor network for future smart homes. *ACS Nano*, *15*(4), 7659–7667. doi:10.1021/acsnano.1c01256 PMID:33871965

Sonika, S. K., Samanta, S., Srivastava, A. K., Biswas, S., Alsharabi, R. M., & Rajput, S. (2022). *Conducting polymer nanocomposite for energy storage and energy harvesting systems*. Review Article, 2266899.

Srinivas, B., Maguluri, L. P., Naidu, K. V., Reddy, L. C. S., Deivakani, M., & Boopathi, S. (2023). Architecture and Framework for Interfacing Cloud-Enabled Robots: In T. Murugan & N. E. (Eds.), Advances in Information Security, Privacy, and Ethics (pp. 542–560). IGI Global. doi:10.4018/978-1-6684-8145-5.ch027

Sultana, S., Salon, D., & Kuby, M. (2021). Transportation sustainability in the urban context: A comprehensive review. *Geographic Perspectives on Urban Sustainability*, 13–42.

Sundaramoorthy, K., Singh, A., Sumathy, G., Maheshwari, A., Arunarani, A. R., & Boopathi, S. (2023). A Study on AI and Blockchain-Powered Smart Parking Models for Urban Mobility. In B. B. Gupta & F. Colace (Eds.), Advances in Computational Intelligence and Robotics. IGI Global. doi:10.4018/978-1-6684-9999-3.ch010

Sun, M., Wang, W., Zheng, P., Luo, D., & Zhang, Z. (2021). A novel road energy harvesting system based on a spatial double V-shaped mechanism for near-zero-energy toll stations on expressways. *Sensors and Actuators. A, Physical*, *323*, 112648. doi:10.1016/j.sna.2021.112648

Syamala, M. C. R., K., Pramila, P. V., Dash, S., Meenakshi, S., & Boopathi, S. (2023). Machine Learning-Integrated IoT-Based Smart Home Energy Management System. In P. Swarnalatha & S. Prabu (Eds.), Advances in Computational Intelligence and Robotics (pp. 219–235). IGI Global. doi:10.4018/978-1-6684-8098-4.ch013

Taha, L. Y. (2009). *Design and Modelling of a MEMS Piezoelectric Microgenerator* (Doctoral dissertation, Ph. D. thesis, Universiti Kebangsaan Malaysia).

Compilation of References

Tang, S., Chen, W., Jin, L., Zhang, H., Li, Y., Zhou, Q., & Zen, W. (2020). SWCNTs-based MEMS gas sensor array and its pattern recognition based on deep belief networks of gases detection in oil-immersed transformers. *Sensors and Actuators. B, Chemical*, *312*, 127998. doi:10.1016/j.snb.2020.127998

Tariq, F., Khandaker, M. R. A., Wong, K. K., Imran, M. A., Bennis, M., & Debbah, M. (2020). A speculative study on 6G. *IEEE Wireless Communications*, *27*(4), 118–125. doi:10.1109/MWC.001.1900488

Tataria, H., Shafi, M., Molisch, A. F., Dohler, M., Sjoland, H., & Tufvesson, F. (2021). 6G wireless systems: Vision, requirements, challenges, insights, and opportunities. *Proceedings of the IEEE*, *109*(7), 1166–1199. doi:10.1109/JPROC.2021.3061701

The Engineer. (2019). *Smart infrastructure to harvest energy from roads*. https://www.theengineer.co.uk/content/news/smart-infrastructure-to-harvest-energy-from-roads/

The Times of India. (2023). *Why India-specific EV charging and integrated solutions are the key to success for green mobility in India*. https://timesofindia.indiatimes.com/blogs/voices/why-india-specific-ev-charging-and-integrated-solutions-are-the-key-to-success-for-green-mobility-in-india/

TheNextWeb. (2022). *Can vehicle charging-roads power our electric future?* https://thenextweb.com/news/energy-generating-roads-can-charge-electric-vehicles

Tientrakool, P., & Rakphongphairoj, V. (2017). Real-time traffic data collection using electrostatic sensors. *Transportation Research Procedia*, *25*, 2561–2574. doi:10.1016/j.trpro.2017.05.495

Trubia, S., Severino, A., Curto, S., Arena, F., & Pau, G. (2020). Smart roads: An overview of what future mobility will look like. *Infrastructures*, *5*(12), 107. doi:10.3390/infrastructures5120107

Tsutsumino, T. (2006). CYTOP electret generator for energy harvesting. *Journal of Energy Harvesting and Systems*, *5*(1), 25.4-39.8.

Tsutsumino, T. S., Suzuki, Y., Kasagi, N., Kashiwagi, K., & Morizawa, Y. (2006). Efficiency evaluation of microseismic electret power generator. In *Proceedings of the 23rd Sensor Symposium* (pp. 521-524). Academic Press.

Tutorials, E. (n.d.). *Introduction to Digital Logic Circuits*. Retrieved from https://www.electronics-tutorials.ws/io/io_1.html

Tutuncuoglu, K., Varan, B., & Yener, A. (2015). Throughput maximization for two-way relay channels with energy harvesting nodes: The impact of relaying strategies. *IEEE Transactions on Communications*, *63*(6), 2081–2093. doi:10.1109/TCOMM.2015.2427162

Ugandar, R. E., Rahamathunnisa, U., Sajithra, S., Christiana, M. B. V., Palai, B. K., & Boopathi, S. (2023). Hospital Waste Management Using Internet of Things and Deep Learning: Enhanced Efficiency and Sustainability. In M. Arshad (Ed.), Advances in Bioinformatics and Biomedical Engineering. IGI Global. doi:10.4018/978-1-6684-6577-6.ch015

Underwood, I., Bailen, K. L., Dellapenna, F. J., Taha, L. Y., Abdeltawab, H. M., & Anwar, S. (2023, August). Experimental evaluation of piezoelectric Vehicle Speed Sensor for smart highways: A Progress Report. In *2023 12th International Conference on Renewable Energy Research and Applications (ICRERA)* (pp. 576-580). IEEE. 10.1109/ICRERA59003.2023.10269353

Varshney, L. R. (2008). Transporting information and energy simultaneously. *IEEE International Symposium on Information Theory*, (pp. 1612-1616). IEEE.

Vazquez-Canteli, J. R., Dey, S., Henze, G., & Nagy, Z. (2020). CityLearn: Standardizing research in multi-agent reinforcement learning for demand response and urban energy management. *arXiv Preprint arXiv:2012.10504*.

Venkateswaran, N., Kumar, S. S., Diwakar, G., Gnanasangeetha, D., & Boopathi, S. (2023). Synthetic Biology for Waste Water to Energy Conversion: IoT and AI Approaches. In M. Arshad (Ed.), Advances in Bioinformatics and Biomedical Engineering. IGI Global. doi:10.4018/978-1-6684-6577-6.ch017

Venkateswaran, N., Vidhya, K., Ayyannan, M., Chavan, S. M., Sekar, K., & Boopathi, S. (2023). A Study on Smart Energy Management Framework Using Cloud Computing. In P. Ordóñez De Pablos & X. Zhang (Eds.), Practice, Progress, and Proficiency in Sustainability. IGI Global. doi:10.4018/978-1-6684-8634-4.ch009

Vennila, T., Karuna, M. S., Srivastava, B. K., Venugopal, J., Surakasi, R., & B., S. (2023). New Strategies in Treatment and Enzymatic Processes: Ethanol Production From Sugarcane Bagasse. In P. Vasant, R. Rodríguez-Aguilar, I. Litvinchev, & J. A. Marmolejo-Saucedo (Eds.), *Advances in Environmental Engineering and Green Technologies* (pp. 219–240). IGI Global. doi:10.4018/978-1-6684-4118-3.ch011

Vennila, T., Karuna, M., Srivastava, B. K., Venugopal, J., Surakasi, R., & Sampath, B. (2022). New Strategies in Treatment and Enzymatic Processes: Ethanol Production From Sugarcane Bagasse. In Human Agro-Energy Optimization for Business and Industry (pp. 219–240). IGI Global.

Venugopal, P., Shekhar, A., Visser, E., Scheele, N., Mouli, G. R. C., Bauer, P., & Silvester, S. (2018). Roadway to self-healing highways with integrated wireless electric vehicle charging and sustainable energy harvesting technologies. *Applied Energy*, *212*, 1226–1239. doi:10.1016/j.apenergy.2017.12.108

Vijayakumar, G. N. S., Domakonda, V. K., Farooq, S., Kumar, B. S., Pradeep, N., & Boopathi, S. (2023). Sustainable Developments in Nano-Fluid Synthesis for Various Industrial Applications. In A. S. Etim (Ed.), Advances in Human and Social Aspects of Technology. IGI Global. doi:10.4018/978-1-6684-5347-6.ch003

Vullers, R. J. M., Schaijk, R. V., Doms, I., Hoof, C. V., & Mertens, R. (2009). Micropower energy harvesting. *Elsevier Solid-State Circuits*, *53*(7), 684–693.

Wang, C., Li, J., Yang, Y., & Ye, F. (2017). Combining solar energy harvesting with wireless charging for hybrid wireless sensor networks. *IEEE Transactions on Mobile Computing*, *17*(3), 560–576. doi:10.1109/TMC.2017.2732979

Compilation of References

Wang, D., Zhang, R., Cheng, X., & Yang, L. (2017). Capacity-enhancing full-duplex relay networks based on power-splitting (PS) SWIPT. *IEEE Trans. on Vech. Tech*, *66*(6), 5446–5450.

Wang, D., Zhang, R., Cheng, X., Yang, L., & Chen, C. (2017). Relay selection in full-duplex energy harvesting two-way relay networks. *IEEE Transactions on Green Communications and Networking*, *1*(2), 182–191. doi:10.1109/TGCN.2017.2686325

Wang, W., Li, X., Zhang, M., Cumanan, K., Ng, D. W. K., Zhang, G., Tang, J., & Dobre, O. A. (2020). Energy-constrained UAV-assisted secure communications with position optimization and cooperative jamming. *IEEE Transactions on Communications*, *68*(7), 4476–4489. doi:10.1109/TCOMM.2020.2989462

Weisberg, S. (2005). *Applied linear regression* (Vol. 528). John Wiley & Sons. doi:10.1002/0471704091

White, R. M., & Sessler, G. M. (1999). Handbook of Sensors and Actuators: Electrostatic Actuators and Sensors. Elsevier.

Wolf, S., & Tauber, R. N. (2018). *Silicon Processing for the VLSI Era: Process Integration* (Vol. 1). Lattice Press.

Wu, Y., Zhang, J., Ravey, A., Chrenko, D., & Miraoui, A. (2020). Real-time energy management of photovoltaic-assisted electric vehicle charging station by markov decision process. *Journal of Power Sources*, *476*, 228504. doi:10.1016/j.jpowsour.2020.228504

Xiong, K., Fan, P., Zhang, C., & Letaief, K. B. (2015). Wireless information and energy transfer for two-hop non-regenerative MIMO-OFDM relay networks. *IEEE Journal on Selected Areas in Communications*, *33*(8), 1595–1611. doi:10.1109/JSAC.2015.2391931

Xu, Y., Shen, C., Ding, Z., Sun, X., Yan, S., Zhu, G., & Zhong, Z. (2017). Joint beamforming and power-splitting control in downlink cooperative SWIPT NOMA systems. *IEEE Transactions on Signal Processing*, *65*(18), 4874–4886. doi:10.1109/TSP.2017.2715008

Yang, C. H., Song, Y., Woo, M. S., Eom, J. H., Song, G. J., Kim, J. H., Kim, J., Lee, T. H., Choi, J. Y., & Sung, T. H. (2017). Feasibility study of impact-based piezoelectric road energy harvester for wireless sensor networks in smart highways. *Sensors and Actuators. A, Physical*, *261*, 317–324. doi:10.1016/j.sna.2017.04.025

Yang, Z., Ding, Z., Fan, P., & Dhahir, N. A. (2017). The impact of power allocation on cooperative non-orthogonal multiple access networks with SWIPT. *IEEE Transactions on Wireless Communications*, *16*(7), 4332–4343. doi:10.1109/TWC.2017.2697380

Yap, K. Y., Chin, H. H., & Klemeš, J. J. (2022). Solar Energy-Powered Battery Electric Vehicle charging stations: Current development and future prospect review. *Renewable & Sustainable Energy Reviews*, *169*, 112862. doi:10.1016/j.rser.2022.112862

Zekrifa, D. M. S., Kulkarni, M., Bhagyalakshmi, A., Devireddy, N., Gupta, S., & Boopathi, S. (2023). Integrating Machine Learning and AI for Improved Hydrological Modeling and Water Resource Management. In V. Shikuku (Ed.), Advances in Environmental Engineering and Green Technologies. IGI Global. doi:10.4018/978-1-6684-6791-6.ch003

Zeng, Y., Chen, H., & Zhang, R. (2016). Bidirectional wireless information and power transfer with a helping relay. *IEEE Communications Letters*, *20*(5), 862–865. doi:10.1109/LCOMM.2016.2549515

Zeng, Y., & Zhang, R. (2015). Full-duplex wireless-powered relay with self-energy recycling. *IEEE Wireless Communications Letters*, *4*(2), 201–204. doi:10.1109/LWC.2015.2396516

Zhang, C., Shen, S., Huang, H., & Wang, L. (2021). Estimation of the vehicle speed using cross-correlation algorithms and MEMS wireless sensors. *Sensors (Basel)*, *21*(5), 1721. doi:10.3390/s21051721 PMID:33801400

Zhang, J. (2018). Structural health monitoring of bridges using electrostatic sensors. *Structural Health Monitoring*, *17*(1), 88–100.

ZhangM.ZhuL.GaoS.YuanH.LiuT. (n.d.). Energy harvesting of novel smart concrete based on nanotechnology: Experimental and Numerical. *Available at* SSRN 4549309.

Zhang, R., & Ho, C. K. (2013). MIMO broadcasting for simultaneous wireless information and power transfer. *IEEE Transactions on Wireless Communications*, *12*(5), 31989–32001. doi:10.1109/TWC.2013.031813.120224

Zhang, Y., & Cai, L. (2018). Dynamic charging scheduling for EV parking lots with photovoltaic power system. *IEEE Access: Practical Innovations, Open Solutions*, *6*, 56995–57005. doi:10.1109/ACCESS.2018.2873286

Zhang, Z., Zhao, T., Ao, X., & Yuan, H. (2017). A vehicle speed estimation algorithm based on dynamic time warping approach. *IEEE Sensors Journal*, *17*(8), 2456–2463. doi:10.1109/JSEN.2017.2672735

Zhao, X., Ke, Y., Zuo, J., Xiong, W., & Wu, P. (2020). Evaluation of sustainable transport research in 2000–2019. *Journal of Cleaner Production*, *256*, 120404. doi:10.1016/j.jclepro.2020.120404

Zhong, C., Suraweera, H., Zheng, G., Krikidis, I., & Zhang, Z. (2014). Wireless information and power transfer with full duplex relaying. *IEEE Transactions on Communications*, *62*(10), 3447–3461. doi:10.1109/TCOMM.2014.2357423

Zhou, J., Baker, M., & Hu, Y. (2008). Energy harvesting for pervasive computing. *IEEE Pervasive Computing*, *7*(4), 14–21. doi:10.1109/MPRV.2008.78

Zhu, D., Bai, Y., Li, Y., & Qin, Y. (2021). Recent progress and perspectives on electrostatic energy harvesting: From fundamentals to applications. *Nano Energy*, *86*, 106075.

Zou, H. X., Zhao, L. C., Gao, Q. H., Zuo, L., Liu, F. R., Tan, T., Wei, K. X., & Zhang, W. M. (2019). Mechanical modulations for enhancing energy harvesting: Principles, methods and applications. *Applied Energy*, *255*, 113871. doi:10.1016/j.apenergy.2019.113871

Zou, H. X., Zhu, Q. W., He, J. Y., Zhao, L. C., Wei, K. X., Zhang, W. M., Du, R. H., & Liu, S. (2024). Energy harvesting floor using sustained-release regulation mechanism for self-powered traffic management. *Applied Energy*, *353*, 122082. doi:10.1016/j.apenergy.2023.122082

About the Contributors

Luay Taha received the B.Sc. degree in Electrical Engineering from Basra University, the M.Sc. degree in Control and Instrumentation from the University of Technology, and the Ph.D. degree in Electrical Engineering from the University of Windsor, Canada, in 2019. He also received a Ph.D. degree in Micro-Engineering and Nano Electronics from UKM University, in 2009. In 2002, he has joined Emirates Airlines/Emirates Aviation University and in 2009, he was promoted to the Assistant Professor rank and appointed as a Program Manager of Electronics Engineering while in Jan. 2014, he was promoted to the Associate Professor rank. In 2017, he was appointed as a sessional instructor at the University of Windsor, to teach different undergraduate and postgraduate courses, and in 2019, he was appointed as a postgraduate fellow to develop new algorithms in the biomedical signal separation of ECG signals using blind source separation techniques. Currently, Dr. Taha is an assistant teaching professor at Penn State Altoona. His research interests include programmable instrumentation, energy harvesting using piezoelectric and electrostatic approaches, and blind source separation of ECG and speech signals. Dr. Taha is a senior member of the IEEE, a member of IEEE signal processing society, a member of the IEEE Circuits and Systems Society, and a member of the IEEE Educational Activities.

* * *

Sampath Boopathi is an accomplished individual with a strong academic background and extensive research experience. He completed his undergraduate studies in Mechanical Engineering and pursued his postgraduate studies in the field of Computer-Aided Design. Dr. Boopathi obtained his Ph.D. from Anna University, focusing his research on Manufacturing and optimization. Throughout his career, Dr. Boopathi has made significant contributions to the field of engineering. He has authored and published over 220 research articles in internationally peer-reviewed journals, highlighting his expertise and dedication to advancing knowledge in his area of specialization. In addition to his research publications, Dr. Boopathi has also

About the Contributors

been granted two patents and has five published patents to his name. With 17 years of academic and research experience, Dr. Boopathi has enriched the engineering community through his teaching and mentorship roles.

Md. Irfan Khan is working as Regional Manager-ASEAN at Supreme & Co. Pvt. Ltd., India. His area of interest includes Automation and Sustainable Development in T & D Sector.

Rajeev Kumar received his Ph.D. and M. Tech. degrees in Dept. of Electronics & Communication Engineering (ECE) from National Institute of Technology (NIT), Silchar, Assam, India in 2019 and 2015, respectively. He obtained B.Tech. Degree in Dept. of ECE from West Bengal University of Technology (WBUT), Kolkata, West Bengal, India in 2012. Currently, he is working as Assistant Professor, in Dept. of ECE, Indian Institute of Information Technology (IIIT) Sri City, Chittoor, A.P. India. His current research interests include Cooperative Communications, Wireless sensor Network, Green Technology, Advanced Protocol design, IoT and applications, 5G Communications and Beyond, transceiver design for 6G Communications, and Optimization Theory. He is Reviewer of IEEE Transactions on Communications, IEEE Access, IET Communications, Wireless Personal Communications and Wireless Network, Springer. Dr. Kumar was also, selected as an Exemplary Reviewer of National Conference on Communication (NCC) at IIT Hyderabad for 2017. He is Member of Technical Program Committee (TPC) of 8th Annual International Conference on Information Technology and Application (ITA), Xiamen, Fujian, China for 2021.

Dharmbir Prasad was born in 1986 at Nalanda, Bihar, India. He received his B.Tech. degree in electrical engineering from Hooghly Engineering & Technology College (under West Bengal University of Technology), Hooghly, India and M.Tech. in Power System from Dr. B.C. Roy Engineering College (under West Bengal University of Technology), Durgapur, India, respectively. Currently, he is working in the capacity of assistant professor in the department of electrical engineering, Asansol Engineering College, Asansol, West Bengal, India. Now, he is pursuing his Ph.D. degree from Indian School of Mines, Dhanbad, Jharkhand, India. His research interest includes economic operation of power system.

Aditya Singh is an Independent Researcher, Editorial Review Board Member in 4 American International Journals (IJSSMET, IJSDS, IJPAEI, IJESGT), Reviewer in 6 American, 3 Singaporean & 1 Indonesian International Journals (IJUPSC, IJCVIP, IRMJ, JTA, IJRLEDM, IJAIBM, JDSIS, AAES, and JCCE, & IJEECS) & Ex-Alumni Mentor (LPU) at present. He had completed his regular B. Tech degree in Civil Engineering from Lovely Professional University, India, in 2020. He is currently affiliated with Amrita Vishwa Vidyapeetham, Coimbatore, India. He has 18 publications under his name, in which he is a single author in 15 of them, whereas in one of them he is the main author. He had also worked as a Peer Reviewer/Sub Reviewer/Ad Hoc Reviewer/ERB for 115 times so far in international conferences, international books and international journals. He had also won 2 International Awards((Thailand & (Cambodia)) as a Reviewer in 2023. He had worked on 32 positions so far in his professional career internationally.

Rudra Pratap Singh was born in 1983 at Chittaranjan, Burdwan, West Bengal, India. He received his B. Tech. and M. Tech degree in electrical engineering from Asansol Engineering (under the West Bengal University of Technology), Asansol, West Bengal, India and MBA in Power Management from the University of Petroleum and Energy Studies, Dehradun, Uttrakhand, India, respectively. He received his Ph.D. degree from the Indian School of Mines, Dhanbad, Jharkhand, India. Currently, he is working in the capacity of assistant professor in the department of electrical engineering, Asansol Engineering College, Asansol, West Bengal, India. His research interest includes the application of state estimation and optimization techniques in various fields of engineering.

Index

3D Printing 98
6G radio 108-109
6G Radio System 109

A

Adaptive Lighting Systems 268, 273
AI-Driven Energy Management 167
Artificial Intelligence 13-14, 109, 160, 167-169, 189
Autonomous Transportation 264, 268

B

balancing supply 168, 189
base station 111-113, 129

C

carbon footprint 14, 145, 153, 155, 160
Charging Infrastructure 42, 44, 46, 54
charging stations 2, 6, 43-44, 49-50, 55-57
closed-form expressions 108, 114, 116, 121
Cloud Connectivity 83, 85
Cold Region Roads 42
consumption patterns 167-169

D

Data Acquisition System 237
data analytics, 76-77, 100
data collection 77-78, 82, 91, 148, 268
data privacy 17, 33-34, 99-100
DC electricity 53
detection systems 76-79, 81-86, 89, 91, 94, 96-100
developing country 7, 17, 34
digital storage 222, 237

E

Eastern Peripheral Expressway 7, 19
eco-friendly practices 137, 143
Electret-Based Systems 251, 273
Electric Vehicle 2, 42-44, 70
electrical energy 144-145, 147, 241, 243-244, 247, 267, 273-274
electrical power 42, 49, 52, 58, 109, 138, 147, 251, 256
electrostatic energy 51, 242, 248-251, 260-261, 268
electrostatic harvester 242, 247, 250, 258-259
Electrostatic Transducers 241-247, 261-264, 267-268, 273
energy constrained 108, 114-115, 117, 121, 123, 126-127, 129
energy conversion 117, 126, 242-244, 247, 249-250, 268, 274
Energy Harvesting 42, 44, 47, 51, 54, 70, 108-109, 116, 136-139, 143-144, 147-149, 151, 153, 155-157, 159-160, 242, 246, 248, 251-252, 256, 259-262, 267-268, 273-274
energy management 43, 54, 166-169, 184-185, 189-190
energy production 2, 43, 49, 54, 58, 167, 180
Energy Savings 268
Energy Storage 43, 48, 53-54, 64
environmental concerns 137-139

environmental effect 48, 54
Ergodic Capacity 108, 113-114, 116, 119-121, 124-127, 129
estimation algorithm 198, 210-211, 224, 231, 233, 237
EV charging 45-48, 50-51, 54-57, 59, 61-69

F

fixed position 126
Full-Duplex 108-109, 111, 114, 116, 121

G

gas detection 76-79, 81-86, 88-89, 91, 93-94, 96-100
gas emissions 48, 54, 137
gas leakage 85-89, 91
greenhouse gas 48, 137

H

Hazardous Gas Detection 76-77
Highway projects 1, 13, 17-18, 21, 33-34
highway systems 139, 155-157, 160

I

incompatible technologies 17, 33-34
Intelligent gas 76-77, 82, 85-86, 89, 96-98
internal combustion 48, 51
Internet of Things (IoT) 2, 15-16, 76, 81, 100, 148, 274
IoT Integration 77, 82, 100

J

job creation 153, 160

L

Learning models 86-91, 167, 169
LPG Leakage 76, 78

M

Machine Learning 76-77, 79, 82, 85-91, 99-100, 167-169, 174, 177
Machine Learning Algorithms 82, 89, 100, 177
MATLAB 198, 206, 211, 218, 222-224, 237, 259
mechanical energy 144, 241, 243-244, 247, 251, 267, 273
MEMS Devices 255, 274
Microcontroller 197

N

Nanomaterials 100, 138, 142, 151, 160
Nanotechnology 98, 136-143, 149-153, 155-157, 159-160, 267
neural networks 169

P

performance gains 110-112, 115, 126-127
Piezoelectric 49, 51, 137-138, 144-145, 151, 197-198, 200-206, 209-215, 218-219, 222, 230-231, 236
Piezoelectric Phenomenon 201
population growth 137
Power-Splitting Protocol 115-116
Predictive Maintenance 82, 99, 167-170, 174
PZT Material 203, 236

R

real-time data 82, 139, 148-149, 267-268
real-time monitoring 76-78, 81-82, 91, 93, 99-100, 170
Renewable Energy 2, 43, 46, 48-49, 52, 55-58, 70, 137, 139, 145, 160, 167-168, 178-181, 189, 267
Renewable Energy Integration 167, 180
renewable sources 174
residual self-interference 112, 115, 127
road network 17, 34, 138

S

self-energy recycling 111, 114
self-healing properties 137-138, 142-143,

Index

160
SIMULINK 198, 218, 221-222, 237
Smart grids 167-168, 170, 174, 181-183
Smart Highway Projects 1, 13, 17-18, 21, 33-34
Smart Highways 1-3, 5, 7, 9, 13-14, 17, 31, 33-34, 70, 136-139, 148-151, 153, 155, 197, 262
Smart Infrastructure 49, 268
Smart Roads 1-2, 7, 15, 17, 34, 44, 51, 70, 242, 262-264, 267, 274
SOLIDWORKS 214-215, 237
sophisticated components 17, 34
spectral efficiency 110-113
speed estimation 198, 200, 209, 211, 223-224, 229, 231, 233, 237
Structural Health Monitoring 267, 274
Sustainable Transportation 44, 70, 136-138, 143, 158, 160
SWIPT 108, 110-111, 113, 116, 121, 129-130

T

Time-Switching Protocol 121
Traffic Engineering 1
Transportation Engineering 13

transportation infrastructure 136-139, 143-145, 147-149, 151-153, 155-157, 159-160
transportation systems 50, 137-138, 143, 149, 153, 262, 268
two-way FD 111, 114, 116, 121, 124

V

Variable Capacitor Structures 248-249, 268, 274
Variable Capacitors 241, 248-249, 259, 267, 273
Vehicle-to-Infrastructure Communication 274

W

wireless communication 83, 111-112, 114, 129
Wireless Sensor Networks 51, 82-83

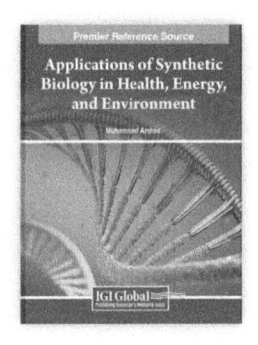

Ensure Quality Research is Introduced to the Academic Community

Become a Reviewer for IGI Global Authored Book Projects

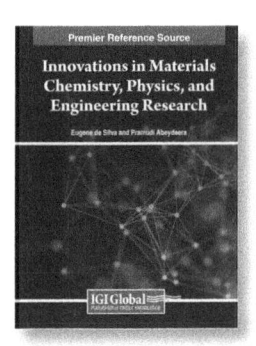

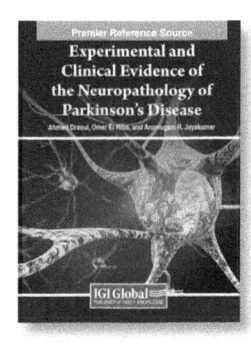

The overall success of an authored book project is dependent on quality and timely manuscript evaluations.

Applications and Inquiries may be sent to:
development@igi-global.com

Applicants must have a doctorate (or equivalent degree) as well as publishing, research, and reviewing experience. Authored Book Evaluators are appointed for one-year terms and are expected to complete at least three evaluations per term. Upon successful completion of this term, evaluators can be considered for an additional term.

If you have a colleague that may be interested in this opportunity, we encourage you to share this information with them.

IGI Global's Open Access Journal Program

Publishing Tomorrow's Research Today

Including Nearly 200 Peer-Reviewed, Gold (Full) Open Access Journals across IGI Global's Three Academic Subject Areas: Business & Management; Scientific, Technical, and Medical (STM); and Education

Consider Submitting Your Manuscript to One of These Nearly 200 Open Access Journals for to Increase Their Discoverability & Citation Impact

| Web of Science Impact Factor | 6.5 | Web of Science Impact Factor | 4.7 | Web of Science Impact Factor | 3.2 | Web of Science Impact Factor | 2.6 |

- JOURNAL OF Organizational and End User Computing
- JOURNAL OF Global Information Management
- INTERNATIONAL JOURNAL ON Semantic Web and Information Systems
- JOURNAL OF Database Management

Choosing IGI Global's Open Access Journal Program Can Greatly Increase the Reach of Your Research

Higher Usage
Open access papers are 2-3 times more likely to be read than non-open access papers.

Higher Download Rates
Open access papers benefit from 89% higher download rates than non-open access papers.

Higher Citation Rates
Open access papers are 47% more likely to be cited than non-open access papers.

Submitting an article to a journal offers an invaluable opportunity for you to share your work with the broader academic community, fostering knowledge dissemination and constructive feedback.

Submit an Article and Browse the IGI Global Call for Papers Pages

We can work with you to find the journal most well-suited for your next research manuscript.
For open access publishing support, contact: journaleditor@igi-global.com

Publishing Tomorrow's Research Today
IGI Global
e-Book Collection

Including Essential Reference Books Within Three Fundamental Academic Areas

Business & Management
Scientific, Technical, & Medical (STM)
Education

- Acquisition options include Perpetual, Subscription, and Read & Publish
- No Additional Charge for Multi-User Licensing
- No Maintenance, Hosting, or Archiving Fees
- Continually Enhanced Accessibility Compliance Features (WCAG)

| Over 150,000+ Chapters | Contributions From 200,000+ Scholars Worldwide | More Than 1,000,000+ Citations | Majority of e-Books Indexed in Web of Science & Scopus | Consists of Tomorrow's Research Available Today! |

Recommended Titles from our e-Book Collection

Innovation Capabilities and Entrepreneurial Opportunities of Smart Working
ISBN: 9781799887973

Advanced Applications of Generative AI and Natural Language Processing Models
ISBN: 9798369305027

Using Influencer Marketing as a Digital Business Strategy
ISBN: 9798369305515

Human-Centered Approaches in Industry 5.0
ISBN: 9798369326473

Modeling and Monitoring Extreme Hydrometeorological Events
ISBN: 9781668487716

Data-Driven Intelligent Business Sustainability
ISBN: 9798369300497

Information Logistics for Organizational Empowerment and Effective Supply Chain Management
ISBN: 9798369301593

Data Envelopment Analysis (DEA) Methods for Maximizing Efficiency
ISBN: 9798369302552

Request More Information, or Recommend the IGI Global e-Book Collection to Your Institution's Librarian

For More Information or to Request a Free Trial, Contact IGI Global's e-Collections Team: eresources@igi-global.com | 1-866-342-6657 ext. 100 | 717-533-8845 ext. 100